Development of Environmental

Development of lights the dynamic nature of environ ... India between the judiciary, the executive and the parliament. This has led to the creation of a wide range of environmental institutions and bodies with varied roles and responsibilities. The book contains a large volume of materials from the late 1990s, which show a marked shift in the nature of environmental governance in India. These materials offer an understanding of the contemporary debates in environment law in the context of India's economic liberalisation. The materials are thematically organised and presented in an accessible manner. The chapters contain definitions and specific clauses from the legal instruments and refer to court orders and judgments on these themes.

Kanchi Kohli is Senior Researcher at the Centre for Policy Research, New Delhi. She works on environment, forest and biodiversity governance in India. Her research explores the links between law, industrialisation and environment justice. She has authored various publications, research papers and popular articles. She is the co-editor of *Business Interests and the Environmental Crisis*, published in 2016. She has taught at universities and law schools on subjects related to biodiversity, environment and community development.

Manju Menon is Senior Fellow at the Centre for Policy Research, New Delhi. Her main areas of work are resource politics, environmental law and regulatory decision-making. She has published popular articles and papers on environment, law and development for over two decades. She is the co-editor of *Business Interests and the Environmental Crisis*, published in 2016.

Development of Environmental Laws in India

Kanchi Kohli

Manju Menon

CAMBRIDGE
UNIVERSITY PRESS

CAMBRIDGE
UNIVERSITY PRESS

University Printing House, Cambridge CB2 8BS, United Kingdom

One Liberty Plaza, 20th Floor, New York, NY 10006, USA

477 Williamstown Road, Port Melbourne, vic 3207, Australia

314 to 321, 3rd Floor, Plot No.3, Splendor Forum, Jasola District Centre, New Delhi 110025, India

103 Penang Road, #05-06/07, Visioncrest Commercial, Singapore 238467

Cambridge University Press is part of the University of Cambridge.

It furthers the University's mission by disseminating knowledge in the pursuit of
education, learning and research at the highest international levels of excellence.

www.cambridge.org
Information on this title: www.cambridge.org/9781108490498

© Kanchi Kohli and Manju Menon 2021

This publication is in copyright. Subject to statutory exception
and to the provisions of relevant collective licensing agreements,
no reproduction of any part may take place without the written
permission of Cambridge University Press.

First published 2021

Printed in India by Avantika Printers Pvt. Ltd.

A catalogue record for this publication is available from the British Library

Library of Congress Cataloging-in-Publication Data

Names: Kohli, Kanchi, author. | Menon, Manju, author.

Title: Development of environmental laws in India / Kanchi Kohli, Manju Menon.

Description: New York : Cambridge University Press, 2021. | Includes bibliographical references and index.

Identifiers: LCCN 2021005343 (print) | LCCN 2021005344 (ebook) | ISBN 9781108490498
(hardcover) | ISBN 9781108748490 (paperback) | ISBN 9781108781053 (ebook)

Subjects: LCSH: Environmental law--India. | Forest conservation—Law and legislation— India. |
Wildlife conservation—Law and legislation—India. | Pollution—Law and legislation—India. | Climatic
changes—Law and legislation—India. | Environmental policy—India

Classification: LCC KNS1507 .K63 2021 (print) | LCC KNS1507 (ebook) | DDC 344.5404/6--dc23

LC record available at https://lccn.loc.gov/2021005343

LC ebook record available at https://lccn.loc.gov/2021005344

ISBN 978-1-108-49049-8 Hardback

ISBN 978-1-108-74849-0 Paperback

Cambridge University Press has no responsibility for the persistence or accuracy
of URLs for external or third-party internet websites referred to in this publication,
and does not guarantee that any content on such websites is, or will remain,
accurate or appropriate.

Contents

Tables

Preface

Indian environmental laws are meant to help governments protect, conserve, extract and acquire natural resources in the processes of development. How are Indian environmental laws designed to perform these roles and how do they actually perform them? This volume lays out the legal frameworks of over two dozen Indian parliament-made environmental laws and nearly 20 executive-made environmental laws. Tracing developmental trajectories of laws requires the understanding that there are no single origins or sources and no predetermined linear pathways but only complex and intersecting contexts within which their development takes place. The preambular texts of laws, elaborate court papers and policy documents that carry varying definitions, meanings and interpretations present the possibilities and scope of these laws. When legal texts are opened up and reassembled, as we have tried to do in this volume, it is possible to see the overlaps, contradictions, duplication, fragmentation and confusions that are part of Indian environmental law.

Our motivations to put together this volume are threefold:

First, environmental laws have mostly been seen and studied as the territory of the judiciary. Most of the existing publications on environmental law provide readers with statutes and court cases. This gives the understanding that environmental laws are developed in courts. This is true to some extent if we look at the decades of the 1980s and the 1990s when higher courts were busy with environmental public interest litigations (PILs). This book can be seen as a complementary volume to those existing publications as it also focusses on the powerful role of the executive in the development of environmental laws. Since the 1990s, the central government, especially the newly formed Environment Ministry has had an unparalleled influence on the generation of legal frameworks to regulate the environmental impacts of all economic sectors. This volume seeks to put in perspective this influence so that those with an interest in the state of India's development can pay closer attention to the actions of India's large environmental bureaucracy.

A chapter dedicated to environmental institutions in this book underscores the enormous spread of regulatory institutions that are involved in the everyday operations of environmental law and their importance in today's environmental governance. The forms, capacities and motivations of governmental departments, regulators and expert agencies to realise environmental and social outcomes deserve much attention from scholars of different disciplines and civil society. While some regulatory themes and institutions have been studied well, we hope this volume sparks interest in studies that investigate the many others as well.

Second, this volume of environmental laws and case materials is a product of its time. Many publications on Indian environmental law refer to laws and cases that are now well established in the field. This is specially the case with the laws that were promulgated until the 1990s and case law from the PIL era when the Supreme Court passed landmark judgments. This volume has a selection of newer laws and cases that are helpful to understand the continuities and departures of India's post-liberalisation environmental laws from earlier legal instruments. This volume contains chapters on the biodiversity law, the evolving discourses on water laws and several new rules related to the management of the types of 'waste' generated by our economy. It also has a short chapter on India's climate policies and emerging international and domestic climate case law. The case materials presented in this volume refer to more recent significant cases in the courts; some of which made headlines and others that drag along slowly and silently. The volume contains many cases heard by the benches of the National Green Tribunal, the newest forum for environmental litigation. There is also a chapter outlining the government's arguments for 'reforms' in environmental laws including those carried out after political power changed hands at the centre in 2014. We hope that these new materials provide a comprehensive view of this dynamic field and generate renewed interest and new analyses of the scope and outcomes of environmental law in India.

Third, most publications on environmental laws are for legal professionals and by legal professionals. They are primarily directed at students and scholars of the law and are written by lawyers and law teachers who draw upon their rich experience and jurisprudence on environmental law. In recent years, environmental subjects, including environmental law, have become popular among students of the social sciences, economics, development studies, journalism, business sustainability, engineering and technology, public policy, social work and environmental sciences. But there are few publications that cater to the curiosity of this very large and growing community of non-legal professionals as well as self-trained environmental activists and public-spirited individuals who engage actively with environmental laws. There are also a large number of grassroots and mid-level government officers and environmental managers engaged with the private sector who deal with environmental issues and legal challenges on a daily basis. We hope that this volume, which does not comply with the rules of legalese, will help to make environmental laws accessible to this growing community interested in the state of our environment.

In the interest of accessibility, the chapters in this volume are arranged as issues because most people interested in environmental matters approach the field through questions about what is being done or to be done rather than through a specific law. The volume contains a list of governmental policies, notifications and orders discussed in the chapters. These have been selected to explain the legal frameworks, so the list is not a comprehensive compilation of all notifications related to environmental laws. The ones discussed here can be seen as signposts to understand the development of environmental laws discussed in this volume. These legal instruments are also amended frequently by bureaucratic processes. So, while we have made an effort to provide updated information about the instruments referred to in this publication, some of them may have undergone changes while this book was at the printers.

Acknowledgements

This book would not have been possible without the help of our colleagues and friends. Our sincere thanks to Prayank Jain, Ahmad Ammar, Debayan Gupta, Kush Tanwani, Sampada Nayak and Sahil Dabas for their research support.

Thanks to our colleagues at the Centre for Policy Research, New Delhi, for providing a scholarly environment and helping us to make connections across disciplinary boundaries. Thanks to our dear colleagues in the Environmental Justice Program who took on bigger responsibilities to let us have time off from project work, planning meetings and report writing.

And most of all, thanks to our families for their unconditional support, enthusiasm and encouragement.

Introduction

Environmental Laws and Development in India

Environmental laws are largely understood as a body of legal codes, statutes, case law, regulations and principles that are used to mediate the relationship between humans and nature. In India, environmental laws govern the protection, management and distribution of natural resources between what are framed as competing realms, such as between the economic and the sociocultural, between ecological needs and livelihood requirements, between the human and non-human and between present and future uses. Indian environmental laws are framed with competition and conflict at their core. Most times, the protective, managerial and distributive aspects of environmental laws are treated as technical matters to be decided based on costs and benefits. The laws fail, for most part, to note that nature is the material basis for the survival and well-being of all.

This volume unpacks environmental laws in India and outlines their development. It tries to understand the development of environmental laws in India by locating them within broader local, national and international sociopolitical and economic influences. More specifically, it seeks to establish the relationships between the last five decades of environmental law-making and practice in India and the institutional ideologies of developmentalism that have held sway during this period. This volume shows that the shifts in the political economy of the Indian state are reflected or even supported by the development of Indian environmental laws.

This book explores that the development and practice of environmental law in India can be understood as a domain of power through which actions of individuals and societies are controlled towards certain ends which are contingent, multiple and fluid. This field of power is dominated by four major actors who have exerted tremendous influence on Indian society through the exercise of environmental laws. These are the governments, courts, international environmental institutions and expert-based domestic regulatory institutions. The control exerted by these institutionalised actors is possible to trace as it is done through the use of statutes and governmental notifications, case law, legal principles and well-documented decisions—all of which are part of Indian environmental law. The chapters in this volume map the development and use of these legal instruments to regulate different kinds of environmental use.

The entities in this field of power are also acted upon. The forms and ends to which power is exercised by the above four sets of actors are also informed by the influence exerted on this field by economic, especially corporate entities and the demands for specific outcomes by social movements and affected communities. The influences exerted by them on institutional actors are difficult to capture as their efforts of lobbying and advocacy tend to be more diffused, episodic and not fully documented. These corporate and social actors are able to shape the workings of the formal institutions because the latter themselves seek to check each other's dominance in the field of environmental law. The understanding of Indian environmental law as a domain of power is best exemplified by the emergence of highly contentious and contradictory legal frameworks to manage India's lands, waters and forests. For example, the development of legal decisions and legislations on forest use and forest rights has involved courts, governments, experts, commercial entities and forest dwellers. These contradictions inbuilt into the legal framework make it nearly impossible to obtain any systematic outcomes.

The roles and functions of Indian environmental law are better understood when they are seen as political tools and not only as legal and regulatory instruments. The endgame in Indian environmental law is political resource distribution and as environment laws and principles are stretched in many directions by powerful stakeholders or interest groups. This puts environmental laws in the realm of resource politics. As most people who work with these laws know, environmental laws can be useful to protect community rights over the environment but all too often they have legalised resource grabs from the poor whose access to the law is very limited. The nub of environmental politics in India is captured in two lines written by Jairam Ramesh, the ex-environment minister, in his comprehensive record of state environmentalism under Indira Gandhi. He says, 'Environmentalists worry about a single objective. Political leaders have to balance competing demands and choose, if not the most desirable, then at least the least undesirable option from among them.'[1] He wrote this in the context of Indira Gandhi's decision to fell trees in Delhi to widen roads for the 'Asian Exhibition' in 1972. Years later, Ramesh himself offered this explanation for several decisions that he took as an environment minister.[2] This shows that our environmental regulations are meant to generate options.

The study and practice of environmental laws can benefit greatly if it is seen in relation to power relations, resource politics, and national and international political economy. The view of environmental laws as tools in power contests and negotiation rather than settled interpretation is critical to evaluate the possibilities and limits of working with them. This understanding helps to go beyond the narratives of institutional apathy or ignorance to explain legal or regulatory actions and inactions. This introduction provides a thematic review of the development of the political, legal and regulatory aspects of India's environmental laws. It outlines some of the debates of the field and hopes to provoke many more.

ENVIRONMENTAL LAWS AND POLITICAL LEGACIES

Postcolonial India's history of development is intertwined with the formulation, enactment and implementation of laws made specifically for the protection, management and regulated use of the environment in the process of economic growth. At the time of India's independence, there existed several state and central laws, such as for forest reservation and land acquisition. Enacted by the British or by the princely states, the intent and stated objectives of these laws were not necessarily the safeguarding of the environment. In some instances, it was to allow for the extraction of revenue or timber or minerals. In other cases, it was to generate community labour to manage natural resources or to keep

---

[1] Jairam Ramesh, *Indira Gandhi: A Life in Nature* (New Delhi: Simon & Schuster, 2017), 144.

[2] Jairam Ramesh, *Green Signals: Ecology Growth and Democracy in India* (New Delhi: Oxford University Press).

communities from using specific areas or resources.[3] The enactment of these
laws often led to retaliatory actions such as protests or civil disobedience.[4]
These laws imposed on certain sections or entire communities caused immense
impoverishment, land alienation, social conflict and environmental impacts.

The central government under the first Prime Minister Jawaharlal Nehru
set up a command economy. Economic development was the objective of the
government. According to Benjamin Zachariah, the independence movement
had coalesced around the theme of India's freedom to develop economically.[5]
The constituent assembly debates had also voted for a strong centre to not only
keep more partitions at bay but to realise this national economic ambition.
During this period, between the 1950s and the 1970s, the central government
was confronted by wars with Pakistan and China, famines and droughts,
and high levels of poverty. To achieve national economic development, the
government chose to bring more land under food production by constructing
massive irrigation projects and taking up industrialisation. The postcolonial
Indian state continued to use many laws that existed in the colonial period.
The many land acquisition laws, of which the central law of 1894 (Chapter 8)
was the prototype, were among them. These laws were deployed in selected
regions to coercively take over private property for energy, industry, agriculture
and irrigation projects. Commenting on the government priorities of this
period, Professor Sivaramakrishnan observes that most policies upheld the
needs of the national economy at the expense of the rural economy.[6]

Indira Gandhi became the prime minister in 1966. Her governments
from 1966 to 1977 and again from 1980 to 1984 can be seen as the period
of ecological modernisation in India, where economic development was

---

[3] M. Gadgil and R. Guha, *This Fissured Land: An Ecological History of India* (New
Delhi: Oxford University Press, 1992); Arupjyoti Saikia, *Forests and Ecological History
of Assam, 1826–2000* (New Delhi: Oxford University Press, 2011); S. Lele and A.
Menon, eds., *Democratizing Forest Governance in India* (New Delhi: Oxford University
Press, 2014).

[4] A. Baviskar, *Contested Grounds: Essays on Nature, Culture and Power* (New Delhi:
Oxford University Press, 2008); A. Agrawal, *Environmentality: Technologies of
Government and the Making of Subjects* (Duke University Press, 2005).

[5] B. Zachariah, *Developing India: An Intellectual and Social History, c. 1930–50* (New
Delhi: Oxford University Press, 2012).

[6] K. Sivaramakrishnan, 'Environment, Law and Democracy in India', *Journal of Asian
Studies* 70, no. 4 (2011): 905–928.

sought to be tempered by efforts in ecological protection. This period saw the enactment of two sets of laws through the parliament: the 'green' laws, such as the Wild Life Protection Act, 1972 (Chapter 6), and the Forest Conservation Act, 1980 (Chapter 3), and the 'brown' laws for the control of water pollution (1974) and the collection of water cess (1977) and air pollution (1981) (Chapter 4). The green laws were instrumental in her ability to take control of the sprawling estates of the erstwhile princely states and support the creation of protected areas (PAs), [7] the symbol of ecological high modernism. The brown laws, which soon gained greater significance than the older legal provisions on 'nuisance' in the Indian Penal Code, were the government's technocratic response to the problems caused by growing industrialisation and urbanisation. These laws signified that state environmentalism included urban areas and not just forest regions. Enacting these central laws through the parliament was no easy feat. States demanded their constitutional right to manage land and water. As Jairam Ramesh notes, the successful passing of these laws is a statement about Indira Gandhi's skills of persuasion and political negotiation. By 1972, the Stockholm Declaration[8] had brought about a global shift amongst the international community for environmental conservation. Indira Gandhi was the only prime minister besides the host to attend the conference in Sweden.[9] Her speech that presented poverty and pollution as two sides of the same coin could have been a trenchant critique of the Western economic models of wealth creation that produced two *public bads*. Instead, it became a call to fight *underdevelopment* through population control and the use of technology.

There are two lasting institutional legacies of Indira Gandhi that go beyond the environmental statutes mentioned above. The constitution as it was adopted in 1949 did not mention the term 'environment'. The constitution was amended in 1976 to introduce Articles 48A and 51A to the Directive Principles of State Policy under Part IV that obligated both the state and the citizens to protect and improve the environment. Several scholars have lauded this move as one of the earliest cases of environmental protection being

---

[7] M. K. Ranjitsinh, *A Life with Wildlife: From Princely India to the Present* (New Delhi: HarperCollins, 2017); Ramesh, *Indira Gandhi*, 125–127.

[8] United Nations General Assembly, United Nations Conference on the Human Environment, 15 December 1972, A/RES/2994.

[9] Ramesh, *Indira Gandhi*, 4, 114.

included in the text of a constitution. But what bears remembering is that this was done in the tumultuous period of the Emergency under the leadership of Indira Gandhi, a period when political rights were suspended and government ruled by administrative fiat. This adoption of the environmental concern by the constitution under these political conditions was a step towards the securitisation of the environmental challenges as urgent and needing centralised and undemocratic measures.[10]

The second institutional legacy of Indira Gandhi's environmental era is the creation of a department of environment and forests that became a full-fledged ministry by 1984. The department was first carved within the central Ministry of Agriculture. The ministry was fashioned as the expert, technical body, comprising bureaucrats and scientists and seen as eligible and equipped to take action on environmental matters (Chapter 2). The Bhopal gas tragedy of 1984 compelled the enactment of the Environment Protection Act (EPA), 1986. As discussed in Chapter 5, the clauses of the EPA vested this new ministry with powers to take the required measures 'to protect and improve the environment'. This includes regulating industrial activities, delegating responsibilities to specialised agencies and creating standards for managing environmental quality. The EPA gave powers to the ministry to draft executive law in the form of notifications, circulars and orders. These do not require any parliamentary approval. Since then, the ministry has used its delegated powers widely to regulate almost all sectors of the economy.

By the 1980s, state-led development was slowly giving way to a new era of privatisation in India. Government control of production functions had created discontent among industrialists and social activists. Those who were in favour of liberalisation caricatured the economic system as the 'license raj'. On the other hand, the government's focus on extractive development with high social costs attracted opposition from Adivasis and other local communities. Large social movements emerged in response to the problems of land acquisition and poor or no rehabilitation of families evicted by dams and mines. Projects built in the earlier decades were revealing their environmental impacts, for example, irrigated lands of Punjab were already waterlogged and saline.[11]

---

[10] N. L. Peluso and M. Watts, *Violent Environments* (Ithaca: Cornell University Press, 2001).

[11] Shekhar Singh, 'Social and Environmental Impacts of Large Dams in India', *Ecologist Asia* 1, no. 1 (2003): 61–70.

By 1991, several mainstream production sectors were opened up to private entities and the government expected to step up from being a developmental agency to being a regulator of development. Privatisation of production sectors involved, on the one hand, the repealing of several laws for labour welfare and protection and replacing them with the Public Liability Insurance Act, 1991, which provides for compensation and relief for anyone affected by industrial injury and accidents. On the other hand, it involved the enactment of environmental regulations to purportedly manage the use of the environment by the growing private industry. Environmental laws became more elaborate to monitor and control industries and projects and manage their growing environmental fallout. Given that social movements were already active, it was imperative for the ruling government to take public concerns related to environmental and social impacts of projects seriously. A focus on the well-being of the environment or in a wider sense, of the people and the natural resources that they depended on, seemed urgent and necessary to sustain the legitimacy of the project of economic development. A slew of new environmental notifications and circulars offered this legitimacy. These legal instruments that targeted specific 'high impact' projects and particular geographies are discussed in Chapter 5.

By the late 1980s, the narrative of nature's limits had been established and the concept of 'sustainable development' had emerged from international conventions related to the environment (Chapter 2). International forums and international aid institutions initiated and supported the development of an environment regulatory regime in India and many other liberalising countries to control the effects of private sector-led development. Scholarly work shows how international institutions such as the World Bank offered 'packaged laws' to many countries as part of their grants and loans for economic development.[12] The World Bank worked closely with the environment ministry to set up the laws for regulating environmental impacts.[13]

---

[12] A. Hironaka, 'The Globalisation of Environmental Protection: The Case of Environmental Impact Assessment', *International Journal of Comparative Sociology* 43, no. 1 (2002): 65–78; M. Goldman, *Imperial Nature: The World Bank and Struggles for Justice in the Age of Globalization* (Yale University Press, 2005).

[13] M. Menon and K. Kohli, 'The World Bank and Environment Policy Reform', in *The World Bank in India: Undermining Sovereignty, Distorting Development*, ed. M. Kelley and D. D'Souza (New Delhi: Orient BlackSwan, 2010), 413–431.

Since the 1990s, we have seen the development of a number of environmental regulations. These regulations that are discussed in this chapter form the body of India's environmental law that is mostly used today. Even as the government developed environmental regulations as a means of resource management, environmental activism internationally and in India during this period crystallised around the discourses of legal rights over the environment itself. This articulation of rights over environmental resources was different from the right to a clean environment as understood by the extension of Article 21 or the constitutional right to life (Chapter 1). The populist discourses of resource rights shaped India's laws on biodiversity (2002), forest rights (2006) and right to fair compensation and transparency in land acquisition (2013). These laws were passed by the Indian Parliament and included the language of prior informed consent and benefit sharing. While the biodiversity law was passed as a requirement under the Convention on Biological Diversity (CBD) (Chapters 2 and 6), the laws for forest rights (Chapter 3) and fair compensation for land acquisition (Chapter 8) were drafted and passed under pressure from social movements. A water law (Chapter 7), framed within the water rights discourses that were developed through courts, is long due.[14] It remains to be seen if this law will fill the gap by providing a single overarching framework for managing the multiple institutions and laws governing conservation and distribution of water. This articulation of individualised rights over a shared resource gained importance over rights to decentralised local governance of the environment at the levels of the *gram* panchayats, *nagarpalika*s or user groups.

During this period, the global discourses on governance of Protected Areas (PAs) took a 'community turn'. This was due to the work of thousands of wildlife ecologists, environmentalists and indigenous rights activists and the International Union for Conservation of Nature and CBD programmes on PAs. Indigenous communities demanded the space for land stewardship within national legal frameworks, and there was openness in the conservation community to accept that landscape-level conservation needed local people to take charge.[15] It was also a time for innovative experiments to demilitarise

---

[14] Shreehari Paliath, 'Without De-bureaucratisation, We Cannot Solve India's Water Problem', *IndiaSpend*, 13 July 2019.

[15] Ashish Kothari, Neena Singh and Saloni Suri, eds., *People and Protected Areas in India: Towards Participatory Conservation* (New Delhi: Sage Publications, 1996); J. Brown, A. Kothari and M. Menon, eds., 'Special Issue on Local Communities and Protected Areas', *PARKS* 12, no. 2 (2002).

lands and set up 'peace parks'. Conservation projects were breaking out of their exclusionary models and aiming to be participatory. These efforts also touched Indian conservation laws in small but significant ways. The Wild Life Protection Act, 1972, got two new categories, which had limited and designated roles for local communities in their management (Chapter 6). But this openness to communities in wildlife area management was affected by the enactment of the Scheduled Tribes and Other Traditional Forest Dwellers (Recognition of Forest Rights) Act, 2006 (FRA). The passing of this law to grant legal resource use rights to communities, most of who had lived in forests as 'encroachers' for years, polarised almost the entire conservation community in India.

A group of conservationists filed a petition for the constitutional review of the FRA in the Supreme Court in 2008. In an exceptional hearing of this case in February 2019, the Court first ordered that all those whose rights have been rejected under the law must be evicted. When it was clarified to the Court that there were inaccuracies in the rejected claims, the Court ordered all state governments to present the status of implementation of the FRA to the Court. More importantly, FRA does not have provisions for evicting forest dwellers if claims are rejected.[16] The FRA and other instruments that legalise community rights to resources came through populist grassroots struggles, but they face the constant risk of being diluted. Scholars have referred to this as a process of negotiating a 'compromise equilibrium' between political elites and the struggling masses.[17] For example, the central and state governments have successfully diluted the new land acquisition law to restrict benefits and consent procedures, as shown in Chapter 8. The forest department and the nodal ministry for the implementation of the FRA, the Ministry of Tribal Affairs, have been caught up in a decade-long 'statute based' tussle over the grant of forest ownership and use titles to individuals and communities residing in forest areas. This is discussed in Chapter 3.

Since 2005, various environmental statues have been amended, and this is, in part, the justification for this new publication. These amendments and reforms have been designed to achieve three objectives: (*a*) to incorporate the

---

[16] Kanchi Kohli, *Historical Injustice and 'Bogus' Claims: Large Infrastructure, Conservation and Forest Rights in India* (New Delhi: Heinrich Boll Stiftung, 2019).

[17] Kenneth Bo Nielsen and Alf Gunvald Nilsen, 'Law Struggles and Hegemonic Processes in Neoliberal India: Gramscian Reflections on Land Acquisition Legislation', *Globalizations* 12, no. 2 (2015): 203–216.

changed economic and technological contexts in which environmental laws operate, especially the changes caused by intensive private-led development in certain regions and sectors, (*b*) to factor in the effects and consequences of litigation and judicial pronouncements on environmental impacts and (*c*) to respond to the prevalent legal discourses on community rights over nature.

ENVIRONMENTAL LAW IN THE PUBLIC INTEREST LITIGATION ERA

The era of judicial activism started in India in the 1980s. Justice S. Muralidhar observes that the general breakdown of rule of law during the Emergency years of 1975–1977 provoked the judiciary out of its usual ways. The post-Emergency years were a period of political instability at the centre, and the higher judiciary felt it important to step up its role of protecting the constitution.[18] Articles 32 and 226 of the Indian constitution that allows citizens to directly approach the high courts and the Supreme Court through the writ jurisdiction is understood to be a flexible form of legal access to higher courts. However, since the 1980s, the courts innovated the judicial system radically by hearing a large number of cases on matters related to socio-economic rights through the route of the PIL (Chapter 2). The PIL emerged as a special class of writs. Divan and Rosencranz provide a useful classification of PILs and how they differ from other legal disputes between private parties.[19]

Through PILs, the higher courts allowed petitions to be filed by individuals who were not directly aggrieved but represented the aggrieved public or by citizens seeking remedies against governmental actions, policies or abuse of power. Public interest litigants did not have to fit into the conventional definitions of *locus standi* of a petitioner. This was allowed with the view that justice was to be served to those classes or groups of society who could not approach the judiciary. The judiciary's enthusiasm to admit PILs and its openness to turn complaints into writs and take up cases *suo moto* brought it much attention. The enactment of the Legal Services Authorities Act, 1987, and setting up of the National Legal Services Authority (NALSA) to provide

---

[18] S. Muralidhar, 'The Expectations and Challenges of Judicial Enforcement of Social Rights', in *Social Rights Jurisprudence: Emerging Trends in International and Comparative Law*, ed. M. Langford (Cambridge: Cambridge University Press, 2008), 102–124.

[19] Shyam Divan and Armin Rosencranz, *Environment Law and Policy in India: Cases, Material and Statutes*, 2nd edn (New Delhi: Oxford University Press, 2002), 133.

free legal aid to the poor[20] also supported the rise of PILs in India. As Senior Advocate Sanjay Parikh observes, these practices of the higher judiciary led to a 'new jurisprudence of social action'.[21]

The PIL, states Justice Muralidhar, 'is an entirely judge led and judge dominated movement' and it helped the Court to easily reach those who needed the protection of the constitution. The higher judiciary in India has been complimented for its 'judicial activism' during these years. While the defining moment of the PIL may have been well intentioned, over the decades, this judicial activism may have turned into a self-limiting approach. Today, the higher courts are clogged with cases and the standards set by the judgments are hard to achieve without effective institutions. Senior Advocate Shyam Divan argues that this activism may have even impeded the development of government structures for effective governance.[22]

The PIL was also used to bring the Court's attention to government failures on environmental matters. Through an environmental PIL, any person could file a case related to environmental problems even if the person was not directly affected. It became the main legal instrument that supported the courts to play an active role in shaping India's environmental and social conditions. 'Green benches' set up in some high courts and the Supreme Court brought focus on a variety of issues centred around rule of law, judicial review and meaningful realisation of rights with respect to the environment. Through the environmental PILs, the Indian courts became known for setting high legal standards and principles of environment laws and for raising significant questions about the character of environmental rights and their corresponding obligations. In environmental PILs, the higher judiciary went beyond adjudicating specific cases and expanded its influence to the areas of law and policymaking, institution building and monitoring and enforcement.[23] As noted by Divan and Rosencranz, members of the higher judiciary and the bar

---

[20] See https://nalsa.gov.in/about-us/introduction, accessed on 1 August 2020.

[21] Sanjay Parikh, 'Foreword', in *Environmental Jurisprudence and the Supreme Court*, ed. Geetanjoy Sahu (New Delhi: Orient BlackSwan, 2014), ix–xiii.

[22] Shyam Divan, 'A Mistake of Judgment', *Down to Earth*, 28 June 2015, available at https://www.downtoearth.org.in/blog/a-mistake-of---judgment-14470, accessed on 10 March 2020.

[23] Geetanjoy Sahu, *Environmental Jurisprudence and the Supreme Court* (New Delhi: Orient BlackSwan).

not only undertook litigation tasks but also conducted fact-finding exercises, served on court-appointed committees and helped the courts as 'amicus curiae' on environmental PIL matters.[24] They observe that the higher judiciary plays the role of a 'super agency' of the environment through environmental cases.

Geetanjoy Sahu undertook an assessment of 191 environmental cases heard by the Supreme Court between 1980 and 2010 to understand their role in the advancement of environmental jurisprudence in India. Based on this study, Sahu observes that these cases created space for environment issues in the highest court of law. The Supreme Court directed changes in policies and laws and pushed environmental institutions to fulfil their constitutional obligation to protect the environment. While the PIL had already earned the Supreme Court the reputation of being 'the final savior of democracy',[25] the environmental cases earned the courts the recognition of being 'green' and 'the sole dispensers of environmental justice'.[26] In cases of environmental conflicts and grievances, the higher courts interrogated the environmental actions of the government to such an extent that courts have been understood as the prime actor in the development of India's environmental law. The many environmental law textbooks whose contents focus on the jurisprudence that developed through environmental cases in the higher judiciary are testimony to this.

As observed by many legal scholars, the higher courts innovated with judicial procedures and principles to deal with environmental cases. For example, the linking of the right to life as inclusive of the right to a clean environment and the development of the absolute liability principle (both discussed in Chapter 1) to deal with industries causing physical harm to workers and communities were done in cases heard soon after the Bhopal gas disaster. These were important cases and the Court's attention to the plight of affected people was necessary. But a critique of green courts also developed over these years, especially regarding the continuing role of the courts in environmental matters through the PIL. This route to environmental matters escalated every environmental case to the level of an emergency, and this came in the way of the Court's ability to assess all aspects and implications of the cases before passing decisions or dictating policies. The activism displayed by

---

[24] Divan and Rozencranz, *Environment Law and Policy in India*, 22.

[25] Parikh, 'Foreword'.

[26] Sahu, *Environmental Jurisprudence and the Supreme Court*.

the Supreme Court not only transformed the judiciary, but also unsettled the state–judiciary relations. While some saw it as the judiciary performing its constitutional obligation, the Court's interventions in policy matters and in securing the enforcement of judgements were viewed as overreach.[27]

As the Supreme Court became 'a powerful arbiter of claims', its decisions attracted greater scrutiny by environmental and social rights advocates alike. Scholars and activists observe that the Court's treatment of environmental cases and its application of legal principles and interpretations was uneven.[28] This unevenness went beyond being merely a function of different personalities or individual ideologies of the judges, differences in the petitioners or how the cases were argued. The discretionary and wide powers of the higher courts is inbuilt into the PIL approach. The Supreme Court preferred not to rule against large infrastructure projects of the government in the name of not interfering with government policy.[29] Justice Muralidhar notes that the higher courts shared the utilitarian logic of the executive on large projects. For example, the Court's support for large dam projects was not shaken by their environmental violations or by their imposition on the rights of Adivasi communities who would be displaced or by the risks posed by large dams in geologically fragile areas.[30] The Supreme Court's support for high modernist projects is borne out by its support to interlink rivers to solve the water problem in India.[31] Senior Advocate Prashant Bhushan adds that the court's environmentalism comes to the fore when the poor or small livelihood enterprises affect nature but the environment becomes dispensable when corporate interests are involved.[32]

Justice Muralidhar provides a sobering assessment of environmental PILs. He observes that for all the celebration of the court's activism, the cases that

---

[27] Videh Upadhyay, 'Reclaiming the Judicial Ground', *Economic and Political Weekly* 43, no. 33 (16 August 2008): 13–15.

[28] Sahu, *Environmental Jurisprudence and the Supreme Court*; Shibani Ghosh, *Indian Environmental Law: Key Concepts and Principles* (New Delhi: Orient BlackSwan, 2019), 1–17.

[29] Sahu, *Environmental Jurisprudence and the Supreme Court*.

[30] Muralidhar, 'The Expectations and Challenges of Judicial Enforcement of Social Rights'.

[31] Supreme Court judgment dated 27 February 2012 in Writ Petition (Civil) 512 of 2002 (*In RE: Networking of Rivers*) Writ Petition (Civil) 668 of 2002.

[32] Prashant Bhushan, 'Misplaced Priorities and Class Bias of the Judiciary', *Economic and Political Weekly* 44, no. 14 (2009): 32–37.

involved environmental rights and socio-economic rights, for example, the *Bombay Environmental Action Group vs the State of Maharashtra* or the *BEAG* case of 1995, were heard as one-sided cases.[33] Those who were likely to be evicted or lose jobs were hardly heard by the Court. The environmental PIL thus differentiated between the publics with and without access to legal services. If the poor affected by these cases did have access to legal services, then it was upon them to defend their rights against this exclusionary form of environmentalism.[34] Scholars who have researched the Court's environmental PIL rulings through the lens of social justice observe the biases of the Court. Caste and class prejudices of the Court framed the environmental PILs and ruled in favour of the rights of the middle class to a clean environment against the rights of the poor to reside and work in cities. The environmental PILs that involved evictions of slum dwellers and bans on rickshaw pullers provoke the question of who has 'rights to the city' in the face of urban middle-class environmentalism.[35]

The active PIL phase waned by the end of the 1990s. In the words of Justice Muralidhar, as the project of neoliberalism unfolded in India, the demands for socio-economic justice shifted from the field of law to the economy. The environmental PILs were also on the decline. The Supreme Court's role in matters of social and environmental justice was affected by the lack of systems for implementation of its orders and judgements. Despite its hold on environmental matters, the institutions involved were unable to provide a comprehensive and lasting approach to tackle the challenges of development. Sahu notes, the environmental actions of the Court failed to alter the ideologies of the state or shake up the environmental apathy of the bureaucracy. 'At best, it only disrupted the balance of power....'[36] However, environmental litigation continued to have a great influence over the development of environmental laws. Several environmental cases have gone on for over a decade in the higher courts. Examples of these, such as the M.

---

[33] Muralidhar, 'The Expectations and Challenges of Judicial Enforcement of Social Rights'.

[34] Ibid.

[35] Amita Baviskar, 'Cows, Cars and Cycle-Rickshaws: Bourgeois Environmentalists and the Battle for Delhi's Streets', in *Elite and Everyman: The Cultural Politics of the Indian Middle Classes*, ed. Amita Baviskar and Raka Ray (New Delhi: Routledge, 2011), 391–418; Mukul Sharma, *Caste and Nature: Dalits and Indian Environmental Politics* (New Delhi: Oxford University Press, 2018).

[36] Sahu, *Environmental Jurisprudence and the Supreme Court*.

C. Mehta cases on pollution, the forest case, the solid waste management case and the case on genetically modified organisms, are discussed in this volume. Some of these are *continuing mandamus* cases in which the courts monitor the ongoing implementation and enforcement of law.

The rise of the environmental regulatory era where the executive took charge of environmental matters may have also contributed to the decline of environmental PILs. The decline of environmental PILs could also be attributed to the creation of specialised judicial bodies to hear environmental matters, especially the National Green Tribunal (NGT) in 2010. Both these aspects, that is, the expansion of the environmental regulatory landscape and the creation of the NGT, can be understood as part of a trend towards the technicalisation of environmental justice. This involves the narrowing of questions regarding the interpretative issues of environmental and social rights to mostly technical aspects of environmental administration.

The NGT became operational in 2011. It was a case of 'third time lucky' after two attempts with the National Environment Tribunal and National Environment Appellate Authority (Chapter 10). The justification for a tribunal was based on the high number of environmental cases, the technical aspects involved in these cases and judiciary's difficulty to monitor and supervise their implementation effectively.[37] The NGT is set up to deal with environmental matters as described in Chapter 2. Over 30,000 cases were filed in the NGT as of February 2020. India has received accolades for the creative expansion of the judiciary to deal with environmental matters in the most populated democracy with a large economy and rich ecologies. While the NGT has become a legal forum to interrogate the environmental impacts of development, it has also attracted criticisms for its tribunalising effect on environment laws. Critics claim that remedies from the NGT are technological fixes rather than substantive justice, that access to the tribunal is restricted due to the limited benches and most importantly, that the tribunal is not an independent body outside the influence of political power. Since its inception, the NGT has at different points of times operated with high vacancies.[38] Since 2011, several cases involving similar issues are being heard by the NGT and the higher

---

[37] Law Commission of India, *One Hundred Eighty Sixth Report on Proposal to Constitute Environment Courts* (New Delhi: Government of India, 2003).

[38] Debayan Roy, 'NGT Working with Just 6 Members instead of at least 21, Zonal Benches Vacant for 2 Yrs Now', *The Print*, 4 November 2019.

judiciary. Two examples of these discussed in this volume are the wetlands case (Chapter 7) and the case on effluents and emissions from thermal power plants (Chapter 4). The focus of the NGT tends to reduce or limit the complex issues involved in environmental cases to very specific questions while the higher judiciary continues to broaden the scope of environmental cases as they hear them.

Despite all these developments, the overall state of environmental adjudication is ineffective compared to the scale of justice needs. Access to quality and affordable legal services on environmental matters is generally poor. It is far worse for the economically weaker sections of society who also face other forms of injustices. There is no legal aid for communities affected by environmental cases outside of the specific groups who can avail them under the National Legal Services Authority (NALSA). There are only a handful of lawyers who offer *pro bono* services for needy clients, and they are stretched beyond capacity to undertake quality legal research for their cases. Besides inadequate resources, the available resources are also used imprudently. These aspects of environmental litigation are barriers to environmental and social justice.

THE ENVIRONMENTAL REGULATORY ERA

Environmental regulation is understood as the exercise of control over economic entities through specific legal and market mechanisms for better environmental outcomes. Regulation is seen as necessary for a clean environment because the costs of environmental protection are not built into economic production costs. With the opening up of the mainstream economic sectors to private corporate entities, the government refashioned itself to regulate these entities and facilitate the creation of public goods. These reforms changed the conditions of economic development since the 1990s not only in India but in many parts of the world. International Financial Institutions operating in developing countries such as India played an important role in co-creating regulatory frameworks by which national governments could manage the environmental issues arising from large liberalised economies. In India, a new Ministry of Environment and Forests was set up to regulate on environmental matters. The ministry led India into a new regulatory phase of environmental governance.

Since its formation, the ministry has promulgated a series of 'subordinate' or delegated legislations that have dominated the environment law landscape in recent years. These laws could be modified on a routine basis, with or without a public or parliamentary discussion. This has caused an erosion of democratic processes in environmental law-making as elected leaders or members of the public are left out. Even though these laws are not full-fledged statutes, they have a countrywide reach and give powers to the executive to control regions, class of projects and processes. The regulations drafted under these powers and clarified by numerous circulars and office orders of the ministry are narrow and specific and, in some cases, have read down the preambular objectives of statutes passed by the parliament. For example, all the waste management laws in 2016 are based on embracing of management approaches to the problem rather than preventive approaches to waste (Chapter 5).

Other regulations are related to the management of coastal zones and decision-making on infrastructure projects including clear provisions for public participation prior to their approval (Chapter 5). The import of such wide powers, under Section 3 (2) (v) of the EPA, 1986, to the central executive, to make national regulations in a federal democracy can be debated. For example, there has been one constitutional review of an environmental regulation in the Indian Council for Enviro-Legal Action case in 1996 (Chapter 5). This case sought to quash a set of notifications to amend the Coastal Regulation Zone Notification, 1991. There are also hardly any assessments of what the overall results of such notifications are in environmental terms and for environmental jurisprudence.

This national environment regulatory framework also includes the creation of expert technical institutions and quasi-judicial bodies to take regulatory decisions. India's transition into the regulatory era can also be marked as the shift away from the environmental PIL phase discussed in the earlier section. Some of these technical institutions were set up after certain environmental cases ordered their creation. These institutions work under or with the environment ministry and government-appointed regulatory bodies. The expert regulatory institutions discussed in this volume are listed in Chapter 2. The adoption of this system to deal with environmental matters reflects the acceptance that environmental matters are technical issues best mediated by technical experts.

As the chapters in this volume show, the implementation of environmental laws has come to rely almost entirely on a large number of expert institutions. At the same time, the development of environmental laws is replete with examples of institutional crowding and collapses. The creation, working and failures of these institutions, although discussed to some extent in this book, deserve in-depth investigation. Take the Pollution Control Boards (PCBs) for example. State pollution control boards (SPCBs) are mostly headed by political appointees even though there are detailed criteria for such positions.[39] Recently, the NGT stated that the inefficiency of the SPCBs has got to do with its composition (Chapter 4). Even though these statutory bodies exist, the Supreme Court made new bodies to perform similar functions. The Delhi Vehicle Pollution case is one such example that limited the role of the Central Pollution Control Board (CPCB) by creating the Environment Pollution (Prevention and Control) Authority (EPCA) (Chapters 2 and 4). Although the EPCA was set up to look at pollution in Delhi-NCR, the recommendations of this small body have been nationalised for the whole country.[40] There are many examples of overlapping institutions, which not only cause confusion but are wasteful and delay environmental actions.[41]

The regulatory processes set up to control industrial activities and developmental processes through most of the environmental laws discussed in this volume include five elements: zoning, expert decision-making, monitoring of compliance, penalties and compensations, and institutionalising costs of environmental damages.

ZONING

The chapters in this volume have several examples of laws that include zoning in their legal frameworks. These include forest areas, PAs, ecologically sensitive areas (ESAs), coastal zones, biodiversity heritage sites and critically polluted areas to name a few. These regions are notionally ring-fenced from surrounding lands (and even water in case of marine PAs) even though they may not be very dissimilar. They are administered through special regulations, notifications or management plans made by agencies that are responsible

---

[39] IANS (Indo-Asian News Service), 'No Political Appointments in Pollution Control Boards: NGT', *Business Standard*, 24 August 2016, available at https://www.business-standard.com/article/current-affairs/no-political-appointments-in-pollution-control-boards-ngt-116082400830_1.html, accessed on 8 March 2020.

[40] Ritwick Dutta, 'EPCA Was Specially Empowered to Deal with Air Pollution Crises: Is It Working?' *The Wire*, 8 November 2018.

[41] Centre for Science and Environment, 'Turn Around: Reform Agenda for India's Environmental Regulators', CSE, New Delhi, 2009, 8–15.

for these jurisdictions. These have been done without adequate attention to prioritisation or differentiation and form a poor basis for environmental governance. As a result, the zoning of areas has been subject to arbitrary and frequent boundary changes.

The regulatory ring-fencing of geographical territories by environment laws have led to discriminatory environmental practices such as treating residents within forest villages as encroachers, locating of environmentally damaging projects such as sewage treatment plants in fishing villages in the coastal No Development Zone and the use of panchayat lands for dumping of municipal solid waste. It has also led to the top-down and undemocratic decisions on what kinds of developments or activities to allow within reserved forests, ESAs and PAs. For example, mining and high-value tourist facilities have been permitted in PAs but village settlements are seen as a threat. While the mapping of ecological spaces such as coastal zones and biodiversity areas are left pending for years, approvals for large-scale, commercial extraction of resources from these spaces are permitted using regulatory procedures. These forms of discriminatory practices legalised by environment laws have faced many legal challenges and public protests. Zoning as a tool to protect the environment faces local backlash as it burdens the people in some geographies with the responsibilities of environmental protection rather than treat it as a shared duty. This has provoked large-scale political mobilisation in specific regions against zoning-based environment laws and policies for fostering 'uneven development'.

Expert-Based Decisions

Today, environment regulation is dependent on expert opinions. Environment regulatory frameworks rely on appraisals, recommendations and conditional approvals given by expert bodies, authorities and committees such as the coastal zone management authorities, Central Ground Water Authority and many others discussed in Chapter 2. These bodies are constituted by experts in the broad fields of general administration, management and engineering. The experts use their technical knowledge to assess applications seeking the commercial use of natural resources. The approvals also mandate several conditions that are seen to minimise or mitigate the undesirable impacts such as pollution or loss of forests. But these expert bodies are neither independent from political influences nor accountable to the public for the recommendations they make. Civil society has played an important role in scrutinising the composition and role of institutions in major environmental decisions such as the committees set up to protect the Western Ghats or Expert Appraisal Committees (EACs) to grant approvals to industries and mines.[42] The committees are expected to make decisions quickly and often based on limited information provided by approval seekers. Most regulators do not use any principles of prioritisation in their decision-making, and the experts themselves are in a difficult position to determine the environmental and developmental needs of the country. So, the government's priorities become their sole guidance. The EACs, under the EIA notification, have a track record of recommending over 90 per cent of the projects applying for approvals.[43] In cases where academic scholars and environmental practitioners have been part of institutions, and have gone against the government's priorities, they have faced isolation in their committees, discontinuation of their term and no reappointment.[44] As a result, environmental bodies have attracted the criticism of being the 'rubber stamp' of the government. Courts or governments have not adequately addressed the structures and procedures of these expert bodies to address biases and conflict of interest even though independent and sound expert opinions are crucial to environmental governance. All the above challenges of creating good regulatory institutions can be understood through the debates that took place between 2010 and 2013 over the government's decision to set up a national environment regulator (Chapter 10).[45]

---

[42] Fourth open letter to the Ministry of Environment and Forests (MoEF) issued on 8 April 2005, titled 'Why Are the Expert Committees of Ministry of Environment and Forests Dominated by Ex-bureaucrats, Politicians and Engineers?' signed by 66 individuals and organisations.

[43] Press Trust of India, '90% of Projects Granted Environment Clearance in 2015', *Economic Times*, 28 November 2016, available at https://economictimes.indiatimes.com/news/economy/policy/90-of-projects-granted-environment-clearance-in-2015/articleshow/55668021.cms?from=mdr, accessed on 8 March 2020.

[44] Ashish Kothari, 'We Should Have Gone to Court: Interview with Ashish Fernandes', in 'Large Dams in North East India: Rivers, Forests, People and Power', *Ecologist Asia* 11, no. 1 (January–March 2003): 38–39; Prerna Singh Bindra, 'Book Excerpt: How Government Policies Are Wilfully and Recklessly Threatening Wildlife Reserves', *Scroll.in*, 4 July 2017.

[45] Sharachchandra Lele, Navroz K. Dubash and Shantanu Dixit, 'Structure for Environment Governance: A Perspective', *Economic and Political Weekly* 45, no. 6 (6–12 February 2010): 13–16.

MONITORING

The third focus of environmental regulation is monitoring of activities and projects that could damage the environment. Environmental monitoring has been the weakest part of environmental regulation.[46] Both government and civil society reports observe that environmental regulatory institutions are extremely short-staffed and under-resourced to perform this task at the scale required. However, the unmanageable burden on monitoring is a systemic outcome of the regulatory design. The approval-based regulatory systems that permit nearly all projects seeking approvals and expects very low-performance standards from the approved units cause the greatest challenge to monitoring. The types of projects and the conditions on the basis of which approvals are granted place a heavy load on monitoring. The present approval systems and court orders uphold the building of environmental management infrastructures such as effluent treatment plants; treatment, storage and disposal facilities; retaining walls; garland canals; compensatory plantations and many others. However, their actual role in managing or mitigating environmental problems at scale has been limited.

Monitoring is done by a handful of regulators, and the data collected by them is not available to the public. In recent years, different types of monitoring systems have been used to measure the environmental performance of polluting industries. The data collected by these instruments, such as Continuous Emission Monitoring Systems (CEMS) and Continuous Effluent Quality Monitoring Systems (CEQMS) (Chapter 4), is also not available to the public. There is less transparency about the methodologies used by government agencies to analyse the data collected through these devices.[47] When official data that is obtained through time-consuming Right to Information processes is used to question government agencies, the latter retreat with the justification that the data quality is poor or suffers from flaws. Scholars and activists working on air quality have repeatedly argued that data quality cannot improve without openness and

---

[46] Manju Menon and Kanchi Kohli, 'Regulatory Reforms to Address Environmental Non-compliance', CPR Policy Challenges (2019–2024), Centre for Policy Research, New Delhi, 2019.

[47] Santosh Harish, Shibani Ghosh and Navroz K. Dubash, 'Clearing Our Air of Pollution: A Road Map for the Next Five Years', Policy Challenges (2019–2024), Centre for Policy Research, New Delhi, 2019.

public scrutiny.[48] There are innovations in the arena of low-cost monitoring devices, and the data collected by these inform public education and advocacy. However, government still relies on its few and expensive devices that provide only patchy data.[49] The use of new technologies is crucial to overcome some challenges of monitoring. But it is also important to recognise that devices cannot measure all forms of environmental indicators or impacts and what cannot be measured should not be ignored. For example, the CPCB eliminated health impacts as a criterion for classifying highly polluting industrial sectors on the justification that these impacts are difficult to measure (Chapter 4). Environmental monitoring systems also need to embrace and encourage the active participation of field researchers and affected people who can provide directions for timely and specific interventions.[50]

Even with these deficiencies in monitoring, it is known that there is a high degree of non-compliance to environment laws.[51] These are recorded in the environmental audits by Comptroller and Auditor General and state auditor generals' (AGs') offices.[52] Environmental organisations and researchers have also reported on non-compliance.[53] These high levels of non-compliance to environment regulations make it abundantly clear that although environment regulations are legal systems, they are practised as policies of the government. Since economic development is the foremost concern of most governments,

---

[48] Ishita Jalan and Hem Dholakia, *What Is Polluting Delhi's Air: Understanding Uncertainities in Emissions Inventories* (New Delhi: CEEW [Council on Energy, Environment and Water], 2019).

[49] Sonam Joshi, 'How Low-Cost Tech Can Help India Monitor the Air It Breathes?' *Mashable-India*, 29 July 2016, available at https://mashable.com/2016/07/29/india-air-pollution-monitoring/, accessed on 8 March 2020.

[50] Menon and Kohli, 'Regulatory Reforms to Address Environmental Non-compliance'.

[51] MoEF, *Report of the Committee Constituted to Examine the Issues Related to Monitoring of Projects (J-11013/30/2009-IA.II (I)* (New Delhi: Government of India, 2010).

[52] Comptroller and Auditor General (CAG), *Report of the Comptroller and Auditor General of India on Environmental Clearance and Post Clearance Monitoring (Report No. 39 of 2016 (Performance Audit)* (New Delhi: CAG, Government of India, 2016).

[53] Kohli and Menon, 'Regulatory Reforms to Address Environmental Non-compliance'; Centre for Policy Research, *Closing the Enforcement Gap: A Groundtruthing Study of the Social and Environmental Impacts and Legal Non-compliance of the National Highway Expansion Project In Coastal Uttara Kannada, Karnataka* (New Delhi: CPR-Namati Environment Justice Program, 2019).

regulatory violations are ignored or accommodated through mechanisms like *post facto* approvals and inaction against non-compliance. There is a need for empirical research to show if environment regulations are enforced arbitrarily or if they are selectively enforced in ways that are politically expedient. This selective application of laws by those who have political power can be understood as the 'rule by law' in place of the rule of law.

Penalties and Compensations

As stated in Chapter 1, courts have dealt extensively with the issue of environmental offences. The concept of nuisance, which predates modern environment laws, provides a framework of remedies that has been applied to environmental cases. Since its establishment, the NGT has also provided a wide range of remedies in environmental cases. In addition, environmental laws dealing with particular themes such as wildlife protection, pollution control or access to biological resources also outline specific kinds of punishments for violations.

Environmental offences can be civil or criminal acts depending on how they are addressed by legal frameworks. Pursuing certain actions of individuals or corporations as environmental crimes involves time consuming and expensive procedures. The burden of proof needed in such cases is higher and the court proceedings for such cases may be beyond the skills and capacities of environmental regulators. Due to these reasons, prosecutions in environmental cases tend to be very few, take too long and may often fail. It has also been argued that criminal sanctions may not always be useful to achieve compliance or reverse any existing damage. [54]

Civil and administrative remedies are seen as flexible and more useful to achieve the results of environmental protection, punishment for the violating person and creating a deterrence. Courts can impose fines on entities that are held liable for certain acts that are unlawful. They can also impose injunctions to deter the violators from carrying out environmentally damaging activities. Penalties can also include the revocation or suspension of approvals or licenses.

---

[54] MoEFCC (Ministry of Environment, Forest and Climate Change), *Report High Level Committee to Review Various Acts Administered by Ministry of Environment, Forests and Climate Change* (New Delhi: Government of India, 2014); Centre for Policy Research, *A Framework of Principles for Environmental Regulatory Reform: Submission to the High Level Committee's Review of Environmental Laws* (New Delhi: Centre for Policy Research, 2014).

Civil remedies could be more efficient and timely than criminal prosecutions. However, they could also result in appeals by those fined.

The earlier section shows that the governmental practices of monitoring the implementation of environmental laws in India have normalised high rates of non-compliance. With growing evidence of habitual offenders of environmental laws, the environmental damage caused by their actions, and the development of climate change science, there is a demand for recognising the actions of large corporations involved in fossil fuel extraction and production as environmental crimes.

While penalties and fines are directed at the violators, the environment or individuals affected by violations also deserve attention. Restitution of damages is mostly achieved through courts and through governmental guidelines.[55] Damages may include individual harms, for example, accidents, loss of property or injury, loss of livelihoods due to environmental degradation. The designing of penalties, fines and compensations do not follow any laid down economic standards. These are arrived at on an ad hoc or case-to-case basis. The compensations that have been provided through courts in a few cases are in the form of environment restoration funds to be managed by governments. Communities who have suffered environmental impacts due to the violations have barely influenced the decisions on the use of such funds.

Remedies available in Indian environmental law hardly take into account the need to improve accountability of regulators to the law. Very few officers and regulators have been penalised for their role in aiding or abetting environmental harms, and they enjoy some immunity under administrative law. For example, there are numerous cases where expert committees have been accused of 'non-application of mind' while recommending approvals.[56] Yet they continue to use their powers to influence important decisions affecting the well-being of communities and the economy.

---

[55] Central Pollution Control Board, *Guidelines on Implementing Liabilities for Environmental Damages Due to Handling and Disposal of Hazardous Waste and Penalty* (New Delhi: Government of India, 2016); Forest Development Corporation of Maharashtra, *Guidelines for Management of Human Elephant Conflicts* (Nagpur: Government of Maharashtra), available at http://mahaforest.gov.in/fckimagefile/ Guideline%20for%20Human%20Elephant%20Conflict.pdf, accessed on 14 March 2020.

[56] Supreme Court judgment dated 29 March 2019 in Civil Appeal No. 12251 of 2018 (*Hanuman Laxman Aroskar v. Union of India*).

Institutionalising Costs of Environmental Damage

The practices of calculating the monetary value of 'environmental services' provided by water, forests or land and the costs of such services lost to the public due to environmentally damaging projects have grown in environmental economics. The concept of 'payment for ecosystem services' tries to show the multiple ways in which human communities are supported by environmental functions that are invisible to the economy and, therefore, not valued by it. This field grew out of the reasoning that environmental protection needs an economic justification if it is to succeed.[57]

However, environment regulations in India impose costs in reductive ways as a result of which these practices do not help to prevent damage but legitimise them through economic valuations. India's environmental regulatory framework is based on governmental facilitation of industries and projects. It has resulted in the institutionalisation of 'costs' of environmental damage into environmental laws and decisions. An example of such costs being included into the legal framework of environment regulation is the case of compensatory afforestation and the net present value. The costs charged to economic entities such as mining and dam projects that have consumed large tracts of forest land have accumulated to over ₹520,000 million in the Compensatory Afforestation Fund (Chapter 3).

These costs are also determined arbitrarily through negotiations with economic actors in specific sectors. The District Mineral Fund (DMF) was instituted in 2015 and is made up of local area development costs charged to mining companies. The DMF, under the charge of the district collectors in mining districts, aims to help dystopian mining regions shake off their 'resource curse' by making operational mines in a district provide funds to invest in public infrastructure in the district. The total funds collected and disbursed through DMF is ₹308,190 million,[58] and Keonjhar district in

---

[57] Pawan Sukhdev, 'Put a Value on Nature', transcript of the TED Talks, TED, December 2011, available at http://www.ted.com/talks/pavan_sukhdev_what_s_the_ price_of_nature/transcript, accessed on 28 April 2014.

[58] CSRBox, *District Mineral Foundations (DMFs) in India: An Overview of Funding and Development Priorities* (Ahmedabad: CSRBox, 2019), available at https://csrbox. org/India_CSR_report_District-Mineral-Foundations-(DMFs)-in-India--An-Overview-of-Funding-and-Development-Priorities_67, accessed on 8 March 2020.

Odisha has one of the largest DMFs.[59] Although such funds are governed by legal statutes, in reality, they can be used in unregulated ways. For example, a coal cess or National Clean Energy Fund was instituted during 2010–11 for funding research on clean energy by charging coal-mining companies. Water cess under the Water Cess Act, 1977, was charged to water users (Chapter 4). By its very definition, cess is a tax meant only for the purpose it is collected. Yet, in a controversial decision by the government in 2017, the unspent coal cess collections amounting to a total ₹567,000 million was diverted to compensate states that lost revenues due to the introduction of the new system of Goods and Services Tax.[60] These costs charged routinely for reducing environmental damage show that environmental regulations have fully embraced the monetisation of environmental degradation. This works as a 'pay and pollute' model of governance.

ENVIRONMENTAL REFORMS

From the chapters in this volume, it is amply clear that Indian environmental laws have failed to develop into a medium for fostering greater equity, justice and well-being of people and nature. The limitations of environmental laws are twofold. First, in their design and implementation, environmental laws have targeted the poor as if nature is to be saved from them. Second, the lack of effective environmental governance affects the poor the most. If we are to expect this to change, there are three main challenges that would need to be tackled in the area of environment law reforms. These three areas are the issues arising from the notion of custodianship in a federal structure of governance, the lack of social legitimacy for hyper-technical environmental regulations and the devolution of nature-based rights and conservation opportunities to people in these times of planetary environmental crisis. While some of these terms exist in environmental policy statements, this volume shows that there is hardly any synergy between such policies and environmental laws.

---

[59] Centre for Science and Environment, 'Keonjhar with Rs 2,500 Crore in DMF Funds Holds Enormous Potential to Improve Human Development Indicators and Create Livelihood Security for the Mining-Affected People, Says CSE Report', press release, 28 February 2019, available at https://www.cseindia.org/keonjhar-with-rs-2-500-crore-in-dmf-funds-holds-enormous-potential-to-improve-human-development-9295, accessed on 8 March 2020.

[60] Kumar Sambhav Shrivastava, 'India Diverts Rs 56,700 Crore from the Fight against Climate Change to Goods and Service Tax Regime', *Scroll.in*, 24 July 2017.

The first aspect of legal reforms is to take seriously the notion of custodianship of resources. Presently, the structures and formats of environment laws reinforce the view that the government is the owner of all resources. If the government is only the custodian of resources, then its systems and processes for conserving these resources across generations need careful redesigning. An added challenge is how to plan for this role of governments as custodian in a federal structure such as ours. As this volume shows, the Seventh Schedule of the constitution which provides for power sharing between the centre and states is a site of power tussles for environmental control. The conflicts created by this aspect alone has damaging results for how we use and conserve environmental resources. The provision of water is a prime example. The Supreme Court interpreted the right to clean water in 1991. Yet, after all these years of discussing water sector reforms, we have a governance regime that cannot negotiate the complex federal jurisdiction between states and the centre; between the water, environment and other sectoral ministries. There are still no settled governance principles for water management, no comprehensive law and no overarching institution.[61] Similarly, the conservation of seas, rivers and wetlands, of marine and aquatic fisheries, are affected by similar inter-state and inter-institutional challenges.

Second, environment laws and policies give particular meanings to abstract notions of cleanliness, protection and conservation. In doing so, they are status quoist and propagate existing regimes of human degradation and exploitation. As a result, environmental laws suffer a social legitimation crisis due to the number of people towards whom it is blind and, worse, hostile. There are inherent racial, class, caste and gender biases in the design and enforcement of Indian environmental laws.[62] There is a need for more studies in India to empirically show how nuisance-causing projects are located in areas used by poor migrant and minority communities. These communities are also often seen as the cause of nuisance, and legal actions are taken that affect their basic rights to life and work.

---

[61] Philippe Cullet, 'Governing Water to Foster Equity and Conservation: Need for New Legal Instruments', *Economic and Political Weekly* 51, no. 53 (31 December 2016).

[62] James K. Boyce and Aseem Shrivastava, 'Delhi's Air Pollution Is a Classic Case of Environmental Injustice', *The Guardian*, 9 March 2016; *The Wire*, 'In "Kodaikanal Still Won't," Artists Call Out Unilever's "Environmental Racism"', 29 June 2018.

Environmental laws fail to tackle economic sectors that promote environmentally damaging reforms and projects and which dismantle local modes of production like shifting cultivation and forest collections. In these systems of community labour, it is not property ownership and growing profit margins but sustainability and regeneration that is the basis of success. Environment law reform should aim to provide legal protections to such systems instead of perpetuating the economic rationale of exploiting nature and intrusions on communities who have a better 'balance' with nature. Another area where environment laws should seek to intervene effectively is to free precarious labour from debilitating work. The occupational health of workers in ginning mills, quarries, asbestos manufacture, ship-breaking and many others must become the centre of environment laws.

The third challenge for environment law reforms stems from the need to build a legal framework that devolves environmental rights and duties to communities. As this volume shows, environment laws operate top-down and are largely suited to the political economy of large-scale, non-inclusive corporate-led development. Decades of this orientation of governance models have left not only India or South Asia but even countries in the West with big environmental challenges in their hands. We now experience environmental crises, such as the explosion of plastics and pesticides, air and water pollution, genetically modified organisms, species extinction and climate change, on a planetary scale. Corporate-backed science trumps public health science in the regulation of industries producing environmentally damaging products. Globally, large corporations, with high stakes in the fossil fuel and other climate-damaging sectors, have had a head start with climate litigation. These industries have almost never asked for a sane regulatory system to immunise society and ecology from their operations. As Morgan states, they control regulations rather than be controlled by regulations.[63]

The present trend is to rely on technology to solve environmental problems for us. These are implicit in the way we frame questions about energy options such as nuclear and renewables or invest in 'cleaning up the air'. While technological interventions may be needed for certain challenges, the decisions to deploy technologies at scale are not the outcome of democratic decision-making. For example, the Indian government's 'ambitious' target-

---

[63] Bronwen Morgan, 'The Economization of Politics: Meta-regulation as a Form of Non-judicial Legality', *Social and Legal Studies* 12, no. 4 (2003): 489–563.

based approach to solar energy, that has received good press internationally, is practically through executive fiat. India's pursuit to generate enough renewable energy to meet the commitments of the Paris Agreement necessarily means that fundamental rights of communities who depend on the forest, agricultural and coastal commons, where wind and solar projects are mostly located, are compromised. Emissions focussed climate policy targets are also imposed without ecological thinking. Renewable energy (RE) specifically targets 'the ecological commons' that are labelled as wastelands in government records. Enviro-legal standards for RE projects are low due to its assumed environment friendliness. For example, RE projects are in the 'white category' in PCB approval procedures for regulating pollution. There is very little interaction between other environment laws and climate policies.

This volume has references to participatory forms of resource management that are yet to implemented. There were some efforts to engage in decentralised planning at state and local levels during the 1990s, but these have been kept aside by top-down climate change and environmental policies and targets. Panchayats and *nagarpalikas* have practically no powers to implement environmental laws. Environmental law reforms should aim to build a democratic and devolved system of governance that values the role of communities in crucial decisions on present and future use of environmental resources. Democratic processes, if designed well, could result in new, creative and more useful framings of the problems we now face and generate greater public ownership of the ways to deal with them. Activists and affected communities around the world are not only taking governments to court for their low climate ambitions and 'for not doing enough' but they are rethinking categories of citizenship, work, growth, development and knowledge to accommodate new conditions of living through these planetary ecological and social crises. Environmental law reforms need to tune in to these creativities to be meaningful and relevant.

1

Fundamentals of Environmental Law

This chapter focusses on the fundamental aspects of Indian environmental law. One of the most significant features of Indian environment law is that the right to life enshrined by the Indian Constitution is interpreted as inclusive of the right to a healthy environment. This interpretation evolved in the 1980s and has been used by the Supreme Court and the high courts to adjudicate complicated environmental matters. Another significant feature of Indian environmental law relates to the provisions of universal legal codes, that is, the Indian Penal Code (IPC), Criminal Procedure Code (CrPC) and Civil Procedure Code (CPC), on 'nuisance'. Indian courts have developed a vast body of case law using provisions of nuisance to address pollution, municipal solid waste generation, or obstruction of water and to seek environmental damages. The third aspect relates to international environmental conventions. Indian environmental laws have been framed to comply with India's commitments to the conventions discussed in this chapter or to advance the objectives of these conventions. Courts and governments have relied on declarations and agreements arising out of these conventions to inform and justify their decisions. The fourth aspect is the group of legal concepts and principles that are applied to Indian environmental laws and policies. Indian courts have drawn upon and interpreted the scope of these concepts and principles.

This chapter elaborates these aspects in three sections:

I.   *Constitutional Right to Life and the Environment*
II.   *Nuisance in the Indian Penal Code, Civil Procedure Code and Code of Criminal Procedure*
III. *International Conventions, Principles and Concepts of Environmental Law*

I CONSTITUTIONAL RIGHT TO LIFE AND THE ENVIRONMENT

The Constitution of India guarantees certain fundamental freedoms and rights to all citizens of India. There are a number of constitutional provisions that are central to environmental laws in India. These have been discussed in different chapters in this book. This section discusses in detail Article 21 listed in Part III of the constitution, which is one of the most widely interpreted provisions in Indian environmental law. Article 21 on the right to life states, 'No person shall be deprived of his life or personal liberty except according to procedure established by law.'

Globally, the right to life has been traditionally applied to civil-political[1] and economic-social[2] aspects of human activities. In the 1980s, the idea of a 'right to live in a healthy and clean environment' emerged in India on account of judicial processes and activism. A series of landmark cases pointed to incidents of environmental degradation and public health impacts. This led to the interpretation of the right to live in a healthy environment as being included within the fundamental right to life under the constitution. The judicial interpretation of Article 21 with respect to the environment is discussed further in this chapter. This article has also been referred to in the thematic chapters related to water, air, and the environment in this volume.

The Supreme Court first considered the question of right to live in a healthy environment in *Rural Litigation and Entitlement Kendra (RLEK) v. State of Uttar Pradesh.*[3] This case emerged from a letter of complaint that RLEK, a non-governmental organisation, sent to the Supreme Court against environmental mismanagement and loss due to mining activities. The apex court treated this complaint as a writ petition under Article 32 of the constitution, which relates to the rights of citizens to approach the Supreme Court of India seeking remedies for enforcement of rights. The Court considered the adverse environmental impacts on soil, water and land and held that people's right to live with minimal ecological disturbance shall be protected. It ordered for the

---

[1]   UN General Assembly, International Covenant on Civil and Political Rights, 16 December 1966, United Nations, Treaty Series, vol. 999, p. 171.

[2]   UN General Assembly, International Covenant on Economic, Social and Cultural Rights, 16 December 1966, United Nations, Treaty Series, vol. 993, p. 3.

[3]   Supreme Court judgment dated 12 March 1985 in Writ Petition Nos. 8209 and 8821 of 1983 (*Rural Litigation and Entitlement Kendra [RLEK] v. State of Uttar Pradesh*).

closure and due phasing out of mining activities in Doon Valley that was seen as responsible for the environmental damage. The judgment acknowledged that the mine workers would be adversely affected by the temporary and final closure of the quarries. The Court recommended that these workers 'as far as practicable and in the shortest possible time, be provided employment in the afforestation and soil conservation programme to be taken up in this area'. While the Court highlighted a new aspect to the right to life in this judgment, it did not expressly mention its legal basis or character.

The judicial interpretation of making Article 21 applicable in environmental matters evolved from a series of public interest writ petitions filed in person by environmentalist and lawyer M. C. Mehta in the 1980s. A set of three cases was filed in 1985 concerning the operations of Shriram Foods and Fertilizer Industries in New Delhi.[4] It is important to remember that this was soon after the Bhopal gas leak that took place on the night of 2 December, 1984. The first petition[5] sought the closure and relocation of the caustic soda and sulphuric acid plants of the company, which were operating in an area with high population density. The petition observed that the factory was engaged in the manufacture of hazardous substances that led to the death of an individual and impaired the health of several others. Even as this case was pending, there was an incident of oleum gas leak from the factory. An order for closing down the plant was given under the Factories Act, 1948. In response to the closure orders, the company filed a writ petition challenging the closure and seeking interim reopening of the plant. Another case, also filed by M. C. Mehta on behalf of the Delhi Legal Aid and Advise Board and Delhi Bar Association, sought compensation for victims of the gas leak. The Supreme Court judgment considered issues of safety of workers and nearby residents and stated that the case raises 'some seminal questions concerning the scope and ambit of Articles 21 and 32'. In its February 1986 judgment, the Court asserted that the state could regulate industries in order to protect people's right to live in a healthy environment and specified conditions for industries engaged in the production of hazardous substances. It also ordered a conditional reopening of the factory, the details of which are in the judgment. The Court asked both the petitioner

---

[4]   S. Divan and A. Rosencranz, *Environmental Law and Policy in India*, 2nd edn (New Delhi: Oxford University Press, 2005), 106–108.

[5]   As referred to in the Supreme Court judgment dated 17 February 1986 in Writ Petition No. 12739 of 1985 (*M. C. Mehta v. Union of India*).

and the company to file submissions on the issue of compensation, which were dealt with separately. On 12 December 1986, the Court ordered the 'Delhi Legal Aid and Advice Board to take up the cases of all those who claim to have suffered on account of oleum gas and to file actions on their behalf in the appropriate court for claiming compensation against Shriram'.

This was a major development in the jurisprudence of the right to a clean environment. Based on the interpretation of the right to life under Article 21 and the right to remedy under Article 32, the Court developed the principle of 'Absolute Liability', discussed in the section on principles in this chapter. It was further held that the compensation amount should correspond to the size of the industry in order to have a restraining effect on polluters. However, the Court, in these cases, did not specifically declare the right to environment as a part of Article 21. Article 21 was also interpreted in another case related to pollution caused by the discharge of liquid waste from leather tanneries in the river Ganga. The Court observed that such pollution was affecting the health of people as well as the ecology of the river.[6] While considering the possibility of economic loss, the Court directed the tanneries to set up treatment plants within six months or close operations. This case did not make a specific reference to right to life and environment; however, it broadened the scope of its enforcement to include corrective measures that private players could take to prevent further environmental damage. This case has been discussed in Chapter 7 on ground and surface water in relation to the interventions filed in the case to ensure environmental protection of the Ganga. Another writ petition highlighting severe reduction in the water quality of river Ganga was filed in the Supreme Court. The grievance here was inaction on the part of the municipal authority of Kanpur to tackle such pollution. The judgment[7] ordered the Kanpur Nagar Mahapalika (municipal corporation) to take several corrective and preventive measures, such as shifting the dairies of Kanpur away from the city, laying new sewage lines, increasing the size of existing sewers, constructing public toilets for poor people at no cost and taking action against pollution causing industries. This judgment highlighted the special role that

---

[6] Supreme Court judgment dated 1 January 1988 in Writ Petition No. 3727 of 1985 (*M. C. Mehta v. Union of India*).

[7] Supreme Court judgment dated 12 January 1988 in Writ Petition No. 1115 of 1985 (*M. C. Mehta v. Union of India*).

the state has in protecting the environment and elucidated the broad nature of its obligations. However, in this judgment too, the Court did not make specific reference to the right to life under Article 21.

These public interest litigation (PIL) cases are significant in terms of the Court's jurisprudence and development of environmental rights. However, the Supreme Court did not set a specific legal precedent with a *ratio decedendi*[8] that could be followed by lower courts. It was the high courts that specifically declared the right to live in a clean and healthy environment as a fundamental right on account of it being intrinsic to the right to life under Article 21 of the Constitution.

- In *T. Damodhar Rao v. Special Officer, Municipal Corporation of Hyderabad*,[9] the Andhra Pradesh High Court while hearing a challenge to a proposal of building a residential facility in a designated open space held that 'environmental pollution and spoilation should be regarded as amounting to violation of the Article 21'.

- In *V Lakshmipathy v. State of Karnataka*,[10] the Karnataka High Court cancelled the proposed establishment of industries in residential areas that were contrary under planning laws. Here, the court affirmed that the entitlement to clean environment is a basic human right, which shall be safeguarded to promote public interest, and recorded,

   The right to life inherent in Art. 21 of the Constitution of India does not fall short of the requirements of qualitative life which is possible only in an environment of quality. Where, on account of human agencies, the quality of air and the quality of environment are threatened or affected, the Court would not hesitate to use its innovative power within its epistolary jurisdiction to enforce and safeguard the right to life to promote public interest.

Since then, several cases have expanded the case law related to right to life that are relevant for environmental law in India.

---

[8] *Ratio decidendi* is a Latin phrase that means 'the reason for a decision'. Lower courts are required to adhere to the reasoning of higher courts in past cases.

[9] High Court of Andhra Pradesh judgment dated 20 January 1987 in Writ Petition Appeal No. 8261 of 1984 (*T. Damodhar Rao v. Special Officer, Municipal Corporation of Hyderabad*).

[10] High Court of Karnataka judgment dated 9 April 1991 in Writ Petition No. 23138 of 1980 (*V. Lakshmipathy and Others v. State of Karnataka and Others*).

- The Supreme Court, in *Chhetriye Pradushan Mukti Sangharsh Samti v. State of Uttar Pradesh*,[11] declared that the right to environment is a part of Article 21 of the constitution. This case concerned the environmental pollution caused by the activities of oil mills located near Varanasi in Uttar Pradesh.

- In *Subhash Kumar v. State of Bihar*,[12] the Court while considering the pollution and water degradation caused by discharge of effluents in the Bokaro river stated that 'the right to life enshrined in Article 21, includes the right to enjoyment of pollution free water and air for the full enjoyment of life'. This case is also discussed in Chapter 7 on ground and surface water extraction.

- In the *M. C. Mehta v. Kamal Nath*[13] group of cases concerning illegal mining and construction activity in sensitive ecological areas of Madhya Pradesh, the Supreme Court stated that all forms of environmental disturbance are hazardous to life in the context of Article 21 of the constitution. These are also discussed in this chapter under 'Public Trust Doctrine'.

The right to live in a healthy environment as part of Article 21 has also been invoked in *Indian Council for Enviro-Legal Action v. Union of India* (1996),[14] *Vellore Citizens Welfare Forum v. Union of India* (1996),[15] *Andhra Pradesh Pollution Control Board v. M. V. Nayudu* (1999)[16] and *Narmada Bachao Andolan v. Union of India* (2000).[17] These cases are well established in most environment law publications and, therefore, are not elaborated here.[18] The

---

[11] Supreme Court judgment dated 13 August 1990 in Writ Petition (Civil) No. 577 of 1988 (*Chhetriye Pradushan Mukti Sangharsh Samti v. State of Uttar Pradesh*).

[12] Supreme Court judgment dated 21 September 1991 in Writ Petition (Civil) No. 381 of 1988 (*Subhash Kumar v. State of Bihar*).

[13] Supreme Court judgment dated 13 December 1996 in Writ Petition (Civil) No. 182 of 1996 (*M. C. Mehta v. Kamal Nath & Ors*).

[14] Supreme Court judgment dated 18 April 1996 in Writ Petition No. 664 of 1993 (*Indian Council for Enviro-Legal Action v. Union of India*).

[15] Supreme Court judgment dated 28 August 1996 in Writ Petition (Civil) No. 914 of 991 (*Vellore Citizens Welfare Forum v. Union of India & Ors*).

[16] Supreme Court judgment dated 1 December 2000 in Civil Appeal Nos. 368–71, 372 and 373 of 1999 (*A. P. Pollution Control Board v. Prof. M. V. Nayudu [Retd.] and Ors*).

[17] Supreme Court judgment dated 1 December 2000 in Writ Petition (Civil) No. 319 of 1994 (*Narmada Bachao Andolan v. Union of India*).

[18] Divan and Rosencranz, *Environmental Law and Policy in India*; P. Leelakrishnan, *Environmental Law in India*, 3rd edn (New Delhi: Lexis Nexis, 2008).

thematic chapters in this book discuss more recent cases where Article 21 and its interpretation have been invoked.

II   NUISANCE IN THE INDIAN PENAL CODE, CIVIL PROCEDURE CODE AND CODE OF CRIMINAL PROCEDURE

'Nuisance' refers to an action or consequence, which unlawfully causes hurt, inconvenience, or damage. In Indian law, nuisance includes obstruction, danger, annoyance and risk.[19] The legal provisions related to nuisance came into existence much before specialised environmental legislations and have been available for addressing common occurrences of water, air, land and noise pollution. The role of Indian lower courts in environmental matters is established primarily through the concept of nuisance. Several high courts have also adjudicated on environmental matters interpreting environmental damage or impacts on human health as nuisance. The legal provisions provide the following definitions:

*   *Private nuisance* is an unlawful interference with another person's use of his land or another right in connection with it. It specifically relates to injuries concerning property such as wrongful disturbances, including obstructing light and air or escape of unwanted substances like smoke, fumes, gas or even vegetation and animals into another person's property.

*   *Public nuisance* implies causing interference in the exercise of a common right held by the public in general.[20] A person's conduct becomes public nuisance when the consequences of his actions are no longer confined to his own property but spread across to another person's property.[21]

A 1981 judgment of the Allahabad High Court distinguishes private nuisance from public nuisance and discusses some general principles to include trivial nuisance and substantial nuisance. It says,[22]

---

[19] Section 268 of IPC, Section 133 of CrPC and Section 91 of CPC.

[20] Bombay High Court judgment dated 11 February 1994 in Criminal Appeal No. 133 of 1994 (*Vasant Manga Nikumba v. Baburao Bhikanna Naidu*).

[21] A 1957 judgment from England and Wales Court of Appeal (EWCA) defined the 'sphere of nuisance' to be 'the neighbourhood'. The judgment concluded that it is not necessary for every member of a community to be seriously affected in order to bring a claim for public nuisance. It is sufficient that a representative cross section of the class or public is affected to bring an action. (The England and Wales Court of Appeal on 15 March 1958 in *Attorney General v. PYA Quarries Ltd.* [1958] EWCA Civ 1.)

[22] Allahabad High Court judgment dated 2 September 1981 in Second Appeal No. 690 of 1970 (*Ram Baj Singh [Dr] v. Babulal*).

person is ordinarily entitled to do any thing on his own property provided that doing of such a thing is lawful. His conduct, however, becomes a private nuisance when the consequence of his acts no longer remain confined to his own property but spill over in a substantial manner to the property belonging to another person. However, any thing done by a person on his property, repercussions of which are felt on the neighbour's land, may not always be a nuisance. The consequences of any thing done by the owner of a land on his own land which are also felt over the neighbouring land may be of such a trivial nature that no reasonable person would object to the same. No precise or universal formula has been devised to determine the distinction between a trivial consequence of an act or a consequence which can be termed to be of substantial magnitude. The test which has always found to be useful in distinguishing the two sets of cases is the test of ascertaining the reaction of a reasonable person according to the ordinary usage of mankind living in a particular society in respect of the thing complained of.

The Supreme Court has interpreted the power of the magistrate in matters of public nuisance as a 'public duty'[23]

the public power of the magistrate under the Code is a public duty to the members of the public who are victims of the nuisance, so he shall exercise it when the jurisdictional facts are present.

INDIAN PENAL CODE AND NUISANCE

The IPC is a substantive law that lays down the classification of offences along with their punishment. It is the basis for the criminal procedure machinery to swing into action. Nuisance constitutes a legal injury, or a violation of rights, which attracts several remedies including damages, injunction[24] and others[25] or punishment including fine and imprisonment.

---

[23] Supreme Court judgment dated 22 September 2004 in Appeal (Criminal) 1350 of 2003 (*Kachrulal Bhagirath Agrawal & Ors v. State of Maharashtra & Ors*).

[24] Injunction is a judicial or court order to restrain any person from (*a*) initiating or continuing an action, (*b*) invading the legal right of another and (*c*) compelling or threatening another person to carry out an act.

[25] Other forms of civil remedies for nuisance include decree, declaration, interim orders and abatement.

The provisions concerning nuisance in the IPC deviate from the standard rule of criminal law that both intention (*mens rea*) and act (*actus reus*) are necessary to constitute a crime.[26] This rule is not rigid, as a statute can deal with issues that may not necessarily require a corresponding intention to make an act criminal. This can be understood well in the context of acts amounting to nuisance. For example, a factory which discharges waste in a local stream of water or stores dangerous chemicals in a residential area may not be doing so with any specific intention to damage the health of neighbouring communities or to harm the environment; nevertheless, their action would attract criminal prosecution and punishment.

The criminal provisions concerning nuisance do not require a large or specific number of complainants as a prerequisite for acting on environmental pollution. The continuance of public nuisance after injunction or order is punishable with simple imprisonment for a term, which may extend to six months, or with a fine, or with both (Section 291 of IPC). Chapter 14 of the IPC concerns 'Offences affecting Public Health, Safety, Convenience, Decency and Morals'. Section 268 [27] provides for 'Public Nuisance', which lists three distinct categories of public, at least one of which should be affected by an act amounting to public nuisance:

- the public
- people who dwell or occupy property in the vicinity and
- persons who may use any public right on occasions

This classification accommodates the many ways in which people exercise public rights. An important case for understanding this distinction is *Krishna Gopal v. State of MP*.[28]

---

[26] 'Actus non-facit reum nisi mens sit rea' means 'An act does not make anyone guilty unless there is a criminal intent or a guilty mind', as discussed on http://lawtimesjournal.in/actus-non-facit-reum-nisi-mens-sit-rea/, accessed on 10 February 2020.

[27] Section 268 of IPC states,

A person is guilty of public nuisance who does any act or is guilty of any illegal omission which causes any common injury, danger or annoyance to the public or the people in general who dwell or occupy property in the vicinity, or which much necessarily cause injury, obstruction, danger or annoyance to persons who may have occasion to use any public right. A common nuisance is not excused on the ground that it causes some convenience or advantage.

[28] High Court of Madhya Pradesh judgment dated 9 December 1985 in Civil Revision Petition No. 1326 of 1984 (*Krishna Gopal v. State of MP*).

As per these legal definitions in the IPC, actions committed by anyone, including any company or association, can be held accountable for creating nuisance. Any group of people, including those living in the same residential area, can be understood as public affected by nuisance. Chapter 14 of the IPC also lists other provisions, which impact public health, safety and convenience, which complement Section 268 in penalising different instances of public nuisance. These include

- negligent (Section 269, IPC) and malignant (Section 270, IPC) acts likely to spread infection of disease dangerous to life;

- contaminating water of public springs or reservoirs to render it unfit for ordinary use (Section 277, IPC);

- acts making atmosphere noxious to health (Section 278, IPC);

- acts causing danger or obstruction in public way (Section 283, IPC) and

- negligent conduct with respect to poisonous substances (Section 284, IPC), fire or combustible matter (Section 285, IPC), explosive substances (Section 286, IPC) and machinery (Section 287, IPC) which endanger human life or are likely to cause injury to others.

In 1999, a writ petition was filed in the Kerala High Court[29] seeking action against tobacco smoking in public. The court analysed this issue and observed that public smoking amounted to public nuisance as per the provisions of Section 268 of the IPC and that it satisfied the definition of 'air pollution' as provided in the Air (Prevention and Control of Pollution) Act, 1981 (see Chapter 4). This led to a statewide ban on public smoking in Kerala. Soon after this judgment, the Tamil Nadu Prohibition of Smoking and Spitting Act, 2002, was enacted. This law banned tobacco smoking and chewing in public premises and their advertisements.

CRIMINAL PROCEDURE CODE AND NUISANCE

Part B of Chapter 10 of the CrPC provides the procedure for judicial action to address public nuisance. Sections 133–143 specifically provide detailed instructions and powers to a magistrate[30] for dealing with public nuisance

---

[29] High Court of Kerala judgment dated 12 July 1999 in O.P. No. 24160 of 1998 (*K. Ramakrishnan and Anr. v. State of Kerala and Ors*).

[30] District magistrate or a subdivisional magistrate or any other executive magistrate, specially empowered in this behalf by the state government.

in cases of emergencies which pose an imminent danger to public health and safety. The magistrate is empowered to make a conditional order for the removal of existing public nuisances, including those affecting the local environment such as open drains, lack of public sanitation facilities and harmful fumes from factories, amongst others.

The scope of Section 133 as a tool of pollution control has been made clear through its use by the judiciary. The courts, even prior to India's independence, have applied the provisions of public nuisance to pre-independence cases relating to water[31] and noise[32] pollution, which have broadened and shaped its relevance as a remedy for environmental protection. It is, however, important to note that the determination of substantial injury is always based on the standard of a reasonable and sober mind.[33] In a case regarding a bakery that was releasing smoke from its chimney and affecting the health of people in the vicinity, the magistrate's order under Section 133 had directed the bakery to destroy the oven and the chimney and later required it to close its business on the site. The Supreme Court in its judgment of 1978 upheld this order.[34]

A Supreme Court judgment in 1980 widened the scope of Section 133's jurisdiction and created guidelines for determining accountability of local bodies and private agencies.[35] This is considered to be a landmark decision about the character and scope of Section 133 as a remedy for environmental protection and its applicability on statutory bodies. In this case, the subdivisional magistrate had directed the municipality of Ratlam, under Section 133, to construct drainage and sanitation facilities for the public within six months. The municipality challenged the order on the grounds of financial limitations and approached the Supreme Court. The apex court asserted that the power of the magistrate under Section 133 is a public duty to members of the public who are victims of nuisance and that the municipality cannot take the plea of financial inability to exonerate itself from statutory liability.

---

[31] Patna High Court judgment dated 25 January 1925 in Criminal Revisions Nos. 534 and 540 of 1924 (*Deshi Sugar Mills v. Tupsi Kahar*).

[32] Allahabad High Court judgment dated 16 March 1931 in Criminal Revision No. 828 of 1930 (*Emperor v. Raghunandan Prasad*).

[33] Allahabad High Court judgment dated 2 September 1981 in Criminal Revision No. 828 of 1930 (*Ram Baj Singh [Dr] v. Babulal*).

[34] Supreme Court judgment dated 15 September 1978 in Criminal Appeal No. 59 of 1973 (*Gobind Singh v. Shanti Swaroop*).

[35] Supreme Court judgment dated 29 July 1980 in Special Leave Petition (Criminal) No. 2856 of 1979 (*Municipal Council, Ratlam v. Vardhichand*).

An interesting issue concerning the application of Section 133 of the CrPC is its relationship with the Water (Prevention and Control of Pollution) Act, 1974, and the Air (Prevention and Control of Pollution) Act, 1981 (discussed in Chapter 4). These laws give statutory powers to judicial magistrates to issue fines or direct imprisonment while taking cognizance of offences issued under the Air or Water Acts. In such cases the magistrates do not need to rely on Section 133. Parallel laws for the same or similar subjects can create legal uncertainty regarding their validity and application. In this context, in the 1980s, it was considered that the Water and the Air Acts had a limiting effect on the magistrate's powers under Section 133 to deal with pollution. A 2003 judgment of the Supreme Court of India[36] addressed this issue. The Court ruling was based on three questions:

> (1) Whether there is direct conflict between the two provisions. (2) Whether the Legislature intended to lay down an exhaustive Code in respect of the subject-matter replacing the earlier law. (3) Whether the two laws occupy the same field.

The apex court was of the view that the areas of operation of the CrPC and the pollution laws were different with distinct aims and objectives. While both of these laws addressed nuisance, they were not of an identical nature. The judgment concluded:

> While as noted above the provisions of Section 133 of the Code are in the nature of preventive measures, the provisions contained in the two Acts are not only curative but also preventive and penal. The provisions appear to be mutually exclusive and the question of one replacing the other does not arise. Above being the position, the High Court was not justified in holding that there was any implied repeal of Section 133 of the Code.

Section 144 of the CrPC deals with cases of 'urgent nuisance', which empowers the magistrate to make orders for removing and apprehending dangers of public nuisance. In the interest of providing a quick remedy in case of emergencies, the absolute order under Section 144 can be issued *ex-parte*, that is,

---

[36] Supreme Court judgment dated 19 August 2003 in Appeal (Criminal) No. 151–158 of 1996 (*State of Madhya Pradesh v. Kedia Leather & Liquor Ltd & Ors*).

it does not require both sides to be heard and can be made as per the legal interest of just the aggrieved person. Such orders are temporary in nature and cannot remain in force for more than two months. However, if the state government considers it necessary for preventing nuisance, it may, by notification, direct that the magistrate's order under Section 144 shall remain in force for a further period not exceeding six months from the date it was passed.

CIVIL PROCEDURE CODE AND NUISANCE

Public Nuisance in Section 91 of the CPC is understood to be the same as given in Section 268 of the IPC. It deals with nuisance that affects a considerable number of people. For applying the provisions of this section, mere proof of obstruction, danger or wrongful omission[37] is not sufficient. It should also be proved that such obstruction, danger or omission constitutes a public nuisance.[38] However, the plaintiff does not need to prove that he has suffered any injury.[39]

Section 91 provides the procedure for claiming remedies. The provision in Section 91 of CPC provides for class action[40] against environmental violations; however, it is not widely used. The Section lays down a process where relief can be sought for 'public nuisances and other wrongful acts affecting the public'. Such cases can be brought to court either by the Advocate-General or 'by two or more persons, even though no special damage has been caused to such persons by reason of such public nuisance or other wrongful act'. These cases can lead to remedies in the form of declaration or injunctions. While the Court allows for adding another person to the suit and the subsequent order is binding on all those on whose behalf the suit is instituted, it is not necessary to show that they have the same cause of action. It is enough to show that the persons who are suing have the same interest as others.[41]

---

[37] Punjab and Haryana High Court judgment dated 16 September 1985 in Civil Revision. No. 1898 of 1985 (*Union of India v. Ishwar Pal Attri*).

[38] S. Shanthakumar, *Introduction to Environmental Law*, 2nd edn (New Delhi: Lexis Nexis, 2009), 319.

[39] Supreme Court of India judgment dated 22 September 2004 in Appeal (Criminal.) 1350 of 2003 (*Kachrulal Bhagirath Agrawal & Ors v. State of Maharashtra & Ors*).

[40] Proceedings brought to court by one or more petitioners on behalf of a bigger group.

[41] Leelakrishnan, *Environmental Law in India*, 16.

Public nuisance in the IPC, CPC and CrPC has been used in Indian environmental law. The use of the relevant clauses has created a tested framework of remedies for environmental protection. A person can have a private right of action for special damages against a public nuisance where the damage suffered by him/her is 'particular', beyond what the general public has suffered, 'direct', not a mere consequence of nuisance, and of a 'substantial' character.[42]

III INTERNATIONAL CONVENTIONS, PRINCIPLES AND CONCEPTS IN ENVIRONMENTAL LAW

This section outlines the international environmental conventions that have shaped environmental laws in India. These conventions were forums for the discussion of environmental challenges, to develop conceptual and legal tools to deal with them as well as to share knowledge and resources on building environmental institutions. Indian environment law has been influenced by and has contributed to these international conventions. Most of these conventions have been held under the aegis of the United Nations, the intergovernmental organisation formed in 1945 to discuss common global challenges and manage shared responsibilities.[43] These conventions work through a system of protocols and agreements that focus on specific actions. For instance, the Kyoto Protocol of the United Nations Framework Convention on Climate Change (UNFCCC) is about reducing carbon emissions, while the Nagoya Protocol of the Convention on Biological Diversity (CBD), 1992, is about operationalizing access and benefit sharing. Every convention has a secretariat that works with subsidiary bodies and working groups, which negotiate the commitments and points for collective action before it is presented at a conference of parties (CoP) that is convened every two years.

In this section, principles and concepts that have now become an established part of Indian environmental jurisprudence are also discussed.[44] The legal principles that are discussed in this chapter are precautionary principle, polluter pays principle, strict and absolute liability, and public trust doctrine. This section

---

[42] Shanthakumar, *Introduction to Environmental Law*, 317.

[43] The United Nations has 193 signatory countries including India. More can be available at https://www.un.org/en/.

[44] Shibani Ghosh, *Indian Environmental Law: Key Concepts and Principles* (New Delhi: Orient BlackSwan, 2019), 4.

also discusses three concepts that have emerged from international conventions: sustainable development, public participation, and free, prior and informed consent (FPIC). The judiciary and the executive have used these principles and concepts extensively in the practice of environmental law in India.

INTERNATIONAL CONVENTIONS AND AGREEMENTS

- Ramsar Convention, 1971: This convention is one of the oldest international agreements related to the environment. It deals with reducing the loss and degradation of wetlands and taking proactive steps for their conservation. The convention was negotiated through the 1960s and adopted in Ramsar, Iran, in 1971. It came into force in 1975. India's legal framework for conservation and protection of wetlands, including the 2017 regulations, draw extensively from the Ramsar Convention.

- Stockholm Conference, 1972: The United Nations Conference on Human Environment is also known as the Stockholm Conference. This conference was held in June 1972 in Stockholm, Sweden, and is considered to be a watershed event that sparked environmental law-making in many parts of the world.[45] The conference outcomes include a 9-point declaration and 21 principles, which are used extensively in environmental laws, policies and in courts. The declaration recognised that the protection and improvement of the environment was a major issue which affected both human well-being and economic development. Another significant aspect of this conference was its emphasis on population as a significant contributor to environment degradation. Addressing pollution and toxic contamination was one key aspect of the declaration. The convention is considered to have laid the foundation for the institutionalisation of environmental protection in India. This includes the setting up of the environment ministry and the enactment of several laws related to pollution control and environment protection in the 1970s and 1980s that are discussed in this book.

- Convention on International Trade in Endangered Species (CITES) of Wild Fauna and Flora, 1973: The convention was signed in Washington, DC, United States of America. Since then, it has provided the framework for all signatory countries to regulate 'international trade in specimens of wild animals and plants', reduce threats and take measures for their survival. There are 183 parties to the CITES agreement, including India. India's regulatory framework for preventing wildlife trade is directly influenced

---

[45] Declaration of the United Nations Conference on the Human Environment held in Stockholm, Sweden, from 5 to 16 June 1972.

by the CITES requirement. The convention has three appendices, which lists species that are threatened by international trade, with the objective of giving them special conservation attention and preventing overexploitation. This includes 5,800 species of animals and 30,000 species of plants.

- Basel Convention, 1989: The Basel Convention relates to the control of 'toxic trade' or the transboundary movement of hazardous waste and disposal. It was adopted in March 1989 and came into force in May 1992. This issue was introduced as one of the three priority areas of the United Nations Environment Program (UNEP) in the 1980s. The scope of the convention covers several important issues for India including management of plastic,[46] persistent organic pollutants (POP) waste, mercury and ship-breaking activity. The Hazardous and Other Wastes (Management and Transboundary Movement) Rules, 2016, discussed in Chapter 5 draws on the commitments to this convention while setting regulations on importing and exporting hazardous waste.[47]

- Rio Earth Summit, 1992 and 2002: The United Nations Conference on Environment and Development (UNCED) and its declaration reaffirmed the Stockholm Declaration[48] and called for a 'new and equitable global partnership'. It was here that the concept of sustainable development was introduced to the global environmental discourses. The summit was held at a time when most parts of the world were opening themselves to economic liberalisation. Principle 7 on common but differentiated responsibility (CBDR) and Principle 10 on public participation were two important principles which have influenced the global practice of environmental law and climate change negotiations. In 2002, all parties of the Earth Summit reconvened to take stock of how the summit's intention had translated into practical steps. The most significant outcome of this meeting was the introduction of the 17 Sustainable Development Goals (SDGs) for the international commitments to 'end poverty, protect the planet and ensure that all people enjoy peace and prosperity'.

---

[46] Parul Kumar and Sridhar Lekha, 'Basel Convention's Plastic Ban Amendment: A New Step against Waste Colonialism', *The Wire*, 21 May 2019 available at https://thewire.in/environment/basel-conventions-plastic-ban-amendment-is-a-new-step-against-waste-colonialism, accessed on 30 September 2020.

[47] Convention text and protocols of liability, available at https://www.basel.int/Portals/4/Basel%20Convention/docs/text/BaselConventionText-e.pdf, accessed on 27 August 2019.

[48] The Rio Declaration on Environment and Development, 1992, finalised at the UNCED, held at Rio de Janeiro from 3 to 14 June 1992.

The Rio Summit, 1992, led to three framework conventions: United Nations CBD, UNFCCC and United Nations Convention to Combat Desertification (UNCCD). These are discussed as follows.

India became a signatory to the CBD, 1992, in 1993 and ratified it in February 1994. The enactment of the Biological Diversity (BD) Act, 2002, was part of India's commitment to the CBD. The discussions at the convention led to the development of two international protocols. These are the Nagoya Protocol on Access and Benefit Sharing of biological resources and associated knowledge as well as the Cartagena Protocol on Biosafety related to research and use of genetically modified organisms (GMOs).

The UNFCCC was also signed in 1992. It is the guiding agreement for all parties on actions to be taken for adapting to existing impacts of climate change and mitigating future degradation. The two significant treaties of the UNFCCC are the Kyoto Protocol and the Paris Agreement. The Kyoto Protocol was signed in 1997 and bound all signatories to targeted reduction in greenhouse gas (GHG) emissions responsible for global warming. The Paris Agreement was signed in 2015 and requires all signatories to present nationally determined contributions (NDCs) to keep the global temperature rise 'below 2 degrees Celsius above pre-industrial levels and to pursue efforts to limit the temperature increase even further to 1.5 degrees Celsius'. It aims to increase financial flows, new technology frameworks and capacity building to support this.

The UNCCD, 1994, is a legally binding agreement between parties for 'sustainable land management'. It focuses on arid, semi-arid and dry sub-humid areas together understood as dry lands. India became a signatory in 1994 and finally ratified the convention in December 1996. The focus on dry land and rainfed areas in the National Water Policy and the National Forest Policy (discussed in Chapters 7 and 3) is also a response to the commitments towards this convention. Several other union ministries and state government departments such as rural development and agriculture have also taken measures in response to this convention.

- Convention on Persistent Organic Pollutants, 2001: This convention deals with regulating the release of POPs through industrial processes, in agriculture and in consumer products. POPs are carbon-based chemical substances that once released are toxic to both human beings and wildlife. They can contaminate soil, water, and air and impact health. The discussions to identify and regulate 12 POPs[49] began in 1995. It was only in 2001 that this convention was adopted. It came into force in 2004, after 152 countries ratified it. India is yet to ensure curbs on the use of some of these pesticides or industrial chemicals.[50]

PRINCIPLES AND CONCEPTS USED IN ENVIRONMENT LAW

Environmental principles are a critical aspect of the development of environment laws in India and they have had a significant role in giving shape to environmental rights in India. These are used in concretising the court's rulings, in developing national legal frameworks, developing policy priorities and in ensuring proper implementation of environment laws. Indian courts have used international environmental principles and approaches and evolved new concepts and principles to respond to domestic challenges in environmental policy and legal adjudication. Since 2010, the National Green Tribunal (NGT) is required to apply the precautionary principle, the polluter pays principle, and sustainable development 'while passing any order, decision or making the award'.[51]

This section discusses a total of seven principles and concepts that have been used in environmental law.

Precautionary Principle

The definition of precaution can be traced back to Principle 15 of the Rio Declaration. The declaration provides for an approach considering the following:

> In order to protect the environment, the precautionary approach shall be widely applied by States according to their capabilities. Where there are

---

[49] These POPs are *pesticides* such as aldrin, chlordane, DDT, dieldrin, endrin, heptachlor, hexachlorobenzene, mirex and toxaphene; *industrial chemicals* such as hexachlorobenzene, polychlorinated biphenyls (PCBs); and *byproducts* such as hexachlorobenzene, polychlorinated dibenzo-*p*-dioxins and polychlorinated dibenzofurans (PCDD/PCDF), and PCBs, available at http://chm.pops.int/TheConvention/ThePOPs/The12InitialPOPs/tabid/296/Default.aspx, accessed on 25 August 2019.

[50] 'Toxics Link Report: POPs in India 2018', International Pollutants Elimination Network (IPEN), 2019, available at https://ipen.org/news/toxics-link-report-pops-india-2018, accessed on 25 August 2019.

[51] Section 20 of the National Green Tribunal (NGT) Act, 2010.

threats of serious or irreversible damage, lack of full scientific certainty shall not be used as a reason for postponing cost-effective measures to prevent environmental degradation.

Since then, this principle has been reiterated in several international conventions, interpreted in judgments of the high courts and the Supreme Court and also emphasised in government policy documents.[52] One of the earliest interpretations of this principle has been discussed in a 1996 judgment of the Supreme Court (also mentioned in Chapter 4 on pollution control and prevention). Here, the Court interpreted the precautionary principle to have three components. First, environmental measures by state governments and statutory authorities should 'anticipate and prevent and attack the causes of environment degradation'. Second, lack of scientific certainty should not be used as a reason to postpone measures to prevent serious and irreversible damage. Third, in such cases, the onus of proof is on the actor or a developer 'to show that his action is environmentally benign'.[53]

Several environmental matters decided by the courts have invoked the precautionary principle, including the 2019 case challenging the construction of a coastal road in Mumbai with two components on the northern and southern sides of the city. The project proponents had carried out two separate environment impact assessments for the two parts of the road. In its judgment,[54] the Bombay High Court concluded that, conceptually and factually, the two roads were one project. Therefore, to 'artificially break the project into two components merely because there was a Sea-link in between' would result in a 'truncated environment impact assessment' and not uphold the precautionary principle as developed in environmental jurisprudence.

---

[52] Rajamani Lavanya, 'The Precautionary Principle', in *Indian Environmental Law: Key Concepts and Principles*, ed. Shibani Ghosh (New Delhi: Orient BlackSwan, 2019), 192–229.

[53] Supreme Court of India judgment dated 18 April 1996 in Writ Petition No. 664 of 1993 (*Indian Council for Enviro-Legal Action and the Union of India*).

[54] High Court of Bombay judgment dated 16 July 2019 in Writ Petition (Civil) No. 560 of 2019 (*Worli Koliwada Nakhwa & Anr v. Municipal Corporation of Greater Mumbai & Ors*); Public Interest Litigation (L) No. 39 of 2019 (*Vanashakti & Anr v. National Board for Wildlife & Ors*); Public Interest Litigation (L) No. 44 of 2019 (*Shweta Wagh & Ors v. Municipal Corporation of Greater Mumbai & Ors*).

The National Environment Policy, 2006,[55] views the precautionary approach as one of its guiding principles. However, it limits the realisation of the principle within an economic argument. It suggests the need for economic efficiency as a means for integrating environment into the planning of other sectors of the government such as power or mining.[56]

Polluter Pays Principle

This principle in environmental law practice evolved with the need for measures required to deal with chronic pollution. According to Lovleen Bhullar, the polluter pays principle can be traced back to the economic theory of externalities where pollution, environmental harm or damage is not accounted for while determining the market price of goods and services produced. This principle was defined in 1972 by the Organisation for Economic Cooperation and Development.[57] The definition accepted three aspects: it agreed that a cost needs to be allocated for the harm, damage or contamination; it required that this cost needs to be borne by the polluter and that this cost needs to be calculated by a public authority.[58]

It was subsequently embodied in Principle 16 of the Rio Declaration as follows:

National authorities should endeavour to promote the internalization of environmental costs and the use of economic instruments, taking into account the approach that the polluter should, in principle, bear the cost of pollution, with due regard to the public interest and without distorting international trade and investment.

---

[55] Ministry of Environment and Forests, *National Environment Policy* (New Delhi: Government of India, 2019).

[56] Ministry of Environment, Forests and Climate Change, *Report High Level Committee to Review Various Acts Administered* (New Delhi: Government of India, 2014), available at http://www.indiaenvironmentportal.org.in/files/file/Final_Report_of_HLC_0.pdf, accessed on 4 April 2020.

[57] Lovleen Bhullar, 'The Polluter Pays Principle', in *Indian Environmental Law: Key Concepts and Principles*, ed. Shibani Ghosh (New Delhi: Orient BlackSwan, 2019), 152–191.

[58] European Commission, 'Workshop on EU Legislation: Principles of EU Environmental Law', presentation, available at https://ec.europa.eu/environment/legal/law/pdf/principles/2%20Polluter%20Pays%20Principle_revised.pdf, accessed on 27 August 2019.

This principle has also been used in court judgments. One of the first judicial interpretations of this principle in India was in a Supreme Court judgment of 1996 related to the contamination of water and soil of Bicchri village in Udaipur district of Rajasthan. The judgment recorded the polluter pays principle as 'accepted universally as a sound principle'.[59] The judicial response and remedies related to nuisance discussed in the previous section have also drawn substantially from the polluter pays principle. The principle has also been used in environmental policy-making in India, including in the National Environment Policy, 2006, and the National Water Policy, 2002, where the concept of pricing the use of water was introduced. The formation of various funds for penal afforestation, water cess and bank guarantees that are discussed in different chapters of the book exemplify the government's understanding of the polluter pays principle. The recent reforms on environmental laws also substantially rely on the polluter pays principle for their recommendations. The reforms proposed by a high-level committee, discussed in Chapter 9, suggest the setting up of an Environment Reconstruction Fund comprising of monies collected from polluters and offenders of environment law.[60]

It has also been the basis for the offset and compensatory measures discussed as part of the international negotiations on climate change. Several international policy measures, including carbon pricing, cap and trade mechanism, and reducing emissions from deforestation and degradation are based on the polluter pays principle. These have allowed polluting operations to offset their GHG emissions by supporting afforestation programmes or investing in non-polluting infrastructure and energy generation.[61] While these mechanisms have been reviewed and critiqued,[62] they continue to be part of India's national policies and responses to climate change which are discussed in Chapters 3 and 10.

---

[59] Supreme Court judgment dated 13 February 1996 in Writ Petition (C) Nos. 967 of 1989, 94 of 1990, 824 of 1993 and 76 of 1994 (*Indian Council for Enviro-Legal Action and Ors v. Union of India [UOI] and Ors*).

[60] Ministry of Environment, Forests and Climate Change, *Report High Level Committee*, 82.

[61] 'Carbon Pricing', Carbon Market Watch, available at https://carbonmarketwatch.org/our-work/carbon-pricing/, accessed on 27 August 2019.

[62] Jutta Kill, *Economic Valuation and Payment for Environmental Services: Recognizing Nature's Value of Pricing Nature's Destruction* (Berlin: Heinrich Böll Stiftung, 2015).

Strict Liability and Absolute Liability

The principle of strict liability emerged from Common Law of Torts, in an 1868 judgment of the House of Lords,[63] United Kingdom. The court held that law should not remain static and needs to 'evolve new principles and lay down new norms which would adequately deal with the new problems which arise in a highly industrialised economy'.

This judgment attributed *strict liability* to a person 'when he brings or accumulates on his land something likely to cause harm if it escapes, and damage arises as a natural consequence of its escape'. However, subsequent interpretations read down the scope of strict liability by introducing a number of exceptions such as act of god, sabotage, act of a third party, plaintiff consent, plaintiff's fault and other reasons. These interpretations justified why liability should not rest on a person or activity causing pollution or environmental harm.

In Indian environmental law, the courts found this principle limited while adjudicating on contamination and damage caused by the oleum gas leak in Delhi in December 1985. In its judgment of 1986,[64] made in the aftermath of the Bhopal gas leak, the Supreme Court observed the limitations of the traditional interpretation of strict liability. The apex court held that any hazardous enterprise poses a threat to persons both working in the factory and living in surrounding areas. Therefore, enterprises engaged in a dangerous or hazardous industry have

> absolute non-delegable duty to the community to ensure that if any harm results to anyone, the enterprise must be held to be under an obligation to provide that the hazardous or inherently dangerous activity must be conducted with the highest standards of safety.

Therefore, if any harm occurs, the enterprise 'must be absolutely liable to compensate for such harm'. This is 'irrespective of the fact that the enterprise had taken all reasonable care and that the harm occurred without any negligence on its part'.

---

[63] The United Kingdom House of Lords on 17 July 1868 in *Rylands v. Fletcher* ([1868] UKHL 1).

[64] Supreme Court of India judgment dated 17 February 1986 in Writ Petition No. 12739 of 1985 (*M. C. Mehta v. Union of India*).

The absolute liability principle was referred to in detail in a 2015 judgment[65] of the NGT. This was in a case filed against the closure orders issued by the Haryana State Pollution Control Board seeking the tribunal's intervention in ordering the board to de-seal the petitioner's premises and restore water and electricity connections. The NGT concluded that the industry was a polluting unit and had not installed pollution control devices. Therefore, the principles of strict and absolute liability as well as the polluter pays principle were applicable. It ordered the applicant to pay ₹0.2 million to the pollution control board (PCB) for 'improving the pollution of groundwater, environment around the areas of the applicant'. With this, the company was allowed to operate for two months during which it would install all pollution control devices, and the state and central pollution control board would inspect the premises jointly. The PCBs would prepare a comprehensive report and place it before the tribunal.

Public Trust Doctrine

The history of this principle goes back to Roman civil law codified in what is known as *Corpus Juris Civilis*, a compilation of ancient Roman law and codes. It states, 'By the law of nature these things are common to all mankind, the air, running water, the sea and consequently the shores of the sea.'

Legal scholars have emphasised that this principle laid the grounds for what is understood as the public trust doctrine today.[66] In environmental law, this principle has been interpreted to say that the state or the governments have custodianship of water, air, land and other natural resources in public trust rather than as private property. The Chawla committee[67] on natural resource

---

[65] National Green Tribunal judgment dated 10 November 2015 in Original Application No. 419 of 2015 (*Sterling Home Innovator Private Limited v. Haryana Pollution Control Board*).

[66] David Takacs, 'The Public Trust Doctrine, Environmental Human Rights and the Future of Public Property', *New York University Environment Law Journal* (2008), available at http://www.ielrc.org/content/a0804.pdf, accessed on 27 August 2019; Joseph Chun, 'Reclaiming the Public Trust in Singapore', *Singapore Academy of Law Journal* 17 (2005): 717–746.

[67] Cabinet Secretariat, *Report of the Committee on Allocation of Natural Resources* (New Delhi: Government of India, 2011).

allocation, discussed in Chapter 9, recorded that the decisions on how to use these resources should be taken considering the following:

> Under the 'public trust' doctrine, the state is perceived of, not as owning the water resources of the country, but as holding them in trust for the people (including future generations). As a trustee, the state will of course have to be empowered to legislate, regulate, allocate, manage, and so on, and all this must involve a degree of control.

There are several court cases where the public trust doctrine has been emphasised; several of them are related to ground and surface water. One of the most significant uses can be found in the case that is popularly known as the Plachimada judgment.[68] The issue related to the cancellation of the license of a company to extract groundwater by the Perumatty gram panchayat (village council). This was done due to the excessive drawing and overexploitation of groundwater by the company, which led to the shortage of water for the villagers. The panchayat filed a writ petition when the company did not adhere to the cancellation notice. The High Court of Kerala laid down strict conditions to regulate 'excessive' withdrawal of water, which would be inspected and monitored by the panchayat and the groundwater department. Prior to this conclusion, the judgment held as follows:

> The Public Trust Doctrine primarily rests on the principle that certain resources like; sea waters and the forests have such a great importance to the people as a whole that it would be wholly unjustified to make them a Subject of private ownership. The said resources being a gift of nature, they should be made freely available to everyone irrespective of the status in life. The doctrine enjoins upon the Government to protect the resources for the enjoyment of the general public, rather than to permit their use for private ownership or commercial purposes.

This judgment relied upon a flagship case of the Supreme Court, which gives a comprehensive account of the history and evolution of the public trust doctrine in international environmental law. This 1996 judgment puts together a large body of international case law on public trust to conclude that the public trust doctrine should be 'expanded to include all eco-systems operating in our

---

[68] High Court of Kerala judgment dated 16 December 2003 in Writ Petition (Civil) No. 34292 of 2003 (*Perumatty Grama Panchayat v. State of Kerala*).

natural resources'.[69] Before this judgment, this principle was largely used for ground and surface water related issues.

Sustainable Development

The concept of sustainable development was in circulation in the international environmental discourses since 1969. One of its first iterations is attributed to the International Union for Conservation of Nature (IUCN). It was also an important theme at the Stockholm Conference, 1972, where ideas for balancing economic growth and environmental damage were in question.[70] Even before a definition for sustainable development was presented at the Rio Earth Summit, 1992, there was scholarly literature pointing to the inconsistency in its interpretation and the role of economic growth in environmental sustainability.[71]

The World Commission on Environment and Development reached a consensus definition in 1987 in what is known as the Bruntland Report.[72] Here, sustainable development was defined as 'development that meets the needs of the present without compromising the ability of future generations to meet their own needs'.

In 1992, Principle 1 of the Rio Declaration put sustainable development at the heart of the international discourse on environment and development paving the way for its use in environmental law and policy globally. Principle 1 states: 'Human beings are at the centre of concerns for sustainable development. They are entitled to a healthy and productive life in harmony with nature.' This definition did not resolve the problem for policymakers, who found the concept vague and intangible to enforce.[73] Yet the definition has been included in policies like India's National Environment Policy.

[69] Supreme Court of India judgment dated 13 December 1996 in Writ Petition (Civil) 182 of 1996 (*M. C. Mehta v. Kamal Nath & Ors*).

[70] W. M. Adams, *The Future of Sustainability: Re-thinking Environment and Development in the Twenty-First Century* (Gland: IUCN–World Conservation Union, 2006).

[71] Sharachandra Lele, 'Sustainable Development: A Critical Review', *World Development* 19, no. 6 (2006): 607–621.

[72] Brundtland Commission (formerly World Commission on Environment and Development), *Report of the World Commission on Environment and Development: Our Common Future* (Oxford: Oxford University Press, 1987), available at https://sustainabledevelopment.un.org/content/documents/5987our-common-future.pdf, accessed on 27 August 2019.

[73] Jairam Ramesh, 'The Two Cultures Revisited: The Environment–Development Debate in India', *Economic and Political Weekly* 45, no. 42 (2010): 13–16.

The concept of sustainable development has also been read with the principle of intergenerational equity which has been defined in Principle 3 of the Rio Declaration to say: 'The right to development must be fulfilled so as to equitably meet developmental and environmental needs of present and future generations.'

Since the Rio Summit defined it, sustainable development and intergenerational equity have been used in international conventions and national laws and policies. In fact, UNFCCC relies on intergenerational equity to discuss CBDR of parties to respond to climate change. This has been discussed in Chapter 10. The justifications for developing the 17 UN SDGs at the Rio+20 Conference in 2002 rose from the critique that sustainable development was too broad and had come to mean many things. The SDGs are target oriented, specific and meant to tie in with domestic laws and programmes to achieve the targets.[74]

Courts in India have also interpreted sustainable development with diverse outcomes. One of the most detailed discussions of the concept is in a 1996 judgment of the Supreme Court while referring to the Stockholm Declaration that established: 'The traditional concept that development and ecology are opposed to each of other, is no longer acceptable. "Sustainable Development is the answer."'[75] The apex court in different matters has referred to sustainable development as a constitutional requirement linking it to Article 21.[76] In other cases, courts have ruled in favour of large projects,[77] even when presented with evidence related to their environmental damage and violation of law.

Public Participation

The concept of public participation has been traced back to international civil and political rights discourses. This was with reference to participation in

---

[74] NITI Aayog, *Sustainable Development Goals (SDGs), Targets, CSS, Interventions, Nodal and Other Ministries* (New Delhi: Government of India, 2017).

[75] Supreme Court of India judgment dated 28 August 1996 in Writ Petition (Civil) No. 914 of 1991 (*Vellore Citizens Welfare Forum v. Union Of India & Ors*).

[76] Supreme Court of India judgment dated 4 April 2013 in Writ Petition (Civil) No. 180 of 2011 (*Orissa Mining Corporation v. Ministry of Environment & Ors*).

[77] National Green Tribunal judgment dated 7 May 2015 in Original Application No. 521 of 2014 (*Om Dutt Singh & Anr v. State of Uttar Pradesh & Ors*).

public office or during elections. It also dealt with access to information. These definitions were considered to be limited for environmental decision-making.[78] A few attempts were made between Stockholm Declaration, 1972, and Rio Declaration, 1992, to develop an expanded meaning of public participation. The Principle 23 of the United Nations World Charter for Nature in 1982 states as follows:

> All persons in accordance with their national legislation shall have the opportunity to participate, individually or with others, in the formulation of decisions of direct concern to their environment, and shall have access to means of redress when their environment has suffered damage or degradation.

However, the push for public participation really came from the Rio Declaration, 1992, which established its need in environment protection and sustainable development. It held that it is the responsibility of national governments to ensure that there is transparency and that citizens have access to information. This participation also implies access to judicial redress. Principle 10 of the Declaration says:

> Environmental issues are best handled with the participation of all concerned citizens, at the relevant level. At the national level, each individual shall have appropriate access to information concerning the environment that is held by public authorities, including information on hazardous materials and activities in their communities, and the opportunity to participate in decision-making processes. States shall facilitate and encourage public awareness and participation by making information widely available. Effective access to judicial and administrative proceedings, including redress and remedy, shall be provided.

The United States Environment Protection Agency has elaborated that public participation cannot be considered a single event. It needs to consist of a series of activities and actions to either inform the public or obtain their inputs to

---

[78] Madhuri Parikh, 'Public Participation in Environmental Decision Making in India: A Critique', *IOSR Journal of Humanities and Social Science* (*IOSR-JHSS*) 22, no. 6 (2017): 56–63.

take a decision. It is only through this that those who are likely to benefit or lose are able to 'influence decisions that affect their lives'.[79]

India's environmental laws allow public participation in a very limited manner. The Environment Impact Assessment (EIA) notification has a relatively elaborate procedure for public consultation that includes a public hearing and written comments. These have been discussed in Chapter 5, which deals with environmental protection. Several cases have been filed that challenge the non-adherence of public hearings with regulatory standards. These have been filed by state governments[80] or in public interest[81] or by project-affected people aggrieved by poorly conducted public hearing proceedings.[82]

More recent laws like the BD Act, 2002, and the Right to Fair Compensation and Transparency in Land Acquisition, Rehabilitation and Resettlement Act, 2013, have introduced the requirement of consultations and public hearings prior to approvals for access to local biodiversity or to identify project-affected people. These have been discussed in Chapters 6 and 8.

Free, Prior and Informed Consent

FPIC is relatively new to environmental law in India. Article 10 of the United Nations Declaration on the Rights of Indigenous Peoples adopted in 2007 states:

> Indigenous peoples shall not be forcibly removed from their lands or territories. No relocation shall take place without the free, prior and informed consent of the indigenous peoples concerned and after agreement on just and fair compensation and, where possible, with the option of return.

---

[79] 'Public Participation Guide: Introduction to Public Participation', United States Environmental Protection Agency, available at https://www.epa.gov/international-cooperation/public-participation-guide-introduction-public-participation, accessed on 27 August 2019.

[80] Supreme Court of India order dated 29 November 2018 in 105699/2018 in Original Suit. No. 4/2007 (*State of Orissa v. The State of Andhra Pradesh & Ors*) related to the Polavarm Multipurpose Project.

[81] High Court of Gujarat judgment dated 2 March 2000 in Special Civil Application No. 8529 of 1999 (*Centre for Social Justice [Jan Vikas] v. Union of India*).

[82] NGT judgment dated 13 December 2013 in Appeal No. 9 of 2011 South Zone (*Samata & Anr v. Union of India & Ors*).

The Nagoya Protocol of the CBD requires FPIC to be instituted in national biodiversity legislations such as the BD Act, 2002, in India. India's BD Act only requires a consultation as discussed in chapter 6.

There is a limited body of case law related to FPIC in India. One case relates to the use of forest land for a bauxite mine in Odisha. The mining activity threatened to take over the customary land and sacred hill of the Dongria Kondh tribal community. The mining operation was proposed despite pendency of forest rights claims to be recognised under the Scheduled Tribes and other Traditional Forest Dwellers (Recognition of Forest Rights) Act, 2006. This case is discussed in Chapter 3 in detail. The 2013 judgment[83] of the Supreme Court relied on a 2007 UN declaration along with the provisions of the Forest Rights Act to direct the state government to seek the consent of *gram sabha*s (village assemblies) and submit a report to the central government, based on which the environment ministry could take a final decision on the proposal for forest diversion. All 14 *gram sabha*s likely to be affected by the mining operations rejected the proposal for the mining activity, and the project was not initiated.

---

[83] Supreme Court of India judgment dated 4 April 2013 in Writ Petition (Civil) No. 180 of 2011 (*Orissa Mining Corporation v. Ministry of Environment & Ors*).

2

Institutions Regulating India's Environment

INTRODUCTION

The Seventh Schedule of India's Constitution distributes powers between the central and state governments to enact and implement laws on subjects according to their listing in the Union, State or Concurrent (shared) Lists. Environment as a subject has not been assigned to any of these lists; as a result, the division of roles on environmental matters is a dynamic, shifting space between the centre and state governments. As the chapters in this volume show, specific natural resources such as land, water and forests are assigned to one of the three lists and are regulated by corresponding institutions. But this distribution of power to govern natural resources is contested. The distribution of roles also makes it necessary for cooperation between the centre and states in the management and regulation of these resources.

Even though the powers to enact laws to manage specific resources are distributed between the centre and states, in effect, the drafting, implementation and enforcement of environment laws require substantial coordination between them. For instance, Entry 14 in the Union List is 'Entering into treaties and agreements with foreign countries and implementing of treaties, agreements and conventions with foreign countries'. Using this entry, the central government has enacted several laws such as Environment Protection Act (EPA), 1986, and the Biological Diversity (BD) Act, 2002, as commitments under specific treaties and conventions. While the EPA, 1986, is a centrally administered law, the centre has delegated regulatory powers to several state-level institutions. These institutions are set up by the central government in consultation with state governments, and their functioning depends on the administrative and financial support from the state governments.

In this chapter, we discuss the main institutions involved in environmental governance in India at the central and state levels. Indian environment laws have also evolved through complex interactions between the parliament, the executive and the judiciary, and these interactions are reflected in the institutions set up to perform legal, policymaking and regulatory functions. This separate chapter on environmental bodies outlines the vast institutional landscape of India's environmental laws. It shows the plethora of institutions that have emerged in recent decades and highlights the immense power and crucial roles ascribed to these institutions in environmental governance.

Many of the institutions described in this chapter are referred to in specific, thematic chapters in this book. Their structures, roles and critiques regarding their design and functions are also discussed in the following chapters. Several governmental audit processes and parliamentary committees have reviewed the performance of environmental institutions. Independent researchers and civil society organisations concerned with environmental governance and social justice have also assessed the scope and limits of environmental institutions and regulatory bodies to deliver environmental sustainability.

There are also several planning bodies and committees that have shaped the design and functions of environmental regulation. This chapter does not include such institutions that may have influenced the functioning of environmental regulatory bodies from outside the arena of environmental laws. The book refers to the role of several sectoral ministries, which have a direct bearing on how natural resources are put to developmental uses. While the primary objective of these ministries is not the protection of the environment, their legal and institutional frameworks do overlap with the environmental institutions. These include the Ministry of Tribal Affairs, Ministry of Water Resources and Ganga Rejuvenation,[1] Ministry of Agriculture and Ministry of Rural Development. The roles and functions of these ministries or various bodies under them are not included in this chapter.

This chapter has four sections as follows:

I.   *Judicial Institutions*
II. *Parliamentary Committees*

---

[1] In mid-2019, the Ministry of Water Resources, River Development and Ganga Rejuvenation and the Ministry of Drinking Water and Sanitation were merged to create the Ministry of Jal Shakti.

I JUDICIAL INSTITUTIONS

The judiciary plays an important role in legal adjudication of environmental disputes and providing remedies in case of breach of environment law, maintaining institutional accountability and directing policy actions. Such jurisdiction rests with the Supreme Court, high courts and district courts. Over the years, specialised appellates and tribunals have been set up to address appeals and hear other grievances related to environmental impacts.

Courts

The jurisdiction of courts has been invoked for several environmental matters.

The law of tort[2] or wrongdoing has traditionally been used to bring pollution-related cases to any court. Such civil and criminal matters are brought to court to redress the harm caused by one party to another. Remedies sought have been in the form of fines, injunctions (directions and orders) and damages. More details on the role of courts and the use of the Indian Penal Code (IPC), the Civil Procedure Code (CPC) and the Criminal Procedure Code (CrPC) in environmental matters have been discussed in Chapter 1.

Since the 1980s, one of the most popular approaches of seeking interventions of the high courts and the Supreme Court in environmental matters is through the writ jurisdiction. Writ petitions can be filed before the Supreme Court using Article 32 of the constitution and in the high courts using Article 226.

Article 32 (rights of citizens to approach the Supreme Court of India seeking remedies for enforcement of rights) states,

> The Supreme Court shall have power to issue directions or orders or writs, including writs in the nature of habeas corpus, mandamus, prohibition, quo warranto and certiorari, whichever may be appropriate, for the enforcement of any of the rights conferred by this Part.

---

[2] 'A tort is an act or omission that gives rise to injury or harm to another and amounts to a civil wrong for which courts impose liability. In the context of torts, "injury" describes the invasion of any legal right, whereas "harm" describes a loss or detriment in fact that an individual suffers.' Available at https://www.law.cornell.edu/wex/tort, accessed on 17 August 2019.

Article 226 (rights of citizens to approach the high courts of India seeking remedies for enforcement of rights) states,

> Every High Court shall have power, throughout the territories in relation to which it exercises jurisdiction, to issue to any person or authority, including in appropriate cases, any Government, within those territories directions, orders or writs, including writs in the nature of habeas corpus, mandamus, prohibition, quo warranto and warranto and certiorari, or any of them, for the enforcement of any of the rights conferred by Part III and for any other purpose.

Divan and Rosencranz state that though both these constitutional provisions are powerful, the writ jurisdiction before the Supreme Court is relatively limited[3]. This is because Article 32 allows for the right to seek the apex court's enforcement of fundamental rights, including those related to the environment. Article 226 allows for the high court's jurisdiction to be invoked not only for the enforcement of a fundamental right such as right to freedom but for 'any other purpose', including 'ordinary legal rights' such as trade and commerce.

In several cases, courts have also stepped in on policy matters. Several chapters in this book discuss examples where courts have either *suo moto* or as in response to a grievance extended the remedy to include directions such as the setting up of new institutions,[4] reconstituting regulatory authorities and developing a *continuing mandamus* for the court to monitor enforcement of one or more laws.

Both the Supreme Court and the high courts appoint short-term and long-term committees to assist them with the implementation of specific orders or to play quasi-judicial roles on specific cases. One of these is the Central Empowered Committee appointed by the Supreme Court's green bench in Writ Petition (Civil) in 202 of 1995, popularly known as the Godavarman or green bench. This committee is discussed in Chapter 3.

---

[3] S. Divan and A. Rosencranz, *Environmental Law and Policy in India*, 2nd edn (New Delhi: Oxford University Press, 2005).

[4] The establishment of authorities for compensatory afforestation and coastal and groundwater regulation and management has emerged from judgments of the Supreme Court. These have been discussed in specific, thematic chapters.

NATIONAL GREEN TRIBUNAL

A dedicated judicial body for environmental matters was gazetted in June 2010 following the enactment of the National Green Tribunal Act, 2010. The bench superseded two earlier authorities: the National Environment Appellate Authority (1997) and the National Environment Tribunal. The National Green Tribunal (NGT) was functional from May 2011, once the NGT rules were finalised. This green court is designed to have a principal bench and regional benches. Since its establishment, the NGT has functioned with a capacity up to two principal benches in New Delhi and three regional benches in Kolkata, Chennai and Pune. In addition, there have been rotational benches in Shillong and Jodhpur, to hear a pool of cases from these regions. Since mid-2018, the NGT's principal bench began hearing cases through video conferencing.[5]

The benches are comprised of judicial and technical members. The chairperson of each of the benches is a judicial member who has served as a judge of the Supreme Court of India or Chief Justice of a high court. The expert members are required to have a minimum qualification of a master's degree and 15 years' work experience. The area of expertise defined includes pollution control, management of hazardous substances, environment impact assessment, climate change, biological diversity and forest conservation.

The NGT has powers to hear all applications which involve a 'substantial question relating to environment'. The tribunal can issue directions for the compensation and restitution of damage for actions of environmental negligence. It has powers to requisition public records and documents, summon any person and also review its own decision as vested in a civil court under the CPC, 1908.

It can hear appeals arising out of orders or directions issued under six laws relating to environment protection, air and water pollution, and diversion of forestland for non-forest purposes. These laws have been discussed further in the book. The 2010 Act defines who is the aggrieved person, who can file such cases and the limitation period of appeals and applications that can be filed.[6]

---

[5]   Express News Service, 'National Green Tribunal Plans to Hear Cases Using Video-Conferencing', *Indian Express*, 15 July 2018.

[6]   Analysis of the structure, functioning and cases of the National Green Tribunal (NGT) is available in Ritwick Dutta and Sanjeet Purohit, *Commentary on the National Green Tribunal Act, 2010* (New Delhi: Universal Law Publishing Company, 2015).

II PARLIAMENTARY COMMITTEES

There are four kinds of parliamentary committees that assist both the Lok Sabha (lower house) and the Rajya Sabha (upper house). These are financial committees, department-related standing committees, other parliamentary committees and ad hoc committees.[7] There are three financial committees that are appointed either by the Lok Sabha or jointly by the Lok Sabha and Rajya Sabha. These are the Estimates Committee, the Public Accounts Committee and the Committee on Public Undertakings.

As per the Rajya Sabha website, there are 24 department-related standing committees covering all the ministries/departments of the government. 16 of these are Lok Sabha committees and 8 are housed in the Rajya Sabha. Each of these committees consists of 31 members: 21 from the Lok Sabha and 10 from the Rajya Sabha who are nominated by the speaker of the Lok Sabha and the chairman of the Rajya Sabha. The term of office on these committees is for one year. There are several department-related committees that have a bearing on environmental issues such as the Department-Related Parliamentary Standing Committees on Science, Technology, Environment and Forests and Climate Change; Urban Development; and Coal and Water resources. Other standing committees can influence environmental regulations, for example, the Committee on Subordinate Legislation. This committee has 15 members, appointed with a one-year tenure. The speaker of the Lok Sabha nominates the members.

Ad hoc committees are usually one-time appointments. This includes matters where the parliament has not been able to arrive at a consensus and need detailed deliberations including seeking expert and public inputs. Such parliamentary committees that have been set up on the land acquisition ordinance, compensatory afforestation or pollution control boards have been referred to in specific, thematic chapters.

III GOVERNMENT-LEVEL INSTITUTIONS

Central and state government environmental ministries and departments are designated with the task of regulating the use of environment and nature

---

[7] Lok Sabha document on 'Parliamentary Committees Introduction', available at http://164.100.47.194/loksabha/Committee/Comm_Introductionnew.pdf, accessed on 20 July 2019.

protection. As ecological landscapes, waterscapes and airsheds do not neatly map on to state administrative units, the state-level institutions have to necessarily interact with the centre and with each other. This interaction is also necessary when existing laws are amended or new laws are proposed by the central government. The financial support for functioning of specific institutions is also a subject area of interaction as some committees are constituted by the central government but their costs have to be borne by state.

CENTRAL GOVERNMENT

The process of formation of the environment ministry can be traced back to 1980 when the then Prime Minister Indira Gandhi set up a committee to prepare a report to recommend 'legislative measures and administrative machinery for ensuring environmental protection'. One of the recommendations of this committee was to create a department of environment to function directly under the prime minister. [8]

The Ministry of Environment and Forests (MoEF) was created in 1985.[9] Since then, it has worked as the nodal ministry for 'planning, promotion, co-ordination and overseeing the implementation of India's environmental and forestry policies and programmes'.[10] For these purposes, the ministry presently has three major divisions—environment, forest and wildlife—highlighting the three major themes and related legislations that the ministry governs.

The ministry also acts as the nodal agency for several international environmental conventions and treaties that India is signatory to. A few examples include the United Nations Conference on the Human Environment, 1972; Convention on Biological Diversity, 1992; United Nations Convention to Combat Desertification, 1994; Stockholm Convention on Persistent

[8] Department of Science and Technology, *Report of the Committee for Recommending Legislative Measures and Administrative Machinery for Ensuring Environmental Protection* (New Delhi: Government of India), iii; Jairam Ramesh, *Indira Gandhi: Life in Nature* (New Delhi: Simon & Schuster India, 2017), 297.

[9] Ministry of Environment and Forests, *National Conservation Strategy and Policy Statement on Environment and Development* (New Delhi: Government of India, 1992).

[10] Website of the MoEF, available at http://moef.gov.in/about-the-ministry/introduction-8/, accessed on 14 July 2019.

Organic Pollutants, 2004; and United Nations Framework Convention on Climate Change, 1994. These have been discussed in Chapter 1.

The ministry is made up of officers of the Indian Administrative Service and officers with scientific qualifications and expertise who play an important role in the standard setting, policy formulation, regulation and enforcement functions of the ministry. The main office is located in New Delhi, and there are 10 regional offices in the cities of Bengaluru, Bhubaneshwar, Bhopal, Chandigarh, Chennai, Dehradun, Lucknow, Nagpur, Ranchi and Shillong.

In 2014, the ministry was renamed the Ministry of Environment, Forest and Climate Change (MoEFCC). In 2019, the entire range of functions related to groundwater regulation was shifted to the water resources ministry.[11]

STATE-LEVEL INSTITUTIONS

Every state government has their own departments of environment and forests. In most cases, they function as separate units under the state government. For instance, a state could have a different state forest department and a state environment department.

State Forest Departments

The state forest departments have a precolonial institutional history. Colonial India was one of the first in the world to have a dedicated service for the scientific management of forests.[12] The then British government started the Imperial Forest Department in 1864. Today, these departments are made up of officers of the Indian Forest Service (IFS) or State Forest Service (SFS). The department also recruits field-level officers who are not from the IFS/SFS but are trained specifically by the department to assist in functions of forest management and enforcement.

The state forest departments also have separate divisions for managing forests in areas specially protected for wildlife and those which are recorded as reserved or protected forests.

The laws that are enforced by state forest departments are Indian Forest Act, 1927; Forest Conservation Act (FCA), 1980; Forest Conservation Rules of 1981, 2003 and 2017; and all state forest Acts and related rules.

---

[11] Cabinet secretariat notification no. S.O. 1972(E) dated 14 June 2019.

[12] Website of the IFS, available at http://ifs.nic.in/hist.aspx, accessed on 26 February 2020.

State Environment Departments

Each state government constitutes its own department of environment that have different nomenclature in different states, for example, the Department of Environment and Climate Change or the Department of Science, Technology and Environment.

Most of the state-level institutions set up through central laws, discussed further in this chapter and in the following chapters, are administered by the department of environment, with an official of this department acting as the secretariat for these institutions. These departments undertake specific environmental awareness and research programmes. They are also involved in the preparation and implementation of State Action Plans on Climate Change often housed within a climate change cell within the department.

DISTRICT-LEVEL INSTITUTIONS

The role of the district collectors and district magistrates is significant in several environmental matters. The collector is the head of the revenue administration, and the magistrate adjudicates criminal justice matters at the district level. In some states, the same official may hold a joint jurisdiction of both revenue administration and criminal adjudication.[13] The role of these officials is crucial in enforcing the provisions of CrPC, IPC and CPC related to *nuisance,* as discussed in Chapter 1 of this book. The nuisance clauses have a significant potential to deal with environmental impacts and remedies.

Specific legal frameworks also direct the setting up of institutions for coastal regulation, grant of permits, impact assessment or fund disbursement, which are to be headed by the district collectors. District collectors have an important responsibility in the conduct of public hearings under the Environment Impact Assessment Notification, 2006. Laws such as Right to Fair Compensation and Transparency in Land Acquisition, Rehabilitation and Resettlement Act, 2013, Wild Life Protection Act, 1972, and the Scheduled Tribes and Other Traditional Forest Dwellers (Recognition of Forest Rights) Act, 2006, give substantial responsibilities to the district collectors for preparing rehabilitation plans and recognition and settlement of rights.

---

[13] Ministry of Personnel, Public Grievances and Pensions, *Second Administrative Reforms Commission: Fifteenth Report* (New Delhi: State and District Administration, chaired by Shri Veerappa Moily, Member of Parliament, Government of India, 2009).

Rural and Urban Local Bodies

Both rural and urban local bodies have an important role to play in environment regulation and management[14]. In rural areas, this role vests with the village panchayats, and in urban areas, this role vests with a municipality or municipal corporation depending on the size of the town or city. These bodies have both elected and nominated members whose powers and functions are defined in specific state-level town planning, municipal or panchayat legislations. These laws give a panchayat or urban local body powers to remove encumbrances, clear encroachments and implement schemes in their area of jurisdiction. These have been used for environment protection and securing livelihoods in select instances as discussed in this book.

Environmental laws vest very few powers with panchayats and urban local bodies. Recently, some legal instruments have delegated roles to them for solid waste management or biodiversity conservation. The Solid Waste Management Rules, 2016 requires these bodies to prepare solid waste management plans, ensure door-to-door collection of waste and issue directions in case of littering.

In some other laws, the role of these bodies is limited to constituting local committees or information dissemination. In the case of the BD Act, 2002, the local body is empowered to nominate Biodiversity Management Committees. In Environment Impact Assessment (EIA) Notification, 2006, the government is required to provide summaries of EIA reports and environment clearance letters to the public through the panchayat office.

IV STATUTORY COMMITTEES AND AUTHORITIES

In this section, we look at various expert committees, authorities or administrative boards that draw their jurisdiction from specific statutes or subordinate legislations such as notifications. Each of these has specific objectives and functions and most often a clearly defined composition.

---

[14] The Solid Waste Management Rules, 2016, defines urban local bodies as 'municipal corporation, nagar nigam, municipal council, nagarpalika, nagar Paliparishad, municipal board, nagar panchayat and town panchayat, census towns, notified areas and notified industrial townships with whatever name they are called in different States and union territories in India'.

FOREST ADVISORY COMMITTEE

The Forest Advisory Committee (FAC) draws its mandate from Forest Conservation Rules, 1981 (as amended in 2003).[15]

The director general (DG) of forests, based at the environment ministry, chairs the FAC, which meets in New Delhi. Other members include the additional DG of forests and the additional commissioner (soil conservation), Ministry of Agriculture, along with three 'eminent experts' in forestry and allied disciplines as non-official members. The inspector general (IG) of forests (forest conservation) is also located at the ministry who acts as the member secretary (MS). In instances where the DG of forests is unable to preside, the additional DG of forests would act as the chair. The term of appointment of non-official members is two years.

The FAC is to meet 'whenever considered necessary, but not less than once in a month'. The committee is to have a minimum of three members for any decision to be taken.

The primary role of the FAC is to appraise proposals related to non-forest use of forestland above 40 hectares and advise the environment ministry. The central government holds the power to approve or reject prior permission that is finally decided upon by state governments. State governments are to transfer such proposals to the FAC for review within 90 days of receiving it from a user agency. The recommendations of the FAC are advisory in nature and not binding on the environment ministry. The environment ministry needs to take the decision on the FAC's recommendation within 60 days.

REGIONAL EMPOWERED COMMITTEES

Regional empowered committees (RECs) were constituted in 2004 through an amendment to the Forest Conservation Rules.[16] Every REC is chaired by the head of the regional office of the environment ministry located in 10 cities, as discussed earlier in this chapter. Other official members include senior-most officers in the regional office of the rank of chief conservator of

---

[15] The Forest Conservation Rules were first enacted in 1981 and subsequently amended in 2003, 2004, 2014 and 2017.

[16] Inserted vide notification number G.S.R 94 (E) dated 3 February 2004 was included in the Forest Conservation Rules through an amendment in 2014 (G.S.R. 713 (E), 10 November 2014).

forests or conservator of forests (CF). A forest officer of the rank of deputy CF acts as the MS to the committee. Three non-official members from forestry and allied disciplines are appointed to serve on the committee. Representatives of state and union territory (UT) administration are also to be invited to be on the committee.

The RECs appraise proposals for non-forest use between 5 and 40 hectares of forestland. Proposals under five hectares are processed by the ministry's regional office without appraisal by the RECs, except for proposals related to regularisation of encroachments and mining projects. RECs also review all proposals related to linear projects like highways, pipelines and transmission lines, irrespective of the amount of forestland involved. The proposals need to be decided upon within 30 days of receipt of all documents. These are then forwarded within five days to the central environment ministry.

NATIONAL COMPENSATORY AFFORESTATION FUND MANAGEMENT AND PLANNING AUTHORITY[17]

The national Compensatory Afforestation Fund Management and Planning Authority (CAMPA) was constituted under the Compensatory Afforestation Fund Act, 2016, and corresponding rules of 2018. The primary objective of this authority is to manage and utilise the national fund set up under the Act. These funds comprise monies collected for

> compensatory afforestation, additional compensatory afforestation, penal compensatory afforestation, net present value and all other amounts recovered from such agencies under the Forest (Conservation) Act, 1980.

The authority needs to ensure that this money is utilised for

> undertaking artificial regeneration (plantations), assisted natural regeneration, protection of forests, forest related infrastructure development, Green India Programme, wildlife protection and other related activities[, including matters that are connected to and incidental to these activities].

---

[17] The audit report of the Comptroller and Auditor General (CAG) of India discusses the role of the ad hoc CAMPA and state CAMPA prior to the enactment of the CAF Act, 2016. This document can be referred from Comptroller and Auditor General (CAG) of India, 'Compliance Audit on Compensatory Afforestation', report no.: 21 of 2013, Government of India, New Delhi, 2013.

The national CAMPA is to consist of a governing body and assisted by an executive committee, a monitoring group and administrative support mechanism. The minister of the MoEFCC chairs the governing body of the National CAMPA. All secretaries of ministries and departments dealing with environment, forest and climate change; finance (expenditure); rural development; land resources; agriculture; panchayati raj, tribal development, science and technology; space and earth sciences and chief executive officer, National Institution for Transforming India Aayog, are ex officio members[18] of the authority. Other ex officio members include the IG, DG and additional DG (forests; additional DG (wildlife); director of the National Mission for a Green India and financial advisor to the MoEFCC. In addition, the governing body at any given point of time needs to include five principal chief conservators of forests from different states who are nominated for a period of two years on a rotational basis. The members of the governing body are five experts, an environmentalist, a conservationist, a scientist, an economist and a social scientist, appointed for a period of two years and not more than two consecutive terms. The chief executive officer (CEO) and MS of the national CAMPA is an officer of the rank of additional DG of forests.

This governing body is assisted by an executive committee, which is headed by the DG (forests). The ex officio members are additional DGs of forest conservation and wildlife along with director, National Mission for a Green India (further discussion on background and genesis of the CAMPA is in Chapter 3).

STATE COMPENSATORY AFFORESTATION FUND MANAGEMENT AND PLANNING AUTHORITY

The state CAMPA is set up under Compensatory Afforestation Fund Act, 2016, and corresponding rules of 2018. The environment ministry has the responsibility to appoint the state CAMPA in each state. This authority is responsible for the management and utilisation of the state fund. Just as the national CAMPA, the state CAMPA has a governing body which has to utilise and monitor the funds received from the national CAMPA. In addition, it has a steering committee and an executive committee, which are the functional arms of the authority.

---

[18] An ex officio member is a member of a body (a board, committee, council, and so on) who is not part of the said body in their own capacity. They are ex officio members because they hold the office that is designated to be on the body by the law.

The chief minister of the state or the lieutenant governor of a UT chairs the governing body. The ex officio members include the forest minister, chief secretary and principal secretaries of all departments dealing with environment, finance, planning, rural development, revenue, agriculture, tribal development, panchayati raj, science and technology. Other ex officio members are the principal chief conservator of forests (PCCF) and chief wildlife wardens of the state. The principal secretary (forests) is to act as the MS of this authority.

An officer of the rank of chief conservator of forests (CCF) or above is to be appointed as the CEO and MS of both the steering committee and the executive committee. The steering committee is to be chaired by the chief secretary. The principal secretaries of the departments dealing with forests, environment, finance, planning, rural development, revenue, agriculture, tribal development, panchayati raj, science and technology are ex officio members. Other ex officio members are the PCCF, the chief wildlife warden and the nodal officer who are responsible for the implementation of the FCA, 1980, in the state and head of the concerned regional office of the environment ministry. In addition, the steering committee is required to have an expert on tribal matters or a representative of tribal communities as a member, who is appointed by the state government.

The executive committee of the state CAMPA is chaired by the PCCF of the state who is the head of the forest department in the state. The ex officio members are the chief wildlife warden or a forest officer not below the rank of the CCF and the nodal officer of state forest development agencies. Representatives of the departments of environment, finance, planning, rural development, revenue, agriculture, tribal development, panchayati raj, science and technology are ex officio members. Another ex officio member is the financial controller or financial adviser, nominated by the finance department. Other members of the executive committee include two 'eminent' non-governmental organisation (NGO) representatives, two representatives of the district panchayats and one expert on tribal matters or representative of tribal communities.

CENTRAL POLLUTION CONTROL BOARD

The Central Pollution Control Board (CPCB) is enacted under the Water (Prevention and Control of Pollution) Act (Water Act), 1974. The board is located in New Delhi. It is to have a full-time chairperson appointed by the environment ministry who is either a person having special knowledge or practical experience in environmental protection or in administering institutions that have dealt with such matters. In addition, another five members are to be nominated as representatives of the government. Two members representing the interests of agriculture, fishery, or industry or trade are also to be nominated by the central government. Two additional persons representing companies or corporations owned, controlled or managed by the central government should also be on the CPCB. The board is to be serviced by a full-time MS with qualifications, knowledge and experience of scientific, engineering or management aspects of pollution control. All members of the board have a term of three years.

The CPCB works closely with the central government on issues concerning prevention and control of pollution. It also coordinates the activities of the state boards including dispute resolution, trainings and technical assistance. Another important function of the CPCB is to prepare guidance manuals and codes for pollution regulation or waste management. Chapter 4 discusses the structure and functions of the CPCB in greater detail.

STATE POLLUTION CONTROL BOARD

State pollution control boards (SPCBs) are also enacted through Water (Prevention and Control of Pollution) Act (Water Act), 1974. Each state needs to appoint a SPCB. The chairman of this board needs to be a person having special knowledge or practical experience in environmental protection or in administering institutions that have dealt with such matters. The SPCB chairperson can either be part time or full time as the state government deems appropriate. The membership of the SPCB is similar to that of the CPCB. It includes five additional members who are to be nominated as representatives of the state government. Two members representing the interests of agriculture, fishery, or industry or trade are also to be nominated by the state government. Two additional persons representing companies or corporations owned, controlled or managed by the central government should also be on the SPCB. The MS has to be of similar qualification as in the CPCB. All members of the board have a term of three years.

The state boards work with the respective state governments to assess the suitability and grant consent to specific industries or industrial processes. They have an important role in enforcing the provisions of pollution control and environment protection laws, through site inspections, show-cause notices and closure orders. They also lay down pollution control norms and standards based on advice from the CPCB. The state boards can identify specific air pollution control areas within their jurisdiction. More details on the SPCB is in Chapter 4.

Both the CPCB and SPCBs need to meet every three months. The boards can appoint additional committees to help carry forward the objectives of the Act. More details on CPCB and SPCBs as well as constitution of regional PCBs at state levels and joint boards are given in Chapter 4. Also discussed in that chapter are the various reviews on the capacity and performance of the PCBs, including the judgment of the NGT on PCBs. These are discussed in Chapter 4.

The Air Act, 1981, and Water Act, 1974, provide for the setting up of pollution control appellate mechanism within the SPCBs where appeals against orders of the state and regional offices of the PCB can be filed. More details on this are in Chapter 4.[19]

ENVIRONMENT POLLUTION (PREVENTION AND CONTROL AUTHORITY)

The Environment Pollution (Prevention and Control Authority) (EPCA)[20] was constituted under the EPA, 1986, following the directions of the Supreme Court in 1998,[21] with the 'objective of protecting and improving the quality of the environment and preventing and controlling environmental pollution in the National Capital Region'. The National Capital Region (NCR) is a notified area which includes the whole of the National Capital Territory:

---

[19] For an analysis of history, capacity and functioning of these appellates, see Shibani Ghosh, Sharachchandra Lele and Nakul Heble, 'Appellate Authorities under Pollution Control Laws in India: Powers, Problems and Potential', *Law, Environment and Development Journal* 14, no. 1 (2018): 47–58, available at http://www.lead-journal.org/content/18045.pdf, accessed on 20 February 2020.

[20] MoEF notification no. S.O. 93(E) dated 29 January 1998 (reconstituted in 2016 as per notification no. S.O. 2311(E), in 2018 as per notification no. S.O. 5120(E) and in 2019 as per notification no. S.O. 1651(E)).

[21] Supreme Court order dated 7 January 1998 in Writ Petition (Civil) No. 13029 of 1985 (*M. C. Mehta v. Union of India*).

Delhi, 13 districts of Haryana, 8 districts of Uttar Pradesh and 2 districts of Rajasthan. It covers an area of approximately 55,083 square kilometres.[22]

Since its constitution, Bhure Lal, a former secretary of the central government, has chaired the authority. Another permanent member is Sunita Narain from the Centre for Science and Environment. Both these members were appointed by name by the Supreme Court. The other members include secretary, MoEFCC; chairperson of the New Delhi Municipal Corporation and Commissioners of north, south and east Delhi municipal corporations. It also includes joint commissioner of police (traffic) and the CEO of Delhi Jal Board as members. The EPCA also has technical members from policy think tanks, university departments, private sector and the health sector.

Under Section 5 of the Environment Protection Rules, the EPCA can issue directions including closure, prohibition and regulation of any industry, operation or process in the NCR region. The authority has the powers to set standards for environment quality, discharge of emissions and environmental pollutants. The EPCA can also identify areas in which operations and processes that are harmful to the environment can be restricted. Finally, the authority can lay down procedures and safeguards for prevention of accidents and handling of hazardous substances. The EPCA also implements a Graded Response Action Plan for controlling air pollution in the Delhi-NCR region when the pollution levels are very high. The plan includes

measures such as prohibition on entry of trucks into Delhi; ban on construction activities, introduction of odd and even scheme for private vehicles, shutting of schools, closure of brick kilns, hot mix plants and stone crushers; shutting down of Badarpur power plant, ban on diesel generator sets, garbage burning in landfills and plying of visibly polluting vehicles, etc.[23]

In October 2020, the EPCA was replaced by a commission for air quality management following the enactment of the Commission for Air Quality Management in National Capital Region and Adjoining Areas Ordinance, 2020.

---

[22] Website of the National Capital Region Planning Board, available at http://ncrpb.nic.in/ncrconstituent.html, accessed on 29 August 2019.

[23] MoEFCC, *Graded Action Plan to Reduce Urban Air Pollution* (New Delhi: MoEFCC, Government of India, 2012), available at https://pib.gov.in/newsite/PrintRelease.aspx?relid=159542, accessed on 26 February 2020.

Expert Appraisal Committee

Expert appraisal committees (EAC) are constituted by the environment ministry with the objective of reviewing all 'Category A' project proposals submitted to the ministry for seeking 'environment clearance' under the EIA Notification, 2006. The EACs appraise all projects under the jurisdiction of the EIA notification that are within 5 kilometres of interstate boundaries, protected areas or satisfy other special conditions laid under the law. At present, there are seven thematic EACs that meet at regular intervals in New Delhi to scrutinise applications, approve terms of reference to carry out EIA studies and finally appraise projects along with all objections received during public hearings. Based on all this information, the EACs recommend the grant of project approval with conditions or rejection.

EACs are to consist of 'professionals and experts' as defined in the EIA notification. A professional is primarily defined as one with a combination of a professional degree and work experience. For instance, a person with a minimum qualification of five years of formal university training with a master's degree in arts or sciences can qualify to be in the expert committee. In case of engineering/technology or architectural disciplines, the formal training has to be a minimum of four years. The EIA notification also gives other eligibility criteria.

The definition of an expert includes 15 years of relevant experience. An expert can also be a professional with an advanced degree such as a PhD along with 10 years of work experience. The kinds of expertise listed in the notification include environmental quality, project management, risk assessment, life sciences, forestry and wildlife, and environmental economics.

The EAC members need to be below the age of 70 years. The maximum tenure of a member, including chairperson, is for two terms of three years each. The MS for these committees is a scientific officer from the environment ministry.[24]

---

[24] National Green Tribunal judgment dated 14 July 2014 in Application No. 116 (THC) of 2013 (*Kalpavriksh & Ors v. Union of India & Ors*) relates to various instances where there was conflict of interest in the appointment of the chairpersons of the EACs and other concerns related to the composition of the EACs. The NGT directed environment ministry to take grievances into consideration and include a criterion and specific requirements for the person to be appointed as a chairperson of the EAC/SEAC within one month of the judgment.

State Environment Impact Assessment Authority (SEIAA)

State environment impact assessment authorities (SEIAAs) are constituted under the EIA Notification, 2006. The environment ministry constitutes a SEIAA in consultation with respective state governments or UT governments. It is the state or UT government that proposes names to the environment ministry for the SEIAA, and the SEIAA needs to be appointed within 30 days of the receipt of these names.

The SEIAA is the authority to grant or reject 'environment clearance' to all Category B projects listed in the EIA notification. A SEIAA needs to have three members including chairman and MS who are nominated by the state government. The MS needs to be a serving government officer within the state or UT government, familiar with environmental laws. Both the chairman and the third member of the SEIAA need to be an expert as defined earlier. All members have a fixed term of three years.[25] All decisions of SEIAA need to be unanimous and taken in a meeting.

State Expert Appraisal Committee

State expert appraisal committees (SEACs) are committees with similar requirements as the EACs and are also constituted under the EIA Notification, 2006. These committees work with SEIAAs to scrutinise, appraise, and recommend or reject project approvals for Category B projects under the EIA notification. Unlike EACs, there is one SEAC that is appointed by the environment ministry in consultation with the SEIAA and state government. Unlike multiple EACs that are appointed according to thematic expertise, SEACs scrutinise all types of projects.

National and State Coastal Zone Management Authorities

The first version of the Coastal Regulation Zone (CRZ) notification in 1991 did not include the setting up of national- and state-level coastal zone management authorities (CZMAs). In April 1996, the Supreme Court, in its

---

[25] In recent years, the non-functioning of SEIAA and SEACs have been important concerns. In Delhi, the SEIAA and the SEAC have been non-functional from May 2018 (MoEFCC notification no. S.O. 919(E) dated 1 April 2015).

judgment in the case of *Indian Council for Enviro-Legal Action and the Union of India* (Writ Petition 664 of 1993), directed the central government to set up state coastal zone management authorities (SCZMAs) in each coastal state and a National Coastal Zone Management Authority (NCZMA).[26] The legal framework of the subsequent versions of the CRZ Notification, 2011 and 2019, discusses their composition and functions in detail.

The NCZMA is appointed under the CRZ Notification, 2019, by the environment ministry[27] for several monitoring, coordination and management tasks as prescribed in the notification. These include coordinating the preparation of state-level coastal zone management plans (CZMPs) along with any changes and modifications required. The authority also has the responsibility to review cases of violations of the notification attracting the penal provisions of the EPA, 1986. This can be done *suo moto* by the NCZMA or on the basis of a complaint. The NCZMA is empowered to issue directions.

The NCZMA is required to provide technical assistance and guidance to SCZMAs and union territory coastal zone management authorities or any other institutions with the objective of protecting the coast. They can also be called upon to advise the central government on policy and planning matters and research and development related to coastal zone management.

The authority is chaired by the secretary, MoEFCC, and has as its members the principal secretaries (environment) of all coastal states. It also has representatives of the Central Ground Water Authority (CGWA) and central ministries of earth sciences, urban development, tourism, agriculture and farmers' welfare (particularly department of fisheries), home affairs and defence. The director, National Centre for Sustainable Coastal Management, Chennai, is also on the NCZMA as the centre is involved with tidal and coastal zone mapping exercises as required under the notification.

The NCZMA meets at the environment ministry in New Delhi. The joint secretary dealing with CRZ at the ministry is the MS of the authority.

---

[26] A detailed research report on these authorities is provided in M. Menon, M. Kapoor, P. Venkatram, K. Kohli and S. Kaur, *CZMAs and Coastal Environments: Two Decades of Regulating Land Use Change on India's Coastline* (New Delhi: CPR-Namati Environmental Justice Program, 2015).

[27] MoEFCC notification no. S.O. 3266(E) dated 6 October 2017 on the constitution of the National Coastal Zone Management Authority (NCZMA).

The quorum needed for meeting of the authority is 10 members and in case the quorum is not available, the meeting is to be 'adjourned for 30 minutes be reconvened with the required quorum'.

An SCZMA/UTCZMA is to be set up in every coastal state and union territory where there is a coastal stretch that attracts the provisions of the CRZ notification. This includes the states of Andhra Pradesh, Gujarat, Goa, Karnataka, Kerala, Maharashtra, Odisha, Tamil Nadu and West Bengal. It also covers the UTs of Andaman and Nicobar islands, Daman and Diu, Lakshadweep, and Puducherry.

These authorities are considered primarily responsible for enforcing and monitoring the CRZ notification. This includes reviewing proposals for regulated activities falling within specific coastal zones. In some instances, the SCZMA directly approves or rejects these proposals and, in others, forwards the same to the environment ministry with recommendations. The SCZMAs have to take the lead in preparing and finalising CZMPs until the time they are finally approved by environment ministry. Any future modification of these plans is also reviewed by the CZMAs.

The SCZMAs have the powers to issue directions under the EPA, 1986, in cases of violation of the CRZ notification. They are also responsible for monitoring the compliance of conditions and safeguards issued as part of the CRZ approvals either by environment ministry or the SCZMA.

The composition of the SCZMAs was discussed in a meeting of the NCZMA held on 15 December 2004. As per the decision, 'The composition of SCZMAs should include 1 NGO (by name), 4 experts (by name) and 5–6 ex-officio members from the various Departments viz Department of Environment, Urban Development, Fisheries, Industry, Pollution Control Board, local bodies.' At the same meeting, it was decided that the secretary and director of the department of environment of the concerned state will function as the chairman and MS of the SCZMA, respectively. The MS of the PCB of a state can also be the MS of its SCZMA.[28]

District-Level Coastal Committees

District-level coastal committees (DLCCs) are to be constituted in every district where the CRZ notification is applicable. The SCZMAs or UTCZMAs assist their respective governments to constitute district-level committees under the

---

[28] Menon, Kapoor, Venkatram, Kohli and Kaur, *CZMAs and Coastal Environments.*

district magistrate. A DLCC needs to include at least three representatives of local traditional coastal communities including fisher communities.

ECOLOGICALLY SENSITIVE AREA COMMITTEES

Since the late 1980s, ecologically sensitive areas (ESAs) have been declared under the Environment Protection Act, 1986. Each ESA notification also specifies the roles and functions of a committee that are specifically set up to manage or regulate activities within the ESA. Some committees such as the Dahanu Taluka Environment Protection Authority (DTEPA) have had overarching powers to regulate approval and take enforcement action against violations. Other committees have limited functions of monitoring the implementation of the notification and report any violations to the central government. ESAs have been discussed in detail in Chapter 5.

GROUND WATER AUTHORITIES

The Central Ground Water Authority (CWGA) was established in 1997[29] following orders of the Supreme Court of India.[30] The authority was set up with the explicit purpose of regulating and controlling the use of groundwater.

The Central Ground Water Board (CGWB) is an institution that has been in existence since 1970 (more details in Chapter 7). The chairman of the CGWB is the chairman of the CGWA. In addition, there are four other members of the CGWB from divisions entrusted with the tasks of surveys and assessments, exploration drilling and material management, sustainable management of groundwater, and training and technology transfer. There are two members from the Ministry of Water Resources and Ganga Rejuvenation at the level of joint secretary, administration and finance. Other members include chief engineer, Central Water Commission (CWC) and director/general manager (exploration).

The CGWA can invite specific representatives from ministries of agriculture, urban development and rural development as special invitees. Others who can be invited include representatives from technical institutions

---

[29] Ministry of Water Resources notification no. S.O. 38(E) dated 14 January 1997 under Section 5 of the Environment (Protection) Act, 1986 (No. L-11011/29/96-IA-III).

[30] Supreme Court order dated 10 December 1996 in Writ Petition (Civil) No. 4677 of 1985 (*M. C. Mehta v. Union of India*).

such as the National Institute of Hydrology and National Geo-physical Research Institute.

The CGWA has the authority 'to regulate and control, management and development of ground water in the country and to issue necessary regulatory directions for this purpose'. This is done by holding powers of issuing no objection certificates (NOCs) for groundwater use in 23 states. The Ministry of Water Resources and Ganga Rejuvenation has framed guidelines for the grant of NOC for withdrawal of groundwater, which have been revised from time to time. The latest operational guidelines are discussed in Chapter 7. The authority also has the powers to take penal action as designated in the EPA, 1986, in case of violations. The states that have separate groundwater laws may set up authorities under their statutes.

CENTRAL WATER COMMISSION

The CWC was first set up in 1945 as the Central Waterways, Irrigation and Navigation Commission.[31] Six years later, in 1951, it was renamed as Central Water and Power Commission (CWPC) and, at the same time, merged with the Central Electricity Commission. When the Ministry of Agriculture and Irrigation was restructured in 1974, the water wing of the CWPC was carved out as the CWC. Just as the CGWB, the CWC works as a subordinate or attached office of the water resources ministry and is one of its main technical arms.[32]

The CWC has the responsibility to appraise, initiate and coordinate schemes for 'control, conservation and utilization of water resources throughout the country, for purpose of Flood Control, Irrigation, Navigation, Drinking Water Supply and Water Power Development' in consultation with state governments.[33] The CWC is also responsible for collecting and publishing hydrological and hydrometeorological data related to major rivers in the country. The commission also prepares statistical data on surface water

---

[31] Department of Labour Resolution No. DW 101(2) dated 5 April 1945.

[32] Ministry of Water Resources, River Development and Ganga Rejuvenation, 'A 21st Century Institutional Architecture for India's Water Reforms: Report Submitted by the Committee on Restructuring the CWC and CGWB', Government of India, New Delhi, 2016.

[33] Website of the Central Water Commission, available at http://cwc.gov.in, accessed on 18 August 2019.

resources and their quality. Another function is that of flood forecasting for major flood-prone interstate river basins.[34]

The water resources ministry appoints the chairman of the CWC who is of the rank of an ex officio secretary of the Government of India. The commission has three wings each headed by a full-time member with the status of an ex officio additional secretary. These wings are Design and Research Wing, River Management Wing and Water Planning and Projects Wing.

BIOTECHNOLOGY COMMITTEES

The Rules for the Manufacture, Use, Import, Export and Storage of Hazardous Micro Organisms, Genetically Engineered Organisms or Cells, 1989, requires setting up of committees at both central and state levels. The first one is the Recombinant DNA Advisory Committee which reviews developments in biotechnology in national and international levels to recommend 'suitable and appropriate safety regulations for India in recombinant[35] research, use and applications'.

The second committee is the Review Committee on Genetic Manipulation (RCGM) which is set up to monitor safety of ongoing research projects and activities that involve genetically modified organisms (GMOs)/hazardous microorganisms. For this purpose, the RCGM is required to bring out manuals and guidelines to regulate activities and research related to GMOs. It includes representatives of Department of Biotechnology, Indian Council of Medical Research (ICMR), Indian Council of Agricultural Research (ICAR) and Council of Scientific and Industrial Research (CSIR) along with experts in their individual capacity. There is no limitation on the number of experts that can be appointed.

Both these committees are housed at the Department of Biotechnology in the Ministry of Science and Technology.

Genetic Engineering Appraisal Committee

The Genetic Engineering Appraisal Committee (GEAC) is set up under the Rules for the Manufacture, Use, Import, Export and Storage of Hazardous

---

[34] Ministry of Water Resources, River Development and Ganga Rejuvenation, 'A 21st Century Institutional Architecture for India's Water Reforms'.

[35] New genetic, cellular or organic materials created by combining such existing materials in research laboratories.

Micro Organisms, Genetically Engineered Organisms or Cells, 1989. When the rules were first notified, the GEAC was referred to as an approval committee. The name was changed to appraisal committee in 2010.[36] The GEAC is housed in the environment ministry to appraise applications seeking prior approval for activities involving 'large scale use of hazardous microorganisms and recombinants' and 'release of genetically engineered organisms and products into the environment'. This is for research or industrial production including experimental field trials.

An additional secretary of the environment ministry chairs the GEAC and a representative of the Department of Biotechnology is a co-chair. The other members are representatives of ministry of industrial development, biotechnology and atomic energy. Officials from the ICAR, ICMR, CSIR, CPCB, Directorate of Plant Protection and General Health Services are all the GEAC's members. Three 'outside experts' are appointed in their individual capacity, with the option of co-opting members/experts as and when required. An official of the environment ministry serves as the MS. The committee can take punitive actions under the EPA, 1986.

In addition, the State Biotechnology Coordination Committee and district-level committees also monitor the research and industrial application of GMOs and hazardous microorganisms and recombinants.

STATE OR UNION TERRITORY WETLAND AUTHORITIES

Wetland authorities are state-level bodies constituted by the central government under the Wetlands (Conservation and Management) Rules, 2017. They are chaired by the concerned state minister responsible for wetlands. In addition, there are 22 other members of the authority. The chief secretary of the state acts as the vice chairman. The ex officio members include secretaries from the departments of environment, forests, urban development, rural development, water resources, fisheries, irrigation and flood control, tourism, and revenue.

The director of State Remote Sensing Centre, the chief wildlife warden, member secretaries of the State Biodiversity Board and SPCB are also ex officio members. The additional PCCF of the concerned regional office of the environment ministry is also a part of the authority. In addition, one expert

---

[36] Crop Biotech Update mailing list, 'India's GEAC Renamed as Appraisal Committee', International Service for the Acquisition of Agri-biotech Applications (ISAAA), Crop Biotech Update, 2010), available at http://www.isaaa.org/kc/cropbiotechupdate/article/default.asp?ID=6475, accessed on 10 February 2020.

each from wetland ecology, hydrology, fisheries, landscape planning and socio-economics need to be as members. A senior official above the rank of a director in the department handling wetlands is designated as the MS. The authority can co-opt up to three members.

The rules outline a long list of functions for the authority; one of the primary ones being notifying a list of wetlands within six months of the publication of the Wetlands Rules, 2017. The authority also needs to identify a zone of influence around the wetlands and 'define strategies for conservation and wise use of wetlands'. Responsibilities also include overseeing the preparation of and reviewing of the integrated management plan for specific wetlands. The authority can also issue directions for conservation and sustainable management of wetlands.

Within 90 days of the publication of the Wetland Rules, state wetland authorities were to establish technical committees to review management plans and give inputs. A four-member grievance committee also was to be set up. These committees are to meet once every quarter.

National Board for Wild Life and Its Standing Committee

The National Board for Wild Life (NBWL) and the standing committee of the NBWL were established following an amendment to the Wild Life Protection Act (WLPA), 2002. Prior to this amendment, the national-level committee was known as the Indian Board for Wild Life.

The NBWL is constituted by the central government and is chaired by the prime minister of India. The mandate of this high-level committee is 'to promote the conservation and development of wild life and forests by such measures as it thinks fit'.

The other members are the environment minister and two members of parliament from the Lok Sabha and one from the Rajya Sabha. It also includes the member of Planning Commission in-charge of wildlife, 5 NGO representatives and 10 eminent conservationists, ecologists and environmentalists. The secretaries of the ministries of environment, defence, information and technology, finance, tribal welfare are also members along with the chief of Army. The list of members includes DG (forests), DG (tourism), DG (Indian Council of Forestry Research and Education), and the heads of Wildlife Institute of India (WII), Zoological Survey of India (ZSI), Botanical Survey of India (BSI), Indian Veterinary Research Institute, Central

Zoo Authority and National Institute of Oceanography. The non-official members are nominated for three years. The director of wildlife preservation at the environment ministry is the MS of this board.

This high-level committee is not involved in the everyday implementation of the objectives of the WLPA at the national level. For this, the environment ministry constitutes a 'Standing Committee' of the NBWL. This committee is set up with delegated powers to perform the duties and functions of the NBWL. The standing committee consists of a vice chairperson and MS from the environment ministry and 10 members nominated by the vice chairperson, which includes members of the NBWL.

The NBWL and any delegated standing committee are duty-bound to frame policies and promote wildlife conservation by controlling poaching and illegal trade of wildlife. The bodies make recommendations for the setting up and management of protected areas (PAs) as per the WLPA. Their powers include carrying out impact assessments of various projects and activities that impact wildlife. This may be within or outside PAs. The NBWL and the standing committee review the progress of wildlife conservation at a national level and are to prepare a status report every two years.

STATE BOARD FOR WILD LIFE

Every state government or UT has to set up a state board for wildlife (SBWL) as per the WLPA, 1972. The chief minister of the state, or the chief administrator of the UT, chairs the SBWL. Just as the NBWL, the minister responsible for wildlife, who could be the environment minister or any other, is the vice chairman of the state board. Other members include three members of the state legislature and in case of UTs, it is two members of the legislative assembly. Non-governmental membership includes 3 NGOs and 10 eminent conservationists, ecologists and environmentalists nominated by the state government. Two of them need to belong to ST communities.

The secretaries in charge of forest and wildlife and tribal affairs as well as the PCCF of the state are also members of these boards. Other members include the managing director of the state tourism development corporation, the IG of police and an Armed Forces representative not below the rank of a brigadier. Interdepartmental participation on the SBWL is through the directors of departments of animal husbandry and fisheries. Other institutions that are part of the SBWL are WII, BSI and ZSI. The chief wildlife warden of the state is the MS of the SBWL. There is no prescribed term for non-official members. This is left to each state to determine.

The duty of the SBWL includes the selection and management of PAs and the formulation of policies for protection and conservation of wildlife at the state level. Some SBWL notifications also list responsibilities such as harmonising of the needs of tribals and other forest-dwelling committees with those of wildlife conservation.[37]

NATIONAL BIODIVERSITY AUTHORITY

The environment ministry appoints the National Biodiversity Authority (NBA) under the BD Act, 2002. It is to be chaired by an 'eminent person having adequate knowledge and experience in conservation and sustainable use of biological diversity' along with matters related to equitable sharing of benefits. The members include three ex officio members from the tribal affairs and environment ministry. In addition, there are seven ex officio members from ministries dealing with agricultural research and education, biotechnology, ocean development, agriculture and cooperation, Indian systems of medicine and homoeopathy, science and technology, and scientific and industrial research, as the activities of these ministries involve the use of biological material. There are also five non-official members with specialised and scientific knowledge related to the objectives of the Act including representatives from 'industry, conservers, creators and knowledge holders of biological resources'. The term for the authority members is three years.

The NBA has the primary responsibility to regulate the access to biological material and related people's knowledge by foreign nationals. It also regulates access related to Intellectual Property Rights. The NBA notifies guidelines to carry forward any regulatory or conservation objectives and plays an advisory role to the environment ministry on all issues relating to biological diversity that are the subject matter of the Act, including the declaration of biodiversity heritage sites.

STATE BIODIVERSITY BOARD

As per the BD Act, 2002, every state government needs to appoint state biodiversity boards (SBBs). For UTs, the NBA would act as the SBB. The chair

---

[37] Government of Gujarat resolution no. WLP-1013-S.F.-23-W dated 22 September 2014.

of the biodiversity board requires a similar qualification as with the NBA. The members of the SBB include up to five ex officio members from concerned departments of the states. No specific departments have been mentioned in the law. It also needs to include up to five expert members from areas that are related to the objectives of the BD Act, 2002.

The SBBs are set up to advise the state governments on conservation, sustainable use and benefit sharing of biodiversity and related people's knowledge. These boards also regulate access related to approvals for any Indian individual or entity and have the powers to enter into benefit sharing agreements following a grant of such an approval. The state government can also assign additional functions to these boards as and when required.

Another significant power of an SBB is to restrict any activity that is 'detrimental or contrary' to biodiversity conservation, sustainable use or equitable sharing of benefits. This needs to be recommended following consultation with the local bodies, for example, panchayats, urban local bodies autonomous councils and so on.

BIODIVERSITY MANAGEMENT COMMITTEES

Urban and rural local bodies are to set up biodiversity management committees (BMCs) under the BD Act, 2002. The BD Act requires every 'local body' to constitute a BMC with six members to carry forward the three objectives of the Act including 'preservation of habitats, conservation of land races, folk varieties and cultivars, domesticated stocks and breeds of animals and microorganisms and chronicling of knowledge relating to biological diversity'. The role of the BMCs needs to be read along with the BD Rules, 2004, where their functions have been defined in detail. The chairperson of a BMC needs to be elected from one of the six nominated members. The term of the BMC is set for three years. Local members of legislative assembly/legislative council and the concerned member of parliament are special invitees to the meetings of the BMCs, though no specific decision-making powers have been assigned to them.

The rules elaborate the above functions to include preparation of Peoples' Biodiversity Register to document biological resources and related knowledge. It also allows BMCs to charge fees in case of any access in their area of jurisdiction.

3

Forest Reservation and Conservation

INTRODUCTION

Forest governance in India is a complex interplay of several old and new laws at the state and central levels. The laws regulating forests can be traced back to the precolonial and colonial periods in India when forests were demarcated and managed as hunting grounds, private estates and timber plantations. Colonial laws granted powers to the British government to legally reserve forests and create new forests. After independence, the administration of these regions passed on to the Indian Forest Service. Forest departments attached to state governments adopted the powers and functions to 'secure' forests. Due to these legacies of top-down regulatory control of forests, the questions of historical ownership and uses of forests by common people, especially *adivasis*, have remained highly contested in forest governance.

Forests were introduced into the concurrent list of the Indian constitution through the 42nd amendment in 1976. This allowed both the central and state governments to enact laws on forests. In 1980, the central government passed a law to give itself the powers to grant 'prior permission' for the diversion and dereservation of forest land in India. These legal changes to forest governance have been understood as the central government's establishment of political control over the country's forest resources. This centralisation has also been justified as a means of protecting forests from exploitation at the state level.

While central and state governments claim ownership and control of forest lands and forest produce, tribal and forest-dwelling communities have mounted challenges to these claims. Communities who live in forest regions and are dependent on forests for their socio-economic and cultural purposes have resisted being alienated from forest governance. Since 2006, the regulatory jurisdiction on forests is not limited to the central and state governments alone but includes local self-governments such as *gram* panchayats, autonomous district councils and *gram sabha*s (village assemblies).

The setting of forest boundaries, transfer of forest land ownership or use and compensating forest loss are matters that have attracted intense legal scrutiny. While the Indian Parliament and the executive have shaped forest legislations in India, the judiciary has exerted a strong influence on forest policies and the administration of forests. Since 1995, the Supreme Court has retained a continuing mandamus[1] on forest-related laws and decisions, especially through the Godavarman case which is discussed in detail in Section III of this chapter. This has allowed several regulatory aspects of forest laws to be substantially interpreted through judicial orders in cases where specific administrative directions were challenged or where the Court took *suo moto* cognizance of forest-related matters.

The legal framework on forests serve five functions: creating and managing forest reserves, regulating the access and use of forest produce, penalising forest destruction, regulating forest diversion and recognising forest rights. These are realised through a slew of statutes including the Indian Forest Act (IFA), 1927, and corresponding state forest legislations; the Forest Conservation Act (FCA), 1980; the Compensatory Afforestation Fund Act, 2016; and the Scheduled Tribes and Other Traditional Forest Dwellers (Recognition of Forest Rights) Act, 2006. In addition, there are several urban laws, which protect trees that are outside of officially designated forest areas. Laws enabling forest conservation with the objective of wildlife conservation have been discussed in Chapter 6.

---

[1] Articles 32 and 226 of the Indian Constitution give powers to the Supreme Court and high courts to 'issue writs' in case there is a breach of fundamental rights or 'any other matters'. Such cases allow the judiciary to 'control the administrative actions and prevent any kind of arbitrary use of power and discretion'. One such writ is that of 'mandamus' or 'we command'. Continuing mandamus implies a long-term monitoring of the compliance with judgments and directions of the Supreme Court by the Court itself. Available at http://www.manupatrafast.com/articles/PopOpenArticle. aspx?ID=deaf8251-5a4a-4c50-b8e1-7be4929c7b29&txtsearch=Subject:%20 Constitution, accessed on 20 July 2019.

This chapter is divided into five sections:

I.   *Reservation of Forests and Regulating Forest Produce*
II.   *Regulating Non-forest Use of Forest Land*
III. *Judicial Regulation of Forests*
IV.   *Forest Rights Act and Forest Diversions*
V.   *National Forest Policies and Programmes*

I   RESERVATION OF FORESTS AND REGULATING FOREST PRODUCE

The IFA was enacted in 1927 by the British colonial government with an objective to

> consolidate the law relating to forests, the transit of forest produce and the duty leviable on timber and other forest produce.

Following India's independence in 1947, several state governments enacted their own laws (and corresponding rules) using IFA as a template legislation. In such cases, the IFA stood repealed and the specific state laws are in operation.[2] In all other states, the IFA is applicable. While the provisions of state laws are largely similar to the IFA, governments have included specific changes.[3] Some states have also enacted specific forest laws for areas to be demarcated as private forests in addition to adopting the IFA's provisions.[4]

The IFA or any corresponding state laws demarcate forests into three categories: reserved forests (RFs), protected forests (PFs) and village forests (VFs). As of 2019, India has a notified forest area of 76.79 million hectares.

---

[2]   Response of the environment minister to a starred question in the Lok Sabha (starred question no. 119, answered on 28 June 2019), available at http://164.100.47.194/loksabha/Questions/QResult15.aspx?qref=1222&lsno=17, accessed on 20 July 2019.

[3]   For instance, the Karnataka Forest Act, 1963 (amendment in 1974), included *ramapatra* and *shigakai* within the definition of forest produce. Similarly, two amendments in 1976 and 1983 imposed a forest development tax on all the forest produce disposed of by the state government, which is not part of the IFA.

[4]   The Himachal Pradesh Private Forests Act, 1954, is related to management and exercise of rights in land that is not government land or is not protected or reserved under the IFA. This law gives powers to the state government to demarcate private forests in the state.

This includes 43.47 million hectares of RFs and 21.94 million hectares of PFs. The remaining forests of 11.38 million hectares are categorised as unclassed (or unclassified) forests, where the classification is still pending.[5]

The legal provision for VFs has been the least implemented provisions of the IFA, with few state governments prioritising their declaration. In 2014 and 2015, Maharashtra and Madhya Pradesh[6] enacted specific rules to initiate the declaration of VFs in any RF. This declaration allows forest-dwelling communities to control and manage forests including timber resources.[7]

The IFA and corresponding state laws lay down mechanisms to regulate collection, transit and sale of forest produce, including timber. They also elaborate the types of offences and penalties in instances of breach of the law. In 2019, the central government introduced a proposal to amend the IFA by strengthening the powers of the forest departments to police forest areas against poaching and other criminal activities more effectively[8]. The proposal also recommends the creation of production forests through private sector participation.

Tribal and forest-dwelling communities dependent on forests have argued that this law is an unjust regime which has restricted their rights and displaced them from their homes.[9] This conflict continues to define the relationship between forest departments, forest-dwelling communities and organisations working to bring justice to forest-dependent communities by officially recognising rights that were rendered unlawful by the IFA or corresponding state laws.

---

[5] Department-Related Parliamentary Standing Committee on Science and Technology and Environment and Forests, *Status of Forests in India: Three Hundred Twenty Fourth Report* (New Delhi: Rajya Sabha Secretariat, 2019).

[6] India Forests (Maharashtra) (Regulation of Assignment, Management and Cancellation of Village Forests) Rules, 2014, and the Madhya Pradesh Village Forest Rules, 2015.

[7] Centre for Environment Law (CEL), *New Village State Forest Rules* (New Delhi: World Wide Fund for Nature, 2016).

[8] MoEFCC (Ministry of Environment, Forests and Climate Change), letter, 'Comments/Views of State/UT Government on the Proposed Draft for Comprehensive Amendments to the Indian Forest Act, 1927-Reg' with copy of amendments dated 7 March 2019 (F. No. 2-1/1997-FP [Vol. 6]).

[9] Shankar Gopalakrishnan, 'The Conflict in India's Forests: Will State-Driven Expropriation Continue?' *Economic and Political Weekly* 54, no. 23 (2019), available at https://www.epw.in/node/154505/pdf, accessed on 10 September 2020.

Defining Forest Produce

The IFA does not define 'forest' or a forest area.[10] However, it gives a clear definition for what is included as 'forest produce' that is to be regulated under the law.

Forest produce is defined as

- produce such as 'timber, charcoal, caoutchouc, catechu[11], wood-oil, resin, natural varnish, bark, lac, mahua flowers, mahua seeds [Kuth] and myrabolams', whether or not they are found in or brought from a forest;
- produce which are found in or brought from a forest:
    o   trees and leaves, flowers and fruits, or their parts
    o   plants that are not trees (including grass, creepers, reeds and moss) or their parts
    o   wild animals and skins, tusks, horns, bones, silk, cocoons, honey and wax, and any other parts that animals produce
    o   peat, surface soil, rock and minerals (including limestone, laterite, mineral oils and all products of mines or quarries).

The IFA also defines a 'tree' to include palms, stumps, brush-wood and canes. Until recently, the definition of tree also included bamboo. It was excluded from the purview of the IFA through the Indian Forest Amendment Act, 2017 (No. 5 of 2018).
The IFA defines timber as

> trees when they have fallen or have been felled, and all wood whether cut
> up or fashioned or hollowed out for any purpose or not.

These definitions and the specific clauses they attract are important to understand the vast jurisdiction of the IFA, 1927, or corresponding state-level laws and how these laws regulate the use of forest produce, trees and timber.

Demarcating Reserved Forests

The IFA allows all state governments, that is, state forest departments to create 'Reserved Forests'. Reserved Forests can be created on any 'forest-

---

[10] There is no legal definition of 'forests' in India under any law. The operative definition is from the Supreme Court in Writ Petition (Civil) 202 of 1995 that applies the dictionary meaning of forests for the purposes of regulating forest diversions. This is discussed further in this chapter.

[11] Caoutchouc is natural rubber and catechu is an extract of acacia tree.

land' or 'waste-land' which is the 'property' of the government or where the government has proprietary rights. An area can also be reserved if it contains a particular forest produce that requires protection. This declaration of an RF makes it possible for state governments to stake ownership, restrict rights of use and penalise[12] any contravention of the law.

The forest department needs to issue a notification to declare any area as an RF. A forest settlement officer is assigned to 'inquire into and determine the existence, nature and extent of any rights' related to land ownership, occupational use or forest produce that may be affected by the creation of an RF. Following this enquiry, the settlement officer has to issue a 'proclamation' in the local language highlighting the specific area proposed as RF and the rights that are likely to be affected. Rights holders need to respond to this proclamation within three months.

The IFA does not specify what kinds of land should be included or left out of an RF. However, specific state Acts have listed exclusions in enforcement manuals. The Karnataka Forest Manual states that cultivable lands under tanks and irrigation channels 'should' be excluded from RF notifications. Land already 'occupied' should also be excluded, provided that leaving out such land from the reservation does not lead to 'disfiguring the shape of the forest'.[13]

The IFA also deals with shifting cultivation lands which may fall within proposed RFs. The prescribed process puts all responsibility on the forest settlement officer to record all claims and recommend to the state government whether shifting cultivation should be 'permitted or prohibited wholly or in part'.

The final notification of the RF needs to have clear details of the boundary demarcated on the ground. Any appeal against the process or final notification can be filed before the district collector or the forest court[14] established by the state government.

Any RF can also be dereserved by the state government at any point of time. This also requires issuance of a notification.

---

[12] Chapter IX, 'Penalties and Procedure' (Sections 52–69) of IFA, 1927 (as amended in 1951).

[13] The Karnataka Forest Manual, available at https://www.karnataka.gov.in/forestsecretariat/Downloads/KFD_Manual_1976.pdf, accessed on 20 July 2019.

[14] Chapter II of the IFA provides for the state government to establish forest courts composed of three persons appointed by the state government.

COLLECTION AND TRANSPORTATION OF TIMBER AND FOREST PRODUCE

The state government has the power to specify routes through which forest produce or timber can be transported within and outside the state boundaries. This includes import and export of these commodities for which a transit pass has to be secured. Concerned officials of the forest department grant these passes. The department also monitors human and vehicular movement through its forest checkposts.

The forest department can establish depots for short-term and long-term storage of both timber and forest produce and also facilitate its sale. The IFA and corresponding state laws allow the forest department to levy a duty, fee or royalty as price for extraction or transportation of forest produce and timber.

OFFENCES AND PENALTIES

The IFA and corresponding state laws have strong penalties against the contravention of the law. This includes both fines and imprisonment. For instance, causing harm to trees, removal of forest produce without permission, breaking up or clearing forests for cultivation, fire and any other activity prohibited under IFA or state laws can be recorded as an offence. These activities can attract imprisonment up to six months and a fine of ₹500 or both.

These offences can be tried by the district magistrate. Forest officers can also seize timber and any other forest produce extracted or transported without due permissions.

II   REGULATING NON-FOREST USE OF FOREST LAND

The FCA[15] was enacted in 1980 with the stated purpose of checking the indiscriminate destruction of forest lands for infrastructure, plantations or mining at the state level. Official documents state that prior to the enactment of this law, 0.15 million hectares of forests were diverted to other uses between 1950 and 1980[16]. The enactment of this law gave powers to the

---

[15] Forest Conservation Act, 1980.

[16] Ministry of Environment and Forests (MoEF), *Handbook of Forest (Conservation) Act, 1980 (with Amendments Made in 1988), Forest (Conservation) Rules, 2003 (with Amendments Made in 2004), Guidelines & Clarifications (up to June, 2004)* (New Delhi: Government of India, 2004).

central government to regulate the 'non-forest use' of forest areas through forest diversion, dereservation of RFs and tree felling in forest areas. A 2004 handbook of the environment ministry claims that the diversion rates came down to 0.038 million hectares per annum once the FCA was enacted.[17]

The FCA defines 'non-forest' purpose as

> the breaking up or clearing of any forest land or portion thereof for— but does not include any work relating or ancillary to conservation, development and management of forests and wildlife, namely, the establishment of check-posts, fire lines, wireless communications and construction of fencing, bridges and culverts, dams, waterholes, trench marks, boundary marks, pipelines or other like purposes.[18]

'Non forest' purposes include the cultivation of tea, coffee, spices, rubber, palms, oil-bearing plants, horticultural crops or medicinal plants and for any purpose other than reafforestation.

The centralisation of powers to regulate forest use for development through the FCA created institutionalised practices of 'forest clearance' or the process of approving projects that need access to forest areas. Some questions that have been recently clarified through legal debates are how forests are to be valued by development projects, how forests lost in development projects are to be compensated, who has the final powers for approving forest diversions, under what conditions can trees be felled in a forest area and is the consent of forest-dwelling communities required for non-forest use.

Forest Clearance Process

The FCA appoints a Forest Advisory Committee (FAC) to appraise applications of projects that require to fell trees or use forest areas. This is part of a process that is popularly understood as 'Forest Clearance'. Approval is required for felling of trees in forest areas irrespective of whether or not ownership is transferred to another user agency or whether the felling requires dereservation of the forest land. By its design, this two-page law ensures that no state government can approve any proposals for diversion, dereservation or tree felling in forest areas without a prior approval of the environment ministry.

---

[17] Ibid.

[18] Section 2 (Restriction on the de-reservation of forests or use of forest land for non-forest purpose) of the FCA, 1980.

Project entities referred to as 'User agencies' who seek forest diversions or tree-felling permissions send their applications to the state forest departments. These applications are scrutinised and decided upon by the ministry's central office (if over 40 hectares of land is involved) or regional offices (if less than 40 hectares of land is involved). Both these offices rely on the recommendations of regional and central FACs. The powers of these committees are discussed in the Forest Conservation Rules, 1981 (amended in 2003 and 2004).

The central environment ministry has developed the practice of a two-stage scrutiny of all proposals forwarded by state governments, even though there was no mention of these in the statute or corresponding rules until 2014.[19] These are the in- principle (Stage I) approval, which is conditional to the user agencies fulfilling regulatory formalities, payments or additional studies, and the final (Stage II) approval which is the environment ministry's prior approval to the state governments.

In 1992, a question was posed in the Lok Sabha, the lower house of the Indian Parliament, on whether the central government had received representations from state governments to repeal/amend the FCA, 1980. The response of Minister Shri Arjun Singh was that the FC rules were modified in May 1992 to accommodate this concern of state governments that the Act was causing delays in implementing projects. The minister responded that a decision was taken to decentralise and streamline the examination of forest diversion proposals. In addition, specific guidelines were issued in October 1992 on how proposals for diversion and tree felling should be examined.[20]

The question of whether states or the centre hold the final powers to approve or reject the proposals for forest diversion or tree felling was clarified by a judgment of the National Green Tribunal (NGT) in 2012.[21] The clarification came in a case in which the petitioners challenged the central environment ministry's approval for forest diversion for a hydro-electric project in Chamoli district of Uttarakhand. The judgment concluded that

---

[19] MoEF Notification G.S.R. 185(E) dated 14 March 2014.

[20] Response by Shri Arjun Singh, Minister of Human Resource Development, to starred question (No. 115) raised by Girdhari Lal Bhargav in Session 5 of 10th Lok Sabha, available at https://parliamentofindia.nic.in/ls/lsdeb/ls10/ses5/0301129205.htm, accessed on 6 January 2019.

[21] National Green Tribunal judgment dated 7 November 2012 in Appeal No. 7 of 2012 (*Vimal Bhai & Ors v. Union of India & Ors*).

it was the state governments, that is, state forest departments, that held the authority to issue final directions for diversion, dereservation of forests or tree felling. These powers are under Section 2 of the FCA.

COMPENSATORY AFFORESTATION

The FCA and its corresponding rules maintain that loss of forests due to development should be offset or compensated for. Originally, the compensatory clauses were to ensure that the amount of forest that is with the forest department is not reduced.

The requirement for compensatory afforestation (CA) is considered to be one of the most important conditions accompanying forest clearance approvals. All proposals for forest diversion, dereservation and tree felling are made with a comprehensive scheme for CA. The Forest (Conservation) Rules, 1981 (as amended in 2003),[22] requires the forest department to fill out detailed forms as the department moves the application on behalf of the user agency.

Compensatory afforestation is to be carried out over non-forest land equal to the extent of forest land required by the user agency. For example, for 100 hectares needed by a user agency for non-forest purpose, CA is to be carried out in 100 hectares of non-forest land. In case of non-availability of non-forest land, afforestation needs to be taken up on double the extent of degraded forest land.

CA areas need to be contiguous with or in proximity to an existing RF or PF. This is primarily to enable the forest department to effectively manage the 'newly planted area'. Looking for a distant site for afforestation outside the district or state is allowed only if land in that particular state is not available. The guidelines allow for other category of forests, which are recognised by the IFA, 1927, and on which FCA, 1980, is applicable, to be used for CA. The CA allows the government to bring revenue lands or categories of land such as *zudpi jungle/chhote/bade jhar ka jungle/jungle-jhari* land/civil–*soyam* lands under the jurisdiction of the forest department through transfer and mutation. In case there is shortage of forest land, the user agency can also purchase equal amount of private land close to a reserve forest and use it for CA.[23]

---

[22] Form for seeking prior approval under Section 2 of the proposals by state governments and other authorities; see Rule 4, FC Rules, 2003.

[23] MoEFCC, *Handbook of Forest (Conservation) Act, 1980 and Forest (Conservation) Rules, 2003 (Guidelines & Clarifications)* (New Delhi: Government of India, 2019).

In most instances, the identification of land and transfer of funds for CA results in Stage II (final) approval from the central environment ministry. Two cases that came up before the Godavarman bench of the Supreme Court (discussed in detail later) present the challenges to the implementation of CA. The central environment ministry has issued specific guidelines related to how CA is to be carried out and monitored. [24]

Two companies, the South Eastern Coalfields Limited (SECL)[25] and the National Mineral Development Corporation (NMDC)[26] in Madhya Pradesh, brought certain challenges to the attention of the Supreme Court as far back as in 1998 and 1999. The companies approached the apex court seeking permissions for tree felling which had been restrained across the country following orders of the Court in Writ Petition (Civil) 202 of 1995. As per the 12 December 1996 order in this case, tree felling could be carried out only in accordance with the working plan of the forest department. The NMDC case was related to six mining leases in Bailadila, located in Bastar district of Madhya Pradesh, and the SECL case concerned operations in Chirmiri Colliery and West Chirmiri Colliery in Manendragarh district of the state.[27] The Court sought details of compliance of CA by NMDC, SECL and the state government.

These matters were heard along with Interlocutory Application (I.A.) 566 that was carved out of statements made by Additional Solicitor General K. N. Rawal before the Supreme Court on 17 April 2000 where the poor compliance with CA was presented along with the need to upgrade the mechanism. The apex court recorded these statements as a separate I.A. and issued notices to all the state governments that had shown poor progress in the utilisation of funds and submission of their quarterly progress reports to the environment ministry.[28] The Court heard the matter for two years. During this time, all state

---

[24] Ibid.

[25] Interlocutory Application (I.A.) 574 in Writ Petition (Civil) 202 of 1995 (*T. N. Godavarman Thirumlkpad v. Union of India*).

[26] I.A. No. 419 and 420 in Writ Petition (Civil) 202 of 1995 (*T. N. Godavarman Thirumlkpad v. Union of India*).

[27] Both these areas are now in the state of Chhattisgarh. Madhya Pradesh was bifurcated into Madhya Pradesh and Chhattisgarh in November 2000.

[28] CEC (Central Empowered Committee), 'Recommendations of the Central Empowered Committee in Interlocutory Application No. 566 of 2000 in Writ Petition (Civil) 202 of 1995', 9 August 2002.

governments had to file details of the financial expenditure and compensatory afforestation undertaken by them. The environment ministry was directed to develop a fresh scheme. The ministry submitted a proposed scheme on 22 March 2002 through an affidavit.

In May 2002, the Court directed the appointment of a Central Empowered Committee (CEC)[29] to assist in the proceedings and monitor the implementation of the Court's orders. The CEC submitted its report on 9 August 2002. This report concluded that out of the total funds recovered from user agencies towards CA, less than 60 per cent was spent. And if land area over which CA to be undertaken is considered, then only 61 per cent of the total target area had been covered.[30] The Supreme Court gave directions to establish a centralised Compensatory Afforestation Management and Planning Authority (CAMPA) to deal with these inefficiencies related to CA funds.

The main issues that have created legal tangles in CA are the following:

- *Contestations over existing uses of land identified for CA*: The lands identified for CA have their own unique uses and management practices. Conversion of these lands to CA areas leads to conflicts with existing uses. For instance, in the central Indian state of Chhattisgarh, the Chhattisgarh Land Revenue Code, 1959, is the operational legislation for all forest and revenue land. This code and a 2002 clarification by the state government recognises the rights of anyone who has planted fruit-bearing trees in 'unoccupied land' which also includes revenue forests like *chhote/bade jhar ka jungle* in the state of Madhya Pradesh.[31] The use of this land for CA, which was also encouraged by a 2017 direction of

---

[29] The CEC was set up vide notification number File No. 1-1/CEC/SC/2002 dated 3 June 2002. It was chaired by Shri P. V. Jayakrishnan, the then secretary of the environment ministry. He continued to head the committee even after his retirement.

[30] Supreme Court order dated 3 April 2000 in I.A. No. 419 and 420 in Writ Petition (Civil) 202 of 1995 observes that the environment ministry 'has clearly been remiss in this respect'. The same order concluded that even though state governments had received the money from user agencies, they have not actually spent the money.

[31] ELDF (Environment Law and Development Foundation), *Protecting and Conserving Commons for Common Good: Needs a Fresh Legal Perspective—An Analysis of the State of Chhattisgarh* (New Delhi: ELDF, 2012).

the central environment ministry, brings CA directly in conflict with prevailing rights of managing and harvesting fruit trees.[32]

• *Non availability of land and creation of land banks for CA*: In recent times, the user agencies have rarely been able to identify non-forest or degraded forest land [33]and have often tried to purchase private revenue land to hand over to the forest department for CA. The current practice is to encourage state governments to identify non-forest land and transfer them to a land bank of the forest department, making them readily available for CA at the time approvals for diversion are sought.[34] While state governments are encouraged by this proposal, it has drawn criticism from community-based forest rights organisations and research groups because afforestation activity curtails existing use rights.[35]

• *Non utilisation and misuse of CA funds*: Despite the creation of CAMPA, the non-utilisation and misuse of CA funds has remained a problem. There have been at least three national-level audits of the Comptroller and Auditor General (CAG) of India[36] on the issue of how funds for CA collected from user agencies have been collected and ulitised. In addition, there have also been state-specific CAG audits on compliance with CA procedures.[37] Comptroller and auditor general audits and court cases have drawn attention to the inadequate or improper utilisation of funds for CA.

---

[32] Jayashree Nandi, 'Community Forests, Common Lands Threatened by New Plantation', *Times of India*, 10 November 2017, available at https://timesofindia. indiatimes.com/home/environment/the-good-earth/community-forests-common-lands-threatened-by-new-plantation-guidelines/articleshow/61592323.cms, accessed on 11 September 2020.

[33] Letter from Additional Chief Secretary, Revenue and DM Department, Government of Odisha, to all district collectors dated 4 August 2014 indicating difficulties in carrying out CA and allowing for 'bilateral' purchase of land, (No. 22958-GE (GL)-S-31/2014.

[34] Letter No. F. No. 11-423/2011-FC dated 8 November 2017 from MoEFCC (Forest Conservation Division) to all state governments.

[35] Chitrangadha Choudhury, 'A Rs 56,000-Cr "Afforestation" Fund Threatens India's Indigenous Communities', *IndiaSpend*, 25 June 2019, available at https://www. indiaspend.com/a-rs-56000-cr-afforestation-fund-threatens-indias-indigenous-communities/, accessed on 10 June 2020.

[36] The national-level CAG audits were carried out in 2004, 2007 and 2013.

[37] Comptroller and Auditor General, *Report No. 3 of 2018: Social, General and Economic Sectors (Non-PSU) Government of Delhi* (New Delhi: Government of India, 2018).

Besides the above implementation-related issues, the debates on compensatory or offset mechanisms for forest loss due to development also include critiques of financialisation and monetisation of valuable forest resources and loss of forest ecologies.[38]

VALUATION OF FORESTS

The genesis of the administrative practice of valuation of forests at the time of diversion is in the Supreme Court's hearings in *T. N. Godavarman v. Union of India* (Writ Petition [Civil] 202 of 1995). In this case, the Court scrutinised the poor implementation of CA.[39] The Court also opened up the question of whether this model to offset forest loss was adequate to conserve and protect forests. A report dated 9 August 2002 was filed by the Court-appointed CEC. In this report, the committee outlined the problems in the state-level implementation of CA. The report also stated,

> the States/UTs as well as Ministry of Environment and Forests are of the view that in addition to the funds realized for compensatory afforestation, the Net Present Value of forest land being diverted for non-forestry purposes should also be recovered from the user agencies. The money so recovered could be utilized for undertaking forest protection, other conservation measures and related activities.

In the same report, the CEC also recommended putting guidelines in place for collection of net present value (NPV) from a user agency.

According to the arguments presented in the Supreme Court, the practice of collecting funds from user agencies in the form of NPV, in addition to CA, was already in place in the states of Madhya Pradesh, Chhattisgarh and Bihar. Their experiences were a critical basis for the discussions in the Supreme Court on determining NPV appropriate to the quality and extent of forest loss. According to the report dated 9 August 2002 of the Court-appointed CEC, the underlying principle behind levying NPV was that plantations raised under CA scheme cannot immediately and adequately compensate for

---

[38] K. Kohli and M. Menon, *Business Interests and the Environmental Crisis* (New Delhi: SAGE Publications, 2016).

[39] K. Kohli, M. Menon, V. Samdariya and S. Guptabhaya, *Pocketful of Forests: Legal Debates on Valuating and Compensating Forest Loss in India* (New Delhi: Kalpavriksh and WWF [World Wildlife Fund] India, 2011).

loss of services provided by forests. The NPV was seen as a cost extracted from the user agency to compensate for the loss of these services due to diversion proposals.[40]

The SC in its order dated 26 September 2005[41] defined NPV as 'the present value (PV) of net cash flow from a project, discounted by the cost of capital'. In simple terms, it is arrived at by deducting the cost of investment in generating forests from the present value of all future services. When applied to forest land diversion, NPV is understood as a compensation, in monetary terms, for the loss of tangible as well as intangible future benefits of that forest. The new user of the forest land is expected to bear the cost of these losses by the payment of NPV.

The arguments in this case led to the institutionalised practice of charging NPV to user agencies. This is now a formal part of the approval process for forest diversions. Over the years, several individual project proponents and public utility projects have appealed to the Supreme Court and the environment ministry for exemptions, reductions or review of payments to be made by them for the diversion of forest land.[42]

ESTABLISHMENT OF CAMPA

As stated in the earlier section, the establishment of a central-level authority to manage and administer funds was entirely driven by the discussions in Writ Petition (Civil) 202 of 1995. In October 2002, the Supreme Court directed that a body be constituted to manage funds collected from user agencies for

---

[40] CEC, 'Recommendations of the Central Empowered Committee in Interlocutory Application No. 566 of 2000'.

[41] I.A. No. 566 with I.A. No. 932 in 819–821, 955, 958, 985, 1001.1001a, 1013, 1014, 1016–1018, 1019, 1046, 1047, 1135–1136, 1137, 1164, 1180, 1181 and 1182–1183, 1196, 1208–1209, 1222–1223, 1224–1225, 1229, 1233, 1248–1249, 1253, 1301–1302, 1303–1304, 1312, 1313, 1314, 1315–1316, 1318 and 1319 in Writ Petition (Civil) No. 202 of 1995 (*T. N. Godavarman Thirumlkpad v. Union of India*).

[42] Ministry of Environment and Forests, *Guidelines for Diversion of Forest Land for Non-forestry Purposes under Forest (Conservation) Act, 1980–Guidelines for Collection of Net Present Value (NPV)* (New Delhi: MoEF, Government of India, 2009), 5 February 2009; MoEF letter no. F. No. 5-1/98-FC (Pt-ll) dated 19 December 2005 is related to *Exemption of Certain Government Projects from Payment of Net Present Value (NPV): Guidelines for Collection of NPV under Forest (Conservation) Act, 1980 for Diversion of Forestland for Non-forestry Use.*

the diversion of forest land for non-forest purposes.[43] This body termed as the CAMPA was notified by the central government in April 2004 [44] and given powers to manage the funds collected for CA, NPV and any other funds recoverable under the FCA for non-forest uses of forest land. This order clarified that all parties including the Ministry of Environment and Forests (MoEF) agreed to the creation of a Compensatory Afforestation Fund (CAF) to be managed by the CAMPA.

Most of the state governments did not object to the recovery of CA and NPV funds from user agencies. However, in May 2005, during a detailed hearing in the I.A. No. 826 in I.A. No. 526 in Writ Petition (Civil) 202 of 1995, the recommendations of the CEC report dated 9 August 2002 were discussed. The state government of Kerala questioned the constitutional validity of the CAMPA. Its challenge was specifically to Clause 6.4 of the CAMPA notification that dealt with the disbursement of funds by the CAMPA.[45] The Kerala government argued that forests are the property of the state government and that the funds accruing from such property cannot be held by any person or authority on behalf of the government. The Kerala state government argued that funds collected by the CAMPA should be paid directly to the state government.

The Court's response to this issue is detailed in its judgment dated 26 September 2005.[46] The 2005 judgment of the Supreme Court in this case concluded that the payments collected by the CAMPA are for the protection of the environment and not in relation to any 'proprietary right'. The judgment states that the collection of monies by the CAMPA are for the purposes of conserving a 'national', 'intergenerational', 'public' asset and not government property and that it is a fee that is levied to undertake some activities that are akin to economic and social planning. Rather than a revenue earned by the government for the sale or use of its property, the monies collected by

---

[43] Supreme Court of India order dated 30 October 2002 in I.A. 566 in Writ Petition (Civil) No. 202 of 1995 (*T. N. Godavarman Thirumlkpad v. Union of India*).

[44] MoEF Order No. S.O. 525(E) dated 23 April 2004.

[45] R. Dutta and K. Kohli, eds., *Forest Case Update*, no. 11 (April 2005): 3–4, available at http://forestcaseindia.org/forest-case-updates/updates-2005/issue-11-april-2005/, accessed on 10 September 2020.

[46] Supreme Court judgment dated 29 September 2005 in I.A. No. 826 in I.A. No. 566 in Writ Petition (Civil) No. 202 of 1995 (*T. N. Godavarman Thirumlkpad v. Union of India*).

the CAMPA are to 'carry out the statutory and constitutional obligations', that is, the protection of natural resources. The Court clarified that the clauses in the CAMPA notification laid down measures to ensure 'financial discipline, transparency and accountability' and that the system of double-entry bookkeeping would be used in accounting.

The setting up of the CAMPA proved difficult. The MoEF had issued the notification under the Subsection (3) of Section 3 of the Environment Protection Act, 1986 on 23 April 2004. However, the authority was not set up and made functional. Two factors may have contributed to this delay. First, the composition of the committee was challenged before the Court due to the lack of non-governmental representation. Second, user agencies, especially public sector undertakings, questioned the imposition of the NPV in addition to CA and sought exemptions from the payment.[47]

The Supreme Court on 5 May 2006 took cognisance of this and directed that an ad hoc CAMPA be set up until a fully functional authority could be put in place.[48] All this while, the money paid by user agencies was being deposited with the CEC.[49] The MoEF appointed an ad hoc CAMPA and a chief executive officer (CEO) by 2007 and began the process of storing funds collected from user agencies.

On 5 May 2008, the central government introduced the Compensatory Afforestation Fund (CAF) Bill, 2008, in the Lok Sabha. The union cabinet had approved the same in March 2008.[50] This bill drew its justifications entirely from the directions of the Court orders on the CAMPA. The bill sought to create the CAMPA by an Act of parliament under Entry 17A (Forests) of the Concurrent List of the constitution.

The Lok Sabha passed the bill in one sitting on 23 December 2008. However, the bill was held up in the Rajya Sabha. It was referred to the

---

[47] Kohli, Menon, Samdariya and Guptabhaya, *Pocketful of Forests*.

[48] CEC, *Report of the CEC Regarding Non-utilisation of Funds Received towards Net Present Value (NPV), Compensatory Afforestation etc. and the Proposed Mechanism for Its Utilisation* (New Delhi: CEC, 2008), 10 January.

[49] CEC, *Report of the CEC about Receipts and Utilisation of Funds and Related Issues*, 2nd Report, I.A. 827 (New Delhi: CEC, 2004), 12 October.

[50] Press Trust of India, 'Compensatory Afforestation Fund Management to Be Created', 13 March 2008, *Outlook Magazine*, available at https://www.outlookindia.com/newswire/story/compensatory-afforestation-fund-management-to-be-created/4803, accessed on 10 September 2020.

Department-Related Parliamentary Standing Committee on Science and Technology and Environment and Forests. The committee was highly critical of the CAMPA and how the MoEF had presented its case. The committee stated,

> [T]he Ministry had not presented the real and complete picture of diversion of funds before the Supreme Court or before this Committee, nor did it take pains to suitably defend the case. The Ministry also admitted that at some stage or the other the directions of the Supreme Court should have been contested, but the same had not been contested.[51]

The committee's objections were as follows:

- By creating a super body such as the CAMPA, the government would centralise forest resources, and this was against the federal character of the constitution. The report suggested that the states should be allowed to generate funds and utilise the same.

- The bill legitimised the monetary compensation for diversion of forest land. It was based on the assumption that collection of monetary compensation and tree plantation would achieve forest conservation.

The bill was rejected in the Rajya Sabha, and it lapsed. The government began a process of building consensus with state governments to set up a state-level CAMPA to receive funds from the ad hoc CAMPA.[52] A note from the Prime Minister's Office (PMO) in June 2009 emphasised that a

> consensual approach to reconcile the Supreme Court's direction for setting up a ring fenced fund and the concerns of the Department Related Parliamentary Standing Committee and States with adequate flexibility was worked out at the meeting held with the State Chief Secretaries and draft guidelines formulated in consultation with the Law Secretary have since been circulated among States for setting up of the State Compensatory Afforestation, Green Management and Planning and Management Authority (State CAMPA).

The note also directed the MoEF to expedite this process to approach the Supreme Court for concurrence on this arrangement.

---

[51] Department-Related Parliamentary Standing Committee on Science and Technology and Environment and Forests, *One Hundred and Ninety Fourth Report on the Compensatory Afforestation Fund Bill, 2008* (New Delhi: Rajya Sabha Secretariat, 2008).
[52] Note (PMO ID no. 250/31/C/6/07-ES-II) from the PMO dated 11 June 2009.

In July 2009, the environment ministry finalised guidelines for the setting up of state-level CAMPA. The Supreme Court also approved the CEC's scheme for the disbursement of funds collected in the ad hoc CAMPA to the states. But the Court cautioned that all the funds should not be released at once. In a July 2009 order the Court suggested that the funds could be released in tranches of ₹10 billion per year for the first five years.[53]

The CAF Act was finally enacted in 2016. This was enabled by a change of government at the centre and its concern that an amount of ₹520 billion had to be unlocked from the ad hoc CAMPA. With the supporting rules in place in 2018,[54] a formal CAMPA was finally established at the centre.[55]

REGULATING TREE FELLING IN URBAN AREAS

Tree felling in urban areas takes place on recorded forest land, private lands and land belonging to other government agencies, for example, development authorities, public works departments, revenue department and urban development departments. If such tree felling is being carried out on forest land, then all the provisions of the IFA, 1927; corresponding state laws; and FCA, 1980, will be applicable.

In case the trees outside officially recorded forest areas are to be felled, then the provisions of state-specific laws come into play. Some of these laws have been in existence since the 1960s such as the Saurashtra Felling of Trees (Infliction of Punishment) (Extension and Amendment) Act, 1960; the Maharashtra (Urban Areas) Preservation of Trees Act, 1975; the Karnataka Preservation of Trees Act, 1976; the Goa, Daman and Diu Preservation of Trees Act, 1984; and the Delhi Preservation of Trees Act, 1994 (in short, Tree Acts).

These and several other urban trees-related laws have the objectives of preserving urban trees and regulating its felling. However, some of these

---

[53] As discussed in Supreme Court judgment dated 12 March 2014 in I.A. Nos. 2143 with 2283, 3088, 3461, 3479, 3693 in 2143, 827, 1122, 1337, 1473 and 1620 and 1693 in 1473 and 3618 in Writ Petition (Civil) No. 202 of 1995 (*T. N. Godavarman Thirumlkpad v. Union of India*).

[54] Compensatory Afforestation Rules, 2010, vide Notification No. G.S.R. 766(E) dated 10 August 2018.

[55] Press Information Bureau, 'Release of Compensatory Afforestation Fund to States Will Help in Meeting Nation's INDCs: Dr. Harsh Vardhan (CAF Act Will Come into Force Next Month', MoEFCC, New Delhi, 2018.

laws elaborate these goals much more than the others. For instance, the 1976 Karnataka Tree Act states,

> Industrialisation and pressure of population have resulted in heavy destruction of tree growth in urban areas. Trees which provide shade, mitigate the extremes of climate, render aesthetic beauty, purify the polluted atmosphere, mute the noise, have been one of the first casualties of pressure on space in our cities and towns.

The legal and institutional structure of these laws include the following:

- Prior Permission for Tree Felling: Prior permission from a tree officer is required for the felling of a single tree or multiple trees. This permission is required whether the tree is on private land or public land. The tree officer is an official of the state forest department. The state forest department also suggests special measures required for pruning of trees causing minimum harm and encouraging preservation. Felling and pruning of trees without prior permissions attracts provisions of offences and penalties listed in specific state level laws. Approval for felling of trees is also conditional on compensating for the loss of trees.

- Constitution of a Tree Authority: The Tree Authority is a monitoring and oversight body under some of the earlier-mentioned laws. The authority has the responsibility of carrying out comprehensive tree census to prepare a tree baseline for the state. It also has powers to monitor the implementation of the compensatory plantations carried out in lieu of tree felling permissions granted. The Tree Authority can also undertake critical studies of the proposals of various government departments and private bodies on tree felling.

Recent judgments[56] and ongoing litigation[57] discuss the implementation and efficacy of the Tree Acts and the need to strengthen their institutional structures.

---

[56] Bombay High Court judgment dated 8 March 2018 in Public Interest Litigation Writ Petition. No. 1 of 2018 (*Federation of Rainbow Warriors v. The Deputy Conservator of Forests (North Goa Division) Forest Department & Others*); Bombay High Court judgment dated 4 October 2019 in Writ Petition No. 1487 of 2019 (*Vanashakti & Ors v. Union of India & Ors*); National Green Tribunal judgment dated 23 April 2013 in Application No. 82 of 2013 (*Aditya N. Prasad v. Union of India & Ors*).

[57] Writ Petition No. 17841 of 2018 in Karnataka High Court (*Dattatraya. T. Devare and Other v. State of Karnataka & Ors*).

III JUDICIAL REGULATION OF FORESTS

Forest matters have been discussed in the Supreme Court, the High Courts and the National Green Tribunal. This section discusses the Supreme Court's intervention on forest matters through the Godavarman bench and the NGT's role in forest diversion cases. Since 1995, the Supreme Court of India has played a proactive role in forest regulation and governance. The interest of the Supreme Court in forest issues is best illustrated through the ongoing matter, *T. N. Godavarman Thirumulpad v. Union of India* (Writ Petition [Civil] No. 202 of 1995), popularly known as the Forest case or the Godavarman case. The case can be traced back to the Supreme Court's actions against large-scale illegal felling of timber and destruction of forests in Gudalur *taluk* (block), Tamil Nadu.

One of the initial orders in this case substantially changed the manner in which forest lands are defined, managed and governed. The order of 12 December1996 expanded the meaning of the word 'forests' to be understood as its dictionary meaning rather than an administrative category. This expansive definition made the FCA, 1980, applicable to all standing forests. Following this definition, use of any forests for non-forest purposes requires an approval under the FCA and its corresponding rules. [58]

The following sections outline two major aspects of the judicial regulation of forests. The first is with respect to the identification of deemed forests, and the second is regarding the Court's regulation of wood-based industries. These are only two thematic aspects of the ongoing Godavarman case, some others being decisions on the temporary working permits for mines, the composition of forest advisory committee and violations of forest diversion procedures.

The Court directed state governments to carry out a detailed exercise of identifying 'deemed forests', so as to rest any controversy regarding which areas should be recognised as forest lands. This exercise to identify and record deemed forests has not been completed in several states. This aspect also had a bearing on specific project-related decisions. For example, a May 2010 affidavit of the MoEF in Application No. 1209 of 2009 (*Kudremukh Wildlife Foundation & Others versus Union of India & Ors*) restricted project approvals in the forests of the Western Ghats due to the status of deemed forests. It states,

---

[58] R. Dutta and B. Yadav, *Supreme Court on Forest Conservation* (New Delhi: Universal Law Publishing Co., 2005); ELDF and WWF India, *Protection of Forests in India: The Godavarman Story* (New Delhi: WWF, 2009).

That, until the list of deemed forests in Karnataka gets amended as recommended above, it is necessary that all projects in Western Ghats region involving tourism, wind power and hydro power, which are on government lands which for the description/inter-alia given in para (II) above (irrespective of their classification in the revenue records), that are under consideration or processing by State and/or Central Governments, should be stayed immediately.

A 2013 report of the CEC in this case directed the state government to identify deemed forests in the Western Ghats. Until such time, no forest diversions would be permitted.[59] The Karnataka government submitted its report on deemed forests to the Supreme Court in 2015.[60] Subsequently, the state government initiated the process of removing revenue lands from deemed forest categorisation. The government filed an affidavit in the Supreme Court for the same, citing complaints received by people affected by deemed forest declaration.[61]

Since October 2019, the Supreme Court's Godavarman bench has been hearing a matter related to the setting up of a metro shed in Aarey Colony in Mumbai, Maharashtra.[62] The government of Maharashtra proposed to fell trees for the construction of the shed and the Bombay High Court concluded that tree felling attracted provisions of the Maharashtra (Urban Areas) Preservation of Trees Act, 1975, as these were urban trees outside recorded

---

[59] Report of the CEC in Application No. 1209 filed before it by the Kudremukh Wildlife Foundation and others regarding alleged illegal diversion of forest lands for non-forest uses in the Western Ghats region in Karnataka dated 25 April 2013.

[60] M. Raghava, 'Identification of Deemed Forest Completed in State, Says PCCF', *The Hindu*, 7 November 2015, available at https://www.thehindu.com/news/cities/Mangalore/identification-of-deemed-forest-completed-in-state-says-pccf/article7853779.ece, accessed on 11 September 2020.

[61] *The Hindu*, '26,600 Acres to Be Delisted from Purview of Deemed Forest', 9 December 2017, available at https://www.thehindu.com/news/national/karnataka/26600-acres-to-be-delisted-from-purview-of-deemed-forest/article21358304.ece, accessed on 11 September 2020.

[62] Supreme Court order dated 7 October 2019 in Suo Moto Writ Petition (Civil) No. 2 of 2019 (*Re: Felling of Trees in Aarey Forest (Maharashtra)*). This matter was heard along with Special Leave Petition (Civil) No. 31178 of 2018, Special Leave Petition (Civil) Diary No.14849 of 2019, Writ Petition (Civil) No.1132 of 2019 and Special Leave Petition (Civil) Diary No.36651 of 2019.

forest. A letter by a law student, to the Chief Justice of India argued that the dictionary meaning of forests as contemplated under the 12 December 1996 order of the Godavarman case should be made applicable and the area should be considered a forest. The Supreme Court took up the case suo moto. There were more petitions filed on this issue and these were heard together. The petitioners argued that the felling is illegal if done without the permission for forest diversion under FCA, 1980. The Court restrained further felling of trees in the project area. In the meantime, the Maharashtra state government took a decision to declare 600 acres of Aarey forests as Reserved Forests under Section 4 the IFA, 1927.[63]

Since 1996, the Godavarman case has been one the main judicial forums where matters related to forest laws, policies and governance have been adjudicated. The orders and judgments cover a range of topics, including project-specific permissions, regulating tree felling, operations of sawmills, violations of approvals for forest diversion, dereservation of forests, compensatory afforestation and valuation, and temporary working permits for mines.

As the intervention applications grew in both numbers and complexity, the Supreme Court constituted a CEC to assist in the cases. The CEC was set up under the provisions of Section 3 (3), Environment Protection Act, 1986, which gives powers to the central government to set up special authorities for environment protection. This order was issued on 9 May 2002 in I.A. No. 295 in Writ Petition (Civil) 202 of 1995 with I.A. No. 171 of 1996. In the initial years of its constitution, the CEC had explicit functions of monitoring the implementation of the Supreme Court's orders and was looking into cases of non-compliance, implementation of working plans, monitoring compensatory afforestation and other conservation issues. The CEC convened detailed hearings on specific cases and continued to assist both the Court and state governments in the implementation of the judgments and orders arising out of the case.

The orders in the Godavarman case and the CEC's directions have had far-reaching consequences on development projects and policies. In the following paragraphs is a summary of a case to illustrate how the Court's

---

[63] Badri Chatterjee, 'Aarey Colony Areas Adjacent to Mumbai's Sanjay Gandhi National Park to Be Reserve Forest: Maharashtra Government', *Hindustan Times*, 12 September 2020.

directions influenced the use of forests in India.[64] This case on the regulation of the operations of wood-based industries created a long-standing dispute since 1997. The Court initiated proceedings using its original jurisdiction in the Godavarman case and was concerned that these units operational across India used natural and plantation-based timber that led to the depletion of forest cover. On 12 December 1996, in addition to extending the definition of forests to its dictionary meaning, the Supreme Court also directed state governments to undertake several other measures for forest conservation and to regulate felling of trees. One of them required filing reports before the Supreme Court on

1. the number of saw mills, veneer and plywood mills actually operating within the State, with particulars of their real ownership;
2. the licenced and actual capacity of these mills for stock and sawing;
3. their proximity to the nearest forest;
4. their source of timber.

The first substantive order on this issue came on 4 March 1997[65] when the Court set up a high-level committee to prepare an inventory of all timber in all forms (including timber products). This included timber lying in the forest, transit depots or mill premises. Wherever possible, the inventory was to indicate the origin and source of timber. In the same order, it directed closure of all unlicenced sawmills, veneer mills and plywood industries in Maharashtra and Uttar Pradesh. These state governments were restrained from granting fresh permits or changing any laws related to the grant of such permits. The order clarified that none of the directions would apply to 'minor forest produce, including bamboos, etc.'

Until 2002, the sawmills case was regularly heard by the Supreme Court and specific difficulties due to the restrictions on timber felling were discussed and relief were sought. These were dealt with different *amicus curiae* (friend of the court) in the Godavarman case on a case-to-case basis.

---

[64] Civil original jurisdiction in Writ Petition (Civil) 202 of 1995 related to saw mills and wood-based industries. This matter was heard with Writ Petition (Civil) No. 171 of 1996 and Writ Petition (Civil) No. 897 of 1996.

[65] Supreme Court order dated 4 March 1997 in Writ Petition (Civil) No. 202 of 1995 (*T. N. Godavarman Thirumlkpad v. Union of India*).

In September 2002, the Supreme Court constituted a CEC for a period of five years. The Court extended the scope of the March 1997 order to the entire country. The 30 October 2002 order restrained all state governments on granting fresh permissions without prior approval of the CEC. The order states,

> No State or Union Territory shall permit any unlicensed saw-mills; veneer, plywood industry to operate and they are directed to close all such unlicensed unit forthwith. No State Government or Union Territory will permit the opening of any saw-mills, veneer or plywood industry without prior permission of the Central Empowered Committee. The Chief Secretary of each State will ensure strict compliance of this direction. There shall also be no relaxation of rules with regard to the grant of license without previous concurrence of the Central Empowered Committee.

Two specific solutions emerged from the court proceedings. First was the exemption for wood-based industries operating with imported material, and the second was to create state-specific plans on how the use of domestic timber would be regulated. The CEC thus would only facilitate a process through which the final licensing authority, which is the state government itself, would grant licences to mills.

In 2006, the CEC recommended the categorisation of all wood-based units based on when licences were issued and when licence fees were deposited.[66] On 18 May 2007, based on the CEC's recommendations, the Supreme Court allowed some categories of sawmills to be reopened following the payment of a one-time amount.[67] In this order, the Court observed,

> The CEC has considered the availability of wood for the industries, which was assessed as 43.70 lakh cu.mt. from trees outside forests and 02.00 lakh cu.mt. from Government Forests.

---

[66] The four categories were as follows: Category I—licences up to December 1996 were renewed before 4 March 1997, Category II—licences up to December 1995 were renewed before 4 March 1997, Category III—licences up to December 1994 were renewed before 4 March 1997 and Category IV—licences from 1993 onwards were not renewed before 4 March 1997.

[67] CEC, report (interim) regarding the closure of the saw mills and other wood-based industries in the state of Uttar Pradesh pursuant to the honourable Supreme Court's order dated 1 September 2006 in I.A. No. 1399 and I.A. No. 1569 with I.A. No. 946 dated 10 October 2006.

It has also assessed the units into four categories.

We accept the CEC's recommendations.

The Saw Mills may be permitted on the basis of the recommendations made by the CEC. Licences may be given by the State Level Committees.

One category was made up of all units which were granted licences in 1993 but had not been renewed at the time the March 1997 order was issued,[68] implying that these units were operating illegally.

On 5 October 2015, the Supreme Court while hearing the matters under Writ Petition (Civil) 202 of 1995 issued directions on expediting the disposal of several pending cases. All matters related to wood-based industries were transferred to state-level committees (SLCs) for the purposes of regulating permits and ensuring compliance. It also directed the environment ministry to issue guidelines to SLCs for 'assessment of timber availability for wood-based industries and grant of license/permission'.[69]

Since 2011, the NGT has adjudicated several matters related to the diversion of forest land for non-forest use. The appeal provisions in the NGT Act, 2010, allow aggrieved parties to approach the tribunal in case there are violations of Section 2 of the FCA, 1980. The NGT does not hear cases related to the IFA and State Forest Acts. In most of these cases, the petitioners sought relief for regulatory failure or violation of the law by project authorities. But the tribunal extended beyond specific cases and clarified questions about the administrative practices in decision-making related to forest diversions and broader interpretations of legal clauses.

Since the enactment of the FCA, 1980, governments and project authorities assumed that the central environment ministry is the final approver of forest diversion applications. This assumption was questioned in an appeal before the tribunal challenging the forest diversion approval granted to a hydroelectric power project.[70] The legal issue argued before the NGT was whether this

---

[68] Supreme Court order dated 30 October 2010 in I.A. No. 1137 with 1319 Writ Petition (Civil) No. 202 of 1995 (*T. N. Godavarman Thirumlkpad v. Union of India*).

[69] Supreme Court order dated 5 October 2015 Writ Petition (Civil) No. 202 of 1995 (*T. N. Godavarman Thirumlkpad v. Union of India*).

[70] National Green Tribunal judgment dated 7 November 2012 in Appeal No. 7 of 2012 (*Vimal Bhai & Ors v. Union of India & Ors*).

approval was appealable under the NGT Act, 2010. The Act restricted the jurisdiction of the tribunal to hear appeals only against orders issued under Section 2 by the state government. Section 16 (e) in the Act allows appeals against

> an order or decision made, on or after the commencement of the National Green Tribunal Act, 2010, by the State Government or other authority under section 2 of the Forest (Conservation) Act, 1980.

On hearing the case, the tribunal stated, '[W]e are surprised to find that most of the State Governments do not pass separate orders.' It further observed that this was creating 'an embargo and depriving a person aggrieved from filing an Appeal'. In its judgment dated 7 November 2012, the NGT clarified that approvals for use of forest land for non-forest use will come into effect only after concerned state governments issue orders. The tribunal asked state governments to pass orders with immediate effect, which could then be appealed before the tribunal. This opened up opportunities for several parties aggrieved by the grant of forest approvals to file fresh appeals. Many projects had initiated construction activity without the issuance of state government approval, rendering the construction activity or operations illegal.[71]

IV FOREST RIGHTS ACT AND FOREST DIVERSIONS

Forest land in India has been largely managed and controlled by the forest departments of respective state governments. The discourses of ownership, management and governance of India's forests underwent a significant change with the enactment of the Scheduled Tribes and Other Traditional Forest Dwellers (Recognition of Forest Rights) Act, 2006 (FRA). The Ministry of Tribal Affairs (MoTA) is the nodal ministry for the implementation of this law. The FRA and its rules (2008 and 2012) put in place a clear legal process for the recognition of rights of tribals and non-tribal forest-dwelling communities, including forest workers who have been living in a forest area for 75 years or three generations. This law grants individual and community forest rights.

---

[71] National Green Tribunal judgment dated 26 February 2016 Appeal No. 04 of 2014 Eastern Zone (*Themrei Tuithung & Ors. v. of Manipur & Ors*); National Green Tribunal order dated 24 January 2014 in Original Application No. 123 of 2013 (*Prafulla Samantara v. Union of India & Ors*).

The MoTA has issued a set of guidelines/clarifications to address the ambiguities and multiple interpretations that have emerged during and as a result of the implementation of the FRA.[72] In this section, we focus on the intersections between the FRA and the FCA.

With the enactment of the FRA, decisions related to forest diversion are required to take into consideration the pending or approved claims, including those in dispute. The implementation of these new processes requires the collaboration of agencies involved in the implementation of two separate laws, the FRA and the FCA, which are under two separate ministries, the Ministry of Environment, Forests and Climate Change (MoEFCC) and the MoTA.

The interface of procedures for forest diversions and the FRA is governed by three legal documents.

- An advisory was issued by the environment ministry on 3 August 2009 to all state governments.[73] This advisory states that all applications for forest diversions by state governments should include evidence of having completed the recognition of rights processes under the FRA. The evidence needs to include consent from the *gram sabha*s (village assemblies) for forest diversion.

- In 2012, the Ministry of Tribal Affairs[74] reiterated that the instructions in the previously mentioned circular should be followed.

- The Forest Conservation Rules, 2017, elaborated the previously mentioned requirement under Rule 6 (3).

The new context of the FRA law has resulted in several legal challenges. The following are three such contentious legal issues that are important for the implementation of the FRA and the FCA.

The requirement of gram sabha *consent*: The issue of the gram sabha's consent was explicitly clarified in an April 2013 Supreme Court judgment.[75]

---

[72] Ministry of Tribal Affairs (MoTA) and United Nations Development Program (UNDP), *Compendium on Guidelines, Circulars and Executive Directions on Scheduled Tribes and Other Traditional Forest Dwellers (Recognition of Forest Rights) Act, 2006* (New Delhi: MoTA and UNDP, n.d.).

[73] MoEF letter dated 3 August 2009 to The Chief Secretary/Administrator (All State/UT Governments except J&K) (F. No. 11-9/1998-FC [pt]).

[74] Ministry of Tribal Affairs letter dated 12 July 2012 with 'Guidelines on the implementation of the Scheduled Tribes and Other Traditional Forest Dwellers (Recognition of Forest Rights) Act 2006' (No. 23011/32/2010-FRA [Vol. II {Pt.}]).

[75] Supreme Court judgment dated 18 April 2013 in Writ Petition (Civil) No. 180 of 2011 (*Orissa Mining Corporation v. Union of India*)

This case was filed by a state-owned public sector undertaking against the orders of the environment ministry, which withheld the grant of forest diversion for the bauxite mine in Niyamgiri Hills. One of the reasons for withholding the approval was the disproportionate impact that the project would have on the culture and traditional livelihood rights of the Dongria Kondh tribal community. This judgment upheld the competence of the *gram sabha* to safeguard and preserve the traditions of the community, their cultural identity, community resources and community modes of dispute resolution. It further stated that the *gram sabha* functioning under the FRA read with Section 4 (d) of the Provisions of the Panchayats (Extension to the Scheduled Areas) (PESA) Act 'has an obligation to safeguard and preserve the traditions and customs of the STs and other forest dwellers, their cultural identity, community resources, etc.'

As per the Supreme Court's directions, the matter of forest diversion had to be placed before the *gram sabha*s for them to take a 'decision' on it and communicate it to the MoEFCC through the state government within three months. However, the Supreme Court also put the 'final decision' in the hands of the environment ministry for approving or rejecting the forest diversion. Following this judgment, all the 12 *gram sabha*s resolved against the mining and the environment ministry also upheld their decision.

'Revocation' of forest rights: On 8 January 2016, residents of Ghatbarra village in Sarguja district of the state of Chhattisgarh received a notice from the district collector, the divisional forest officer and the assistant commissioner of the tribal development department stating that since the time villagers had received their community forest rights (CFR)[76] in September 2013, they had caused disturbances to the ongoing coal mining operations.[77] As a result, the district-level committee constituted under the 2006 FRA cancelled their forest titles.

---

[76] The Ghatbarra CFR was issued as per district collector order no. 10378/a-19(1) 2012–13. It recognised three specific rights for the villages: Section 3 (1) (b) community rights such as *nistar*, by whatever name called, including those used in erstwhile princely states, *zamindari* or such intermediary regimes; Section 3 (1) (c) right of ownership, access to collect, use and dispose of minor forest produce which has been traditionally collected within or outside village boundaries; and rights to grazing (both settled or transhumant) as per Section 3 (1) (d) of the FRA, 2006.

[77] Order from district collector office: No/Forest Rights/A.V./No. 42/2015-16/10669 (letter number translated from Hindi).

The letter further justified the decision to cancel their rights by stating that the villagers were granted the CFR titles only after the approval for forest diversion in 2012. However, the rights holders claimed that the company had initiated ground inspection activities related to the coal mining before their claims were recognised under the FRA, 2006. Ghatbarra's village forest committee issued a letter pointing to the government that their forest rights process was incomplete and that forest clearance should not be granted before their rights are determined.[78]

The *gram sabha* of this village and community-based organisations challenged the decision of the government to withdraw the titles. They filed a case in the Chhattisgarh High Court,[79] and a final decision is pending. The legal question that needs to be clarified is whether rights once granted under the FRA can be revoked. This is an important question for the courts to clarify since the framework of the law does not provide for withdrawal of rights.

Liability in case of violation of legal procedures: A 2016 judgment[80] of the NGT relates to an appeal filed against the violation of legal requirements for recognition of forest rights and *gram sabha* consent prior to the grant of approval for forest diversion. The project in question was a hydro electric project in Kinnaur district in the state of Himachal Pradesh.

The NGT judgment dated 4 May 2016 granted relief in favour of the petitioners. It directed the project authorities to seek consent from the four affected villages for the forest diversion. The court also added that the *gram sabha* should take into account community and individual claims, cultural and religious aspects, and other impacts likely to be caused by construction before extending their consent to the forest diversion proposal.

V   NATIONAL FOREST POLICIES AND PROGRAMMES

FOREST POLICIES

The purpose of policy is to give direction to laws, programmes and schemes of the government. The National Forest Policy currently in operation in India is

---

[78] Letter dated 26 February 2016 from the Hasdeo Arand Bachao Samiti to chairman, Chhattisgarh State Level Monitoring Committee constituted under the FRA, 2006.

[79] Writ Petition (Civil) No. 1346 of 2016 (*Forest Right Committee Ghatbarra v. Union of India*) in High Court of Chhattisgarh, Bilaspur.

[80] National Green Tribunal judgment dated 4 May 2016 in Appeal No. 28 of 2013 (*Paryawaran Sanrakshan Sangarsh Samiti Lippa v. Union of India & Ors*).

from 1988. Since 1980, there have been at least three iterations of the National Forest Policy, which has guided government action on forest conservation, wildlife management, production forestry and regulation. A new draft policy is under preparation. Its challenge would be to take into account emerging threats to forest degradation and the role of forests in complex problems such as climate change and air pollution.

Table 3.1 shows how India's forest policies compare on a few themes.

Table 3.1 Thematic Comparison of Forest Policies

	1952	1988	2018 (Draft)
Objective	To evolve a system of balanced and complementary land use	To ensure environmental stability and maintenance of ecological balance	To safeguard the ecological and livelihood security of people, of the present and future generations
Definition of forest	No discussion	No discussion	No discussion
Classification of forests	PFs, NFs, VFs and tree lands	No discussion	No discussion
Forest cover needed	One-third of India's total land area needs to have forest cover. About 60% of it needs to be 'in the Himalayas, the Deccan and other mountainous tracts' that are vulnerable to erosion	Minimum one-third of the total land area of the country needs to be under forest or tree cover	At a national level, one-third of the total land area needs to be under forest and tree cover, except in hills and mountainous regions where it is proposed to be the two-third of the area to prevent soil erosion
Focus of afforestation programmes	Increasing 'Treelands'* in agricultural areas, envisaged the planting of 30 crores of trees in ten years	A massive need-based and time-bound programme of afforestation and tree planting, with particular emphasis on fuelwood and fodder development	Intensive scientific management of forest plantations of commercially important species

(Contd)

(*Contd*)

	1952	1988	2018 (Draft)
Wildlife conservation	Affording protection to all wildlife and particularly to rare species such as lion and one-horned rhinoceros	Provide for 'corridors' linking protected areas	Management of protected areas and specific focus on human–wildlife conflicts
Forest rights	No discussion	No discussion	Harmonisation required with CFR through Community Forest Management Mission
Climate change	No discussion	No discussion	Assimilation of international conventions including climate change into national programmes

Source: Authors.

JOINT FOREST MANAGEMENT

Forest governance and management has also been shaped by a significant programmatic intervention of the government. The Joint Forest Management (JFM) programme was introduced in the 1990s as an effort to encourage participatory forest management. The programme encouraged the joint management of forests by state forest departments and local communities and incentivised local participation through the sharing of revenue from forest harvest. This programme was seen as an inclusive model of forest management that could replace the prevalent model of forest conservation where rural communities were seen as a threat to forests. The programme received financial support from several international donor agencies.

The programme had limited success. Research studies pointed out that the poor performance of JFM can be attributed to

- structural problems in the implementation of the programme where the forest department retained control and maximised profits from plantations,

- restricted coverage of forest lands under the programme and inadequate forest produce rights to communities,

- absence of autonomy in everyday operations related to forest management and lack of transparency and accountability of the forest department and

- uncertainty of tenure as the land under plantations were under the ownership of forest departments.

Scholars have pointed out that in states such as Gujarat, Madhya Pradesh and Andhra Pradesh, where some success had been achieved, JFM efforts are proving to be unsustainable in the long term.[81]

---

[81] Lele and Menon, *Democratizing Forest Governance in India.*

4

Pollution Control and Prevention

INTRODUCTION

Laws for the prevention and control of pollution have been part of centralised environmental law in India since the 1970s. Parliamentary documents state that before the enactment of these laws, pollution control was mostly carried out through the provisions of the Indian Penal Code, the Criminal Procedure Code, the Factories Act, 1948, and the Merchant Shipping Act, 1958. The enactment of the first law specifically dealing with pollution is attributed directly to India's commitment to the Stockholm Conference on Human Environment in June 1972. Reports of high-level committees record, 'it was considered appropriate by Government to have uniform laws all over the country for broad environmental problems endangering the health and safety of the people as well as of the country's flora and fauna'.[1]

The legal framework for pollution control relies on a conditional consent-based system to regulate polluting projects and processes. Through this system, pollution laws regulate emissions, effluents and toxicity of industrial processes and projects within limits set by the law. The permissible pollution standards are either specific to types of industries such as coal power, manufacturing and construction or types of raw materials being used in industrial processes including asbestos, iron ore and mercury. These standards apply across locations and on projects based on their size and production capacity. For instance, pollution norms apply to the construction of mini dam projects as well as large nuclear facilities. Specific norms also apply to the entire life cycle of an industrial process from pre-construction to end of operations.

---

[1] Public Accounts Committee, *Central Pollution Control Board: Audit Review* (New Delhi: Lok Sabha Secretariat, 1994), 1.

Pollution control boards (PCBs) at central, state and regional levels implement pollution laws. These boards set parameters for the permissible levels of pollutants in the air, water or land. Individuals and organisations are expected to invest in pollution control equipment and infrastructure to keep discharges and emissions within the permissible levels. The pollution levels of projects or processes are to be regularly monitored by devices installed for collecting data on quality and quantity of emissions and discharges. In recent years, monitoring data is made available to the PCB regulators through continuous online real-time emission monitoring systems.

The PCB regulators use a range of measures to direct projects towards legal compliance with the permissible standards. These include bank guarantees, warnings, closures and fines imposed on projects and processes to push them to comply with pollution laws. They also undertake spot checks, usually in response to complaints. During such checks, the authorities manually collect relevant material samples that help to detect pollution levels. These are sent for laboratory testing, and based on these reports, directions for further actions may be issued by PCBs. In the last few years, new mechanisms of facilitating compliance to pollution laws have been introduced. These include amnesty schemes to polluters, long duration consents and self-regulation. These reflect broader global trends in environmental regulation.

The regulation of polluting projects and processes through this system has not produced good results. As stated by the Central Pollution Control Board (CPCB), 31 states and union territories (UTs) have rivers not meeting water quality criteria.[2] There are several polluting industrial zones, and these sites and affected communities have been waiting for remedial actions for years. Lack of restoration of damaged areas and habitual legal non-compliance are questions that face pollution regulators today.

Pollution regulation does not emphasise on siting as an important regulatory tool to prevent pollution impacts of projects. While there have been efforts to create area-specific pollution mitigation action plans, they have largely been done after contamination and pollution has crossed manageable limits. There have also been some regulations for locating, for transportation and for handling of hazardous and polluting material, but the law does not require a careful study of the location before the grant of consent to projects or processes.

---

[2]   Central Pollution Control Board (CPCB), *River Stretches for Restoration of Water: Quality; Statewise and Prioritywise* (New Delhi: Government of India, 2018), 2.

Pollution laws have been invoked by governments and by environmental groups and affected communities within and outside courts. The poor implementation of pollution laws has been at the centre of environment and development conflicts. For example, the impacts of unrestrained and illegal mining in Goa led to protests and litigation.[3] The Court directed the shutdown of mining operations in Goa, and the loss of mining jobs led to heated arguments in the Indian parliament and protests.[4] A recent flashpoint was in the case of a copper smelter in Tamil Nadu where the company was allegedly allowed to continue polluting activities for two decades.[5]

While the outcomes in each of these cases are varied, the concern over the economic future of polluting industries has remained a critical argument in court judgments or administrative proceedings against specific operations. In the much-cited pollution matter of 1996, *Vellore Citizens Welfare Forum v. Union Of India & Ors*,[6] on the discharge of untreated effluent into Palar river of Tamil Nadu, the closure of leather tanneries ordered by the pollution control authorities was lifted by the Court in the light of the leather industry being a major foreign exchange earner and the state of Tamil Nadu contributing 80 per cent of the country's export. In the copper smelter case, the PCB, the Madras High Court and the Supreme Court have directed the closure of the plant after two decades of its operations.[7] These cases are important to understand the practical implementation and consequences of the legal and administrative framework for pollution prevention, control and abatement.

---

[3] Supreme Court judgment dated 14 September 2018 in Writ Petition (Civil) No. 435 of 2012 (*Goa Foundation v. Union of India*); Supreme Court judgment dated 7 February 2018 in Civil Appeal No. 32138 of 2015 (*Goa Foundation v. M/s Sesa Sterlite Ltd. & Ors*).

[4] Times News Network, 'SC Agrees to Hear 20-Year-Old Mining Case', *Times of India*, 14 March 2019, available at https://timesofindia.indiatimes.com/city/goa/sc-agrees-to-hear-20-yr-old-mining-case/articleshow/68400256.cms, accessed on 11 September 2020.

[5] Pamela Philipose, 'The People of Tuticorin versus Sterlite Copper', *Himal Southasian*, 29 June 2018, available at https://www.himalmag.com/the-people-of-tuticorin-versus-sterlite-copper/, accessed on 11 September 2020.

[6] Supreme Court judgment dated 28 August 1996 in Writ Petition (Civil) No. 914 of 1991 (*Vellore Citizens Welfare Forum v. Union of India & Ors*).

[7] Supreme Court judgment dated 18 February 2019 in Civil Appeal Nos. 4763–4764 of 2013 (*Tamil Nadu Pollution Control Board v. Ms Sterlite Industries (I) Ltd*).

This chapter is divided into five sections:

I.   *Legal Instruments for Regulating Polluting Activities*
II. *Institutional Set-Up for Pollution Control*
III. *Regulatory Functions of PCBs*
IV. *Action Plans*
V.   *Appeals before Appellates and Tribunal*

I   LEGAL INSTRUMENTS FOR REGULATING POLLUTING ACTIVITIES

This section deals with the legal framework for the regulation of polluting activities. The central laws that have been enacted for the objectives of prevention, control and abatement of pollution in India are the Water (Prevention and Control of Pollution) Act, 1974 (Water Act) and Air Prevention and Control of Pollution Act, 1981 (Air Act). Along with the main Acts, there are other legal and policy instruments for pollution prevention and management of impacts of specific types of pollution such as hazardous waste. There are also special guidelines and notifications for activities such as the monitoring of pollution in the river Ganga, for the utilisation and management of fly ash from coal burning and for emission standards for coal power plants. Among the latest plans related to pollution that has gained a lot of attention is the National Clean Air Action Programme (NCAP).

There are also several important pollution regulation laws that derive their mandate and powers from the Environment Protection Act (EPA), 1986. Prominent among these are several waste management rules such as the Hazardous and Other Wastes (Management and Transboundary Movement) Rules, 2015, and the Solid Waste Management Rules, 2016. The Noise Pollution (Regulation and Control) Rules, 2000, also draw their mandate from the EPA. These instruments are discussed in Chapter 5 on environmental protection.

AIR AND WATER ACTS

The definition of pollution used in the air and water laws is important to understand how the legal framework limits or expands its jurisdiction for tackling pollution.

• The Water Act sees pollution as a contamination of water or alteration of physical, chemical or biological properties of water. This includes the discharge of sewage, trade effluent or any other substance into water that is likely to create nuisance or be harmful to any legitimate use including domestic, commercial, industrial or agricultural. Human, animal or plant and aquatic health should not be hampered by such pollution. It should also not be harmful or injurious to public health or safety.

• The Air Act looks at pollution as the presence of any air pollutant in the atmosphere including solid, liquid or gaseous substance which is or can be injurious to human and other living beings. The injury can also be to property and the environment. The Air Act also includes noise within its ambit by giving it the status of an 'air pollutant'.[8]

The overall mandate of these laws, their scope and limits are tied to the terms 'Prevention', 'Control' and 'Abatement'. While all three broadly imply any actions or practices that reduce, alleviate or eliminate pollution at its source, these actions have to be planned for different stages of project implementation. The government and regulatory authorities have to take necessary steps to avoid or prevent the occurrence of pollution—strictly control the impacts in instances of release of pollutants and to remedy the damage by both eliminating the pollutant and taking other restoration measures.

The enforcement of both these laws rests with the CPCB, state pollution control boards (SPCBs) and regional pollution control boards and is discussed in section II. The regulatory aspects of these laws are discussed in section III.

COLLECTION OF WATER CESS

The Water (Prevention and Control of Pollution) Cess Act, 1977 (Water Cess Act) was enacted to institute the mandatory payment of a cess or fee by industries directly consuming water and by local bodies supplying water for specific uses. The law gave a list of uses and amounts levied. These uses were as follows:

• Industrial cooling, spraying in mine pits or boiler feeds

• Domestic purposes

• Processing that leads to water pollution where pollutants are biodegradable, non-toxic or both

• Processing that leads to water pollution where pollutants are not easily biodegradable, toxic or both

---

[8] Noise pollution was also brought under the ambit of activities of PCBs in 1988. Public Accounts Committee, *Central Pollution Control Board: Audit Review.*

The money collected as cess was to 'augment the resources' of PCBs that were set up to implement the provisions of the Water Act, 1974. This money was to be deposited in the Consolidated Fund of India and transferred to central and state boards thereafter. The water cess was considered to be an important source of revenue for SPCBs despite partial payments by local bodies in most states. There were also delays in the payment as the users challenged the applicability of the law to argue that they were not liable to pay amounts levied by PCBs.[9]

In June 2017, this Act was abolished when the central government rolled out the Goods and Services Tax (GST). In all, 13 taxes including the water cess were discontinued using the Taxation Laws Amendment Act, 2017.[10]

In the last decade, the environment ministry, the central and state boards have issued several norms and guidelines to manage rising, uncontrolled pollution. The following measures illustrate the kinds and scale of these directions.

REAL-TIME EFFLUENT AND EMISSION DATA COLLECTION AND MONITORING

The 2017 guidelines of the CPCB related to emission monitoring make the case for long-distance online monitoring and self-regulation by industries:[11]

> With rapid industrialization, it is becoming a need and necessity to regulate compliance by industries with minimal inspection of industries. Therefore, efforts need to be made to bring discipline in the industries to exercise self-monitoring and compliance.

Pollution control boards have recognised the challenge in ensuring compliance with effluent discharge and emission standards. This is especially the case with highly polluting industries. In 1984, the CPCB initiated the National

---

[9]   G. D. Agarwal, 'Paying for Environmental Abuse', *Down to Earth*, 14 February 1994, available at https://www.downtoearth.org.in/indepth/paying-for-environmental-abuse-29412, accessed on 22 July 2019.

[10] Press Information Bureau, 'The Central Government Abolished Various Cesses in the Last Three Years for Smooth Roll-Out of GST', Government of India, New Delhi, 2017, available at http://www.pibmumbai.gov.in/scripts/detail.asp?releaseId=E2017PR1051, accessed on 11 September 2020.

[11] CPCB, *Guidelines for Continuous Emission Monitoring Systems* (New Delhi: Government of India, 2017).

Ambient Air Quality Monitoring by setting up its own monitoring stations. The board identified three parameters for monitoring, namely, sulphur dioxide (SO2), nitrogen dioxide (NO2) and suspended particulate matter (PM). In 1994, the Public Accounts Committee of the Lok Sabha recorded that CPCB has 'not been able to make the monitoring stations operative and secure the necessary data through the State Boards even after eight years'.[12]

In recent years, the onus to collect data has been shifted to industrial units. They are required to directly feed the effluent and emission discharge data into an online database, which can be accessed by SPCBs and CPCBs for monitoring. The PCBs have put in place two systems by which the above objective can be operationalised. These are the continuous emission monitoring systems (CEMS) and continuous effluent quality monitoring systems (CEQMS).

In 2014, the CPCB issued directions[13] to SPCBs and pollution control committees (PCCs) under Section 18(1) b of the Water and the Air Acts. The directions required 17 categories of highly polluting industries[14] to install online monitoring system to ensure round-the-clock electronic data transmission on effluents and emissions. The same was also required of all common effluent treatment plants (CETPs), sewage treatment plants (STPs), common bio waste and common hazardous waste incinerators across the country. The directions included the installation of systems for regular maintenance and operation of the online monitoring system with tamper-proof mechanism and facilities for online calibration.

The stack emissions are to be monitored through the CEMS method. According to the 2017 guidelines of the CPCB, the CEMS can measure concentration of gaseous emissions and PM concentrations and emissions rate using a computer program.

In 2014, the CPCB also issued directions for installation of real-time water quality monitoring systems to 11 SPCBs in the Ganga river basin.[15] These systems are to be set up in all the industries discharging effluents directly

---

[12] Public Accounts Committee, *Central Pollution Control Board: Audit Review.*

[13] CPCB letter no. B-29016/04/06 PCI-1/5401 dated 2 February 2014.

[14] These industries include pulp and paper, distillery, sugar, tanneries, power plants, iron and steel, cement, oil refineries, fertiliser, chloral alkali plants, dye and dye-intermediate units, pesticides, zinc, copper, aluminum, petrochemicals, and pharma sectors.

[15] CPCB letter no. B-190019/NGRBA/CPCB/2011-12 dated 2 February 2014.

into the river Ganga or its tributaries. The system in place for monitoring these industries is called the CEQMS. The CPCB released revised guidelines on CEQMS in 2018,[16] which presented different technologies that can be deployed to ensure compliance with this system.

The SPCBs use this data to measure compliance with prescribed standards and take necessary enforcement action under the Air and the Water Acts. The CEMS and CEQMS guidelines are illustrative of the growing importance of self-regulation by the polluting units as an acceptable approach in environmental regulation. This data is accessible to only polluting companies and the concerned PCBs or PCCs and is not public information.

UTILIZATION AND MANAGEMENT OF FLY ASH

Fly ash is a residue from burning coal and is generated in vast quantities by coal-based power plants. The utilisation and management of fly ash has been an area of concern for successive governments. In a 2015 meeting at the environment ministry, a senior official had observed:[17]

> [F]ly ash generation is reported to be 172.87 million tons during 2013–14 and it is likely to increase up to 260 million ton by 2020. Unless immediate strict measures are taken to utilize fly ash appropriately, it will become a big problem in near future.

Regulatory bodies first acknowledged the problem of managing fly ash in 1999 when the environment ministry issued a notification laying out the various ways fly ash should be utilised. The push for the notification came from an order of the Delhi High Court[18] in which the central government was directed to publish the notification within a stipulated time. The SPCB and CPCB were given the responsibility of monitoring the utilisation of fly ash on an annual basis.

---

[16] CPCB, *Guidelines for Online Continuous Effluent Monitoring Systems (OCEMS)* (New Delhi: Government of India, 2018).

[17] Minutes of the meeting of an environment ministry held on 7 January 2015, following the directions of the National Green Tribunal (NGT) (in Application No. 102 of 2014). The meeting was to discuss issues related to fly ash utilisation with a range of 'stakeholders'.

[18] High Court of Delhi order dated 25 August 1999 in Writ Petition (Civil) No. 2145 of 1999 (*Centre for Public Interest Litigation, Delhi v. Union of India*).

The mechanism proposed for fly ash management was ash-mixed brick manufacturing, use of other ash-based products in construction activities and use of ash in reclamation of low-lying areas. In order to enable this, local authorities in areas where fly ash was being generated were to make necessary changes in their building bye-laws to ensure that there is mandatory utilisation of fly ash bricks to a specific percentage. For at least 10 years from the time this notification was published, the power plants were to make fly ash available to users without any payment or any other conditions. All this was to be done within 100 kilometres of where the ash was generated.

However, the problem of fly ash generation and management has persisted. The environment ministry has modified its notification several times, including in 2003, 2009 and 2016, but continued to emphasise only the utilisation end of the problem and not the production.[19] Given the scale of coal burning in India, the generation of fly ash far exceeds what can be accommodated in brick kilns, cement plants, construction of roads, backfilling of mines and other uses suggested in the utilisation notifications. The fly ash utilisation norms also do not address the problems caused by the dumping and mismanagement of ash on road sides, river banks and agricultural fields. There have been several complaints filed before the pollution control authorities and courts on these issues.

The 2016 notification requires 100 per cent utilisation of fly ash within a year of the notification, a direction yet to be complied by most generators of ash. This is confirmed by a set of directions issued by the CPCB in 2018 that builds on the environment ministry's notification. A December 2017 report of the Central Electricity Authority states that even though fly ash utilisation of about 63.28 per cent was achieved during 2016–17, a lot more would need to be done to achieve 100 per cent utilisation.

One enabling direction in this notification is related to the cost of transportation of ash to users. These costs were not addressed in the earlier notifications. It is now required that the cost of transportation of fly ash for road construction projects or for manufacturing of ash-based products is shared by both user and generator of the ash if the transportation is beyond 100 kilometres. In the new notification, this utilisation is to be done by industries and processes up to 300 kilometres from where the ash is generated.

---

[19] Ministry of Environment, Forests and Climate Change (MoEFCC) notification S.O. 254(E) dated 25 January 2016.

This clause opened a new set of disputes before the Central Electricity Regulatory Commission (CERC). In one such petition,[20] the National Thermal Power Corporation (NTPC) sought the recovery of additional expenditure incurred by its compliance to the environment ministry's revised notification. It was argued that this new notification should be understood as 'Change of Law' defined in the 2014 tariff regulations of the CERC, as it imposed additional costs on the company.

In the judgment dated 5 November 2018, the CERC interpreted that the additional expenditure incurred towards transportation of fly ash is admissible under 'Change of Law'. However, the revision of payments and related claims were to be on a case-to-case basis and with transparency in bidding and accounting for purchase of fly ash by state government undertakings. This includes disclosing details of actual additional expenditure incurred on ash transportation after 25 January 2016 that is duly certified by auditors.

New Emission Standards for Coal Power Plants

The coal sector in India has been under international and domestic pressure to monitor emissions from coal power plants. While coal plants were already collecting data on the emission of PM, there was no data on the production of Sulphur oxides (SOx) and nitrogen oxides (NOx). From December 2014, the CPCB engaged in studies and expert consultations to determine a new set of emission standards for coal-fired plants.[21]

On 7 December 2015, the environment ministry notified revised water consumption and emission standards for PM, NOx and SOx and mercury emissions from coal-based thermal power plants.[22] The standards were varied for different coal plants depending on their production capacity and age. The age of the power plants mattered because older projects were limited by the extent to which they could retrofit new mitigation technologies without adding to the cost of power production. The notification categorised thermal power

---

[20] Central Electricity Regulatory Commission judgment dated 5 November 2018 in Petition No. 172/MP of 2016 (*NTPC Ltd v. Uttar Pradesh Power Corporation Ltd & Ors*).

[21] Presentation of Priyavrat Bhati of Centre for Science and Environment (CSE) during a panel discussion at Centre for Policy Research, New Delhi, available at https://www.cprindia.org/news/7192, accessed on 22 July 2019.

[22] MoEFCC Notification S. O. 3305(E) dated 7 December2015.

plants in three, namely (*a*) installed before 31 December 2003 (*b*) installed after 2003 up to 31 December 2016 and (*c*) installed after 31 December 2016. All coal power plants were given two years to put systems in place to comply with these new emission and water use standards.

The new emission standards claimed to reduce the pollution load from coal power stacks. The CPCB claimed that the new standards would reduce SOx and NOx by 48 per cent, PM by 40 per cent and mercury by 60 per cent. Researchers have argued that these are significant reductions in air pollution from the coal power sector that generates 75 per cent of India's electricity and whose emissions are not seasonal in nature.[23] Compliance with the notification required power plants to be fitted with flue gas desulphurisation, selective catalytic reduction and electrostatic precipitation systems, which could reduce specific pollutants in the emissions. Project proponents claimed that their compliance to the new norms was challenged by technology, monetary and space constraints.[24] In effect, there was little effort by power producers or by relevant government ministries to prepare for complying with these norms by the deadline of December 2017.

In June 2016, a case was filed before the National Green Tribunal (NGT)[25] to expedite the compliance with 2015 notification. The case continued to be heard, and the NGT sought regular updates from the ministry on the status of compliance with the standards. Even as this case was being heard, on 12 December 2017, the environment ministry took a contradictory stand in the Supreme Court. The ministry filed an affidavit in an ongoing case monitoring pollution issues in Delhi-NCR (National Capital Region) on pollution matters[26] and asked for the implementation of the new norms

---

[23] According to Vinuta Gopal, Asar Social Impact, during a panel discussion on 'Air Pollution from Thermal Power Plants in India', Clearing the Air Seminar Series, available at https://www.cprindia.org/news/7192, accessed on 22 July 2019.

[24] FICCI (Federation of Indian Chambers of Commerce and Industry), *Developing a Roadmap for Implementing New Emission Norms for Thermal Power Plants* (New Delhi: FCCI, n.d.), available at http://ficci.in/events/24114/Add_docs/Background-Note_Workshop-on-New-Emission-Norms-in-Power-Sector.pdf, accessed on 23 July 2019.

[25] National Green Tribunal Original Application No. 315 of 2016 (*Sunil Dahiya v. Union of India*).

[26] Supreme Court Writ Petition (Civil) No. 13029 of 1985 (*M. C. Mehta v. Union of India & Ors*).

to be extended by five years.[27] It cited the lack of locally proven technology to curb NOx emissions as an important reason for the delay in retrofitting or installing new components on existing plants for seeking the extension. The ministry did not disclose in the affidavit that it was already facing a case on the delay in implementation of the 2015 notification before the NGT. Before the ministry's affidavit was filed, the CPCB had already issued letters to more than 400 thermal power units, allowing them to continue releasing emissions until 2022 in violation of the limits set by the new norms.[28]

On 17 January 2018, once the time frame for compliance had expired, the NGT recorded that the environment ministry had approached the Supreme Court for an extension of the time frame for compliance. In its order, it directed the environment ministry not to grant environment clearance to any new thermal power plant 'unless they show that every mechanism/technique to achieve the standards set out in the Notification dated 07th December, 2015 is in place/adopted'.

In July 2018, the Supreme Court while monitoring the compliance with the 2015 standards observed,

> The affidavit now seems to suggest that positive steps cannot be taken until 2022 but on a closer reading of the affidavit filed by the Union of India, it is clear that the dead line of 2022 cannot be reached by the Union of India and is completely illusory in nature. It, therefore, appears to us quite clear that the Union of India proposes to do nothing substantive in the matter.

In August 2019, the Supreme Court agreed to upwardly revise the NOx emission norms for all coal power plants installed after 2003, that is, from 1 January 2004 to 31 December 2016. This allowed for higher emissions by power plants. The request was made by Ministry of Power (MoP) and the

---

[27] Supreme Court order dated 7 July 2018 in Writ Petition (Civil) No. 13029 of 1985 (*M. C. Mehta v. Union of India & Ors*); Vishwa Mohan, 'Environment Ministry Aligns Itself with Power Ministry, Seeks Relaxed Deadline on Emission Norms for Polluting Power Plants', *Times of India*, 15 December 2017, available at https://timesofindia. indiatimes.com/home/environment/pollution/Environment-ministry-aligns-itself-with-power-ministry-seeking-relaxed-deadline-on-emission-norms-for-polluting-power-plants/articleshow/62088026.cms, accessed on 11 September 2020.

[28] CPCB letter no. B33014/07/2017-18/IPC-II/TPP to M/s Barsingar Thermal Power Station, Rajasthan, dated 11 December 2017.

NTPC and was upheld at a meeting held in July 2019 'where a consensus had been reached' between the environment ministry, the CPCB, the Environment Pollution (Prevention and Control) Authority (EPCA), the NTPC and the MoP.[29]

The case is pending before the NGT and the Supreme Court, with the apex court regularly monitoring the measures being taken to ensure compliance with the standards.

VEHICULAR POLLUTION

The emissions caused by vehicles is a significant source of air pollution in India.[30] According to a 2018 Reference Note of the Lok Sabha: 'Besides substantial Carbon Dioxide (CO_2) emissions, significant quantities of Carbon Monoxide (CO), Hydrocarbon (HC), Nitrogen Oxide (NOx), Suspended Particulate Matter (SPM) and other air toxins are emitted from these motor vehicles in the atmosphere, causing serious environmental and health impacts.'[31] The note also lists reasons for increase in pollution due to vehicles, which includes high density of mostly older, or vintage, vehicles in urban areas. There is also a predominance of private vehicles on the roads, especially cars and two wheelers, due to an inadequate transport system. This causes both increased emissions and traffic congestions. Two other reasons cited are inadequate inspection and maintenance of vehicles and fuel quality and adulteration. Finally, the note considered the increase of high-rise buildings in urban areas as a reason for stagnation and prevention of dispersion of vehicular emissions.

The control of vehicular pollution has been driven by a series of orders issued by the Supreme Court since the early 1990s. It started with a 1985 case[32] (discussed in Chapter 1) in the Supreme Court related to the pollution in Delhi-NCR. Directions in this case have continued to monitor several

---

[29] Supreme Court order dated 5 August 2018 in Writ Petition (Civil) No. 13029 OF 1985 (*M. C. Mehta v. Union of India & Ors*).

[30] Environment Pollution (Prevention & Control) Authority (EPCA) for Delhi and CSE, *Air Pollution Report Card: 2017–18* (New Delhi: CSE, 2018), 13.

[31] Lok Sabha Secretariat Reference Note No. 14/RN/Ref./June/2018 on vehicular pollution in India.

[32] Supreme Court Writ Petition (Civil) No. 13029/1985 (*M. C. Mehta v. UoI & Ors*) relation to pollution in Delhi-NCR.

matters related to pollution control including those of vehicles. In the initial years, the focus of the Court was to find alternate fuels to diesel as a measure to check vehicular pollution. In 1994, the Court observed that government and public transport vehicles should shift to compressed natural gas (CNG). In the 2002 order[33] of the same matter, the Court recalled that

> it was suggested that to begin with of Government vehicles and public undertaking vehicles including public transport vehicles could be equipped with CNG cylinders with necessary modification in the vehicles to avoid pollution which is hazardous to the health of the people living in highly polluted cities like Delhi and the other metros in the country.

In January 1998, the apex court appointed the EPCA (discussed in Chapter 2) to assist the Court in monitoring its directions.[34] Since its appointment, the EPCA has played a central role in standard setting and enforcement of vehicular pollution norms in India. One of the first recommendations of the EPCA was the 'phasing out of non-CNG buses' in the city. This was followed by the restriction on movement of trucks into the city for which orders were issued in 2001. The Court and the EPCA have continued to monitor, update and introduce vehicular pollution standards for Delhi-NCR.[35]

The CPCB has suggested vehicular standards that should be applied at a national scale which draw upon several directions of the Supreme Court in Writ Petition (Civil) No. 13029 of 1985, which have been detailed out in the 2010 guidelines issued by the CPCB.[36]

In comparison to the Supreme Court and the EPCA, the role of the environment ministry and the CPCB has been limited in controlling vehicular pollution in India. A 2010 report of the CPCB lists a legal framework for

---

[33] Supreme Court order dated 5 February 2002 in Writ Petition (Civil) No. 13029 of 1985 (*M. C. Mehta v. Union of India*).

[34] Supreme Court order dated 7 January 1998 in Writ Petition (Civil) No. 13029 of 1985 (*M.C. Mehta v. Union of India*).

[35] EPCA, *Report Status of Implementation of the Hon'ble Supreme Court Order for Bypassing Trucks to the City of Delhi to Control Vehicular Air Pollution (January 2003)* (New Delhi: EPCA, 2013).

[36] CPCB, *Status of the Vehicular Pollution Control Programme in India* (New Delhi: Government of India, 2010).

vehicular pollution in India which discusses the various committees set up by the environment ministry to set standards and present strategies. However, the directions for enforcement on this issue have emerged from the Court.

II INSTITUTIONAL SET-UP FOR POLLUTION CONTROL

MANDATE AND COMPOSITION OF POLLUTION CONTROL BOARDS

The Air and Water Acts are implemented through PCBs, which are set up at national, state and regional levels. These institutions were set up under Sections 3 and 4 of the Water Act, 1974. Since then, the work done by PCBs has generated a long history of implementation. The incremental changes to the regulatory practices of these institutions set up for pollution control have emerged in response to industrial growth, technological changes, and growing environmental and economic challenges.

The CPCB[37] located in New Delhi was established under Section 3 of the Water Act. Their primary mandate is to provide technical assistance and guidance to state boards including by setting standards and guidelines to prevent air and water pollution. It also advises the environment ministry to devise nationwide programmes for regulating air and water pollution and collecting and disseminating information pertaining to air and water pollution. These are implemented through SPCBs whose powers are discussed in the following paragraphs. Some standards that the CPCB has laid out are emission standards for thermal power plants, noise limits for diesel generator sets, effluent discharge in water bodies, transportation, vehicular pollution and handling of minerals. There are also industry-specific air emission and effluent discharge standards. In specific instances, the CPCB also has powers to issue directions in case the environment ministry finds that an SPCB has defaulted on its duties.

SPCBs were established under Section 4 of the Water Act. They function independently at the state level for day-to-day implementation of the pollution control laws and related guidelines. However, the directions

---

[37] In 1974, the CPCB was first known as the 'Central Board for the Prevention and Control of Water Pollution' under the then Ministry of Works and Housing. The board was subsequently brought under the Ministry of Environment and Forests in 1981. It was renamed as the Central Pollution Control Board (CPCB) in October 1988. Public Accounts Committee, *Central Pollution Control Board: Audit Review.*

of the CPCB guide the standard setting and enforcement functions. State boards conduct studies and research projects. They prepare research reports on environmental protection. For instance, comprehensive studies of the water quality of polluted river stretches of Bima, Godavari, Tapi, Krishna and other rivers flowing through Maharashtra were prepared by agencies appointed by the Maharashtra PCB.[38]

The state boards grant conditional consents to establish and operate projects that pollute the environment. This function is described in detail in the next section of this chapter. It also undertakes inspections of industrial facilities and equipment to check their effluent discharge and fugitive emissions. Following such inspections, SPCBs can direct 'closure, prohibition or regulation of any industry, operation or process or the stoppage or regulation of supply of electricity, water or any other service'.

Union Territories have PCCs that function directly under the CPCB. For instance, Daman has PCC Daman. Delhi, which does not have full statehood and is jointly governed by the central and the elected Delhi government, also has a PCC.

Chapter III of the Water Act deals with the creation of joint boards. A joint board can be constituted by an agreement (a) either by two or more governments of contiguous states or (b) by the central government (in respect to one or more UTs) or UTs and participating state or states.[39]

The composition of PCBs is detailed in Chapter II of the Water Act. It requires those steering the mandate of these boards at central and state levels to have special knowledge and expertise on environment protection and administration of institutions. These positions are to be full-time appointments and are to follow detailed guidelines laid out by the environment ministry.

In August 2016, the NGT[40] issued directions to all state governments to ensure that the officials heading the posts of chairman and member secretary of SPCBs 'are competent and eligible with requisite knowledge or practical experience in the field of environment protection and pollution control,

---

[38] Maharashtra Pollution Control Board, *Comprehensive Study Report on Polluted River Stretches* (Mumbai: Maharashtra Pollution Control Board), available at http://mpcb.gov.in/ereports/PoolutedRiverStruche.php, accessed on 27 July 2019.

[39] Ibid., Section 13.

[40] National Green Tribunal judgment dated 26 August 2016 in Original Application No. 318 of 2013 (*Rajendra Singh Bhandari v. State of Uttarakhand and Ors*).

with experience of management'. (The composition and eligibility criteria of SPCBs are discussed in Chapter 2.)

This subject of eligibility was argued in this case dealing with the lack of experienced and full-time officials in the Uttarakhand SPCB. The petitioner argued that poor appointments and lack of competence were directly constraining the required enforcement actions. At first, this case was filed before the Uttarakhand High Court and the petitioner was directed by the court to approach the NGT.[41]

The Principal Bench of the NGT expanded the scope of the case to a review of the appointments of chairpersons and member secretaries of all SPCBs. It observed,

> [W]e are of the considered view that the State Pollution Control Board, with growing industrialization and increasing urbanization, has high responsibility. The functions of the Board are enormous which relates to the essential items like water on which stands the very existence of human being.

The bench asked the environment ministry to submit details of the appointments of chairpersons and member secretaries of all SPCBs and came to the conclusion that there was a gross violation of the Air and Water Acts. The tribunal listed an 11-point mechanism through which the appointments to the posts of PCBs should be made. This included advertising the posts and seeking independent applications. It directed the detailed procedure for appointments to be prescribed through new rules issued under the Water and Air Acts. The judgment directed that suitable infrastructure be created for SPCBs to function as full-time institutions and for the proceedings of the boards to be carried out regularly and with proper record keeping.

INSTITUTIONAL COORDINATION

Section 18 the Water Act addresses the basic coordination mechanism between regulatory institutions. It lays out the hierarchy of these bodies and which institution would have the final word. It focuses on conflict resolution mechanisms if disputes arise between various regulatory institutions. There have also been some cases in high courts that have further defined these relationships.

---

[41] High Court of Uttarakhand at Nainital in Writ Petition (PIL 136/2013).

The four scenarios envisaged by the law are as follows:

- *Disputes between state boards*: The state boards are subject to the directions of state governments, the central board and the central government. In the event of any dispute, the final decision vests with the central government.

- *Disputes between directions of the central board and the state government*: Where there is an inconsistency between the directions given by the central board and the state government, the matter is referred to the environment ministry for its decision.

- *Defaults by state boards*: The central government can direct the central board to perform the functions of the state board in public interest when it is of the opinion that the state board has failed to comply with any direction given by the central board and has caused a grave emergency as a result of such default.

- *Violation of public interest by the central or state boards*: The law empowers the central and state governments to supersede the respective boards if they consistently fail to perform their duties in public interest.

III   REGULATORY FUNCTIONS OF PCBs

CONSENTS TO OPERATE AND ESTABLISH

The Air and Water Acts require any industry, operation or process that is likely to release a pollutant into water or air to take a prior consent from an SPCB or a PCC. This includes manufacturing and processing units, storage and waste disposal facilities, mines and quarries, power plants (including hydro, thermal and gas) as well as building and road construction projects.

PCBs grant two types of consent:

- *Consent to establish (CTE)*: The consent needs to be sought prior to the commencement of any work on constructing or establishing an industrial facility or activity. This is a one-time approval. PCBs have guidelines which prescribe a validity of the CTE. For instance, the Maharashtra PCB grants a CTE for five years. If a project is not established during this time, there needs to be reapplication with fresh fees as per the prescribed norms.[42]

- *Consent to operate (CTO)*: This consent needs to be taken prior to commencement of operation of a unit or a project. This approval comes

---

[42] Maharashtra Pollution Control Board, Circular No. MPCB/AB/88 (guidelines for calculation of consent fees) dated 18 June 2016.

with a set of conditions, which requires the project authorities to comply and report. These consents are time bound and need to be renewed. These consents were granted for one or two years. More recently, PCBs have developed a practice of granting long-term consent, sometimes up to 10 years, as a mechanism to support ease of doing business.[43]

The process of grant of consent is aided by the categorisation of industries into Red, Green, Orange and White, based on the CPCB's calculation of their pollution load. The CPCB traces this back to 1989, when the environment ministry introduced this concept while declaring Doon Valley as an ecologically sensitive area under the Environment Protection Act, 1986. According to the CPCB,

> the application of this concept was extended in other parts of the country not only for the purpose of location of industries, but also for the purpose of Consent management and formulation of norms related to surveillance/inspection of industries.

However, this categorisation was being differentially applied by various SPCBs. The first attempt to harmonise this was in 2012 when the CPCB used its powers under the Air and Water Acts to identify 85 types of industrial sectors as 'Red', 73 industrial sectors as 'Orange' and 86 sectors as 'Green'.[44]

In 2016, this criterion was revised to include White-category industrial sectors that are considered to be 'practically non-polluting'. A total of 36 industrial sectors were added to the White category which would not have to obtain consents. Intimation to the concerned PCB/PCC would suffice. This includes units such as dry cotton and woollen hosiers making, fly ash brick manufacturing, electrical and electronic items assembling, solar power generation through solar photovoltaic cell, wind power and mini hydel projects (less than 25 megawatts).

The CPCB also recategorised the Red, Green and Orange industrial sectors.

---

[43] Karnataka Pollution Control Board, Notification No. KSPCB/798/COC/2016-17/ dated 15 June 2016.

[44] CPCB directions dated 4 June 2012 based on recommendations of the Working Group on criteria of categorisation, formed as per resolution passed during the 57th Conference of the Chairmen and Member Secretaries of the CPCB and SPCBs.

- The Red-category industrial sectors were reduced to 60 and included units such as lead acid manufacturing, automobiles, power generation (except wind, solar and mini hydel), sugar, firecrackers, pulp and paper, oil and gas extraction.

- The Orange category gained another 10 sectors to move up to 83 sectors and included sectors such as coal washeries, bakery and confectionery units, food and food processing units, manufacturing of toothpaste, silica gel, and building and construction projects with more than 20,000 square metres of built-up area.

- Sixty-three industrial sectors such as carpentry and wooden furniture manufacturing, cement products (without using asbestos), cold storage and ice making, cotton spinning, leather footwear and leather products are in the Green category.

On 30 April 2020, the CPCB wrote to all PCBs/PCCs with a revised categorisation of industries. This letter added a new classification for 'Non-Industrial Operations (Activities/Facilities/Infrastructure/Services)' which includes operations such as airports, ports and harbours, hotels, common effluent treatment plants, health care establishments, and so on.[45]

Since 2016,[46] this categorisation is tied to the pollution index score for each sector.[47] The CPCB guidelines suggest that SPCBs/PCCs may issue CTO to the industries in the following manner:

- Red-category industries for 5 years

- Orange-category industries for 10 years

- Green-category industries for 15 years

- No consent needed for White-category industries

---

[45] Central Pollution Control Board, Letter No. F. No. B-2 90 16/ROGW /IPC-VI/2020-21 / dated 30 April 2020.

[46] CPCB, *Final Document on Revised Classification of Industrial Sectors under Red, Orange, Green and White Categories* (New Delhi: Government of India, 2016).

[47] Based on the score of the pollution index, the following categorisation is to be made:

> Type of industries, if scores 60 and above be categorized as Red; Type of industries, if scores from 30 to 59 be categorized as Orange; Type of industries, if scores from 15 to 29 be categorized as Green; and Type of industries, if less than 15 be categorized as White or non-polluting industry.

No Red-category industries are to be permitted in eco-sensitive areas and protected areas declared under the Wildlife Protection Act, 1972.

This was one way by which PCBs also reduced their administrative load to grant and renew consents. Another attempt towards this was merging the grant of CTE with environment clearances. In November 2018, the CPCB reiterated its earlier advisory directing that all projects granted environment clearance under the Environment Impact Assessment Notification, 2006 (discussed in Chapter 4), would not require a CTE. In a letter to all PCBs/PCCs, the central board said,

> For industries requiring EC, issuing of consent by SPCBs/PCCs shall be one- step process and EC will be deemed as CTE. In such cases SPCBs/PCCs shall be involved in the process of granting of EC.[48]

However, this communication was challenged in the High Court of Delhi.[49] In January 2019, the court stayed the operationalisation of this direction on the grounds that the scheme for involvement of SPCBs/PCCs was not ready. Until such time, 'EC cannot be deemed to be considered as CTE'.

The matter was still pending before the Delhi High Court as of January 2020. However, the order dated 2 December 2019 recorded that the environment ministry was in the process of initiating the modalities for creating 'one step process of Consent to Establish and Environmental Clearance'.

The consent granting process of PCBs/PCCs has been based on the type of industry and not on where it is located. There have been attempts by PCBs to build siting into the mechanism of how approvals can be reviewed, but this has not been operationalised as yet.

In 1995, the CPCB started a land use and an environmental planning programme known as 'Zoning Atlas for Siting of Industries'. The stated objective of this programme was to identify and protect environmentally sensitive zones and achieve developmental objectives by informing the decisions on siting of industries with the environmental importance

---

[48] Letter No. F. No. B-29012/MSMEs/IPC-VI/201 7·18/12189-12230 from CPCB to all SPCBs/PCCs dated 2 November 2018.

[49] High Court of Delhi order dated 9 January 2019 in Writ Petition (Civil) 13521 of 2018 & CM No. 52715/2018 (*Social Action for Forest and Environment v. Union of India & Anr*).

of specific locations. It was later extended under the Environmental Management Capacity Building Technical Assistance Project funded by the World Bank. These atlases were prepared at both district and state levels with the tasks of demarcating and classifying the environment in a district, identifying suitable locations for siting of industries and suitable industries for identified sites.

The zoning atlases to be prepared for the districts in a state were to be compiled into an abridged Industrial Siting Atlas for the entire state. A national atlas was envisaged combining siting atlases of various states. The website of the CPCB also introduced this exercise as one that could facilitate the setting up of industries with lesser environmental scrutiny and paperwork once the zoning was completed.[50]

State governments identified priority districts where this exercise could be carried out. This prioritisation was based on three factors:

- Districts which have been declared for rapid industrial development by the government

- Districts that are facing pollution problems that could grow in the future

- Districts which are environmentally sensitive and need to be protected from pollution

By 2017, 55 district environmental atlases across 13 states were pending the CPCB's final nod.[51] The programme is reported to be discontinued in 2008 due to lack of resource allocation.[52]

ENFORCEMENT

Besides granting consents to projects and processes that can lead to pollution, the enforcement of compliance with air and water laws is a significant function of PCBs. There are six main ways in which PCBs implement their enforcement

---

[50] Website of CPCB, http://cpcb.nic.in/environment-planning-introduction/, accessed on 23 July 2019.

[51] Status of district-level atlases updated until 15 July 2017, available at http://cpcb.nic.in/district-level-atlas/, accessed on 23 July 2019.

[52] Sanjeev Kumar Kanchan, 'Newer, Not Better', *Down to Earth*, 20 July 2017, available at https://www.downtoearth.org.in/news/environment/newer-not-better-58292, accessed on 11 September 2020.

functions. These are taken up *suo moto* or in response to a complaint. Both the Air and the Water Acts allow these actions to be taken. They are as follows:

- *Site inspections* or a visit to check and ascertain sources of pollution, to collect samples and verify complaints. Concerned officials usually inform the project authorities. But there are provisions for surprise checks that are mostly carried out by the Vigilance Departments of PCBs.

- *Show cause notices* are issued upon confirming legal violations. The PCBs issue a warning to an alleged violator seeking an explanation on why action should be taken against legal non-compliance.

- *Directions for compliance* are issued in case the response of the violating party is not satisfactory. The PCB issues a direction notice with an attempt to bring a facility into compliance. Such directions are also issued if there is repetitive non-compliance. In such cases, there is no show cause notice.

- *Closure orders* are given by PCBs to a polluting facility or where legal violations have been established. This notice does not always imply that employees cannot enter the premises. It is a direction to shut down or cease a specific operation, part or component which is responsible for the pollution or legal violation. Closure orders can also be recalled.

- *Fines and court action*: A PCB/PCC cannot directly impose penalties under the Air and Water Acts. A criminal case needs to be filed not below the court of a metropolitan magistrate or judicial magistrate of first class. The amount of fine and imprisonment differs upon the nature of offence or violation. Other than the boards, any person can file a case by giving a minimum 60-day notice to the SPCB indicating their intention to approach the court. In such instances, the PCB/PCC would need to provide all documents in their possession to the complainant.[53]

- *Bank guarantees*: The PCB/PCC seeks action plans from specific units that will help to meet the prescribed standards, especially in the cases of non-compliance. This action plan is required to be submitted along with a bank guarantee that can be forfeited against non-compliance with the action plan. This system was recommended by the CPCB in 2003 for 17 highly polluting industries.[54] In 2014, the CPCB also issued directions to 'all industries' in the highly polluting category to submit a bank

---

[53] Shibani Ghosh, 'Reforming the Liability Regime for Air Pollution in India', Working Paper, Centre for Policy Research, New Delhi, 2015.

[54] CPCB, *Charter on Corporate Responsibility for Environmental Protection* (New Delhi: Government of India, 2003).

guarantee of 25 per cent of the cost of online monitoring of effluent or emission systems that they are required to set up.[55] In January 2014, a case related to the imposition of bank guarantees on two projects in Odisha concluded that bank guarantees 'should be liable to be invoked/encashed for environmental compensation and restoration purposes'.[56]

An ongoing study by researchers at the Centre for Policy Research, New Delhi, highlights that PCBs do not follow the same enforcement procedure. For instance, there is no standard practice of issuing notices prior to undertaking site inspection. The study highlights that states such as Karnataka, Chhattisgarh, and Daman and Diu mandate that a notice be given to the industry before conducting an inspection. In Chhattisgarh, this prior intimation can be ignored in urgent situations. In Gujarat and Odisha, notice can be given at the time of entry into the project premises. Prior intimation is not mandatory. The PCBs/PCCs in Daman and Diu, Karnataka, Gujarat, Odisha and Chhattisgarh mandate that the inspection reports should be uploaded online within 24–48 hours.

In a 2014 case, the NGT observed that PCBs lacked an inspection policy. In its judgment, the tribunal directed:

> Thus, we direct that the Boards henceforth shall clearly formulate their inspection policy, which should be fair, transparent and objective.[57]

Supreme Court Case on Effluent Treatment Plants (ETPs)

In 2012, a writ of mandamus[58] was filed seeking the Supreme Court's intervention

---

[55] CPCB letter no. B-29016/04/06/PCI-1/5401 dated 4 February 2014.

[56] National Green Tribunal judgment dated 19 January 2014 in Appeal No. 68 of 2012 (*State Pollution Control Board, Odisha v. M/s Swastik Ispat Pvt. Ltd. & Ors*) and Appeal No. 69 of 2012 (*State Pollution Control Board, Odisha & Ors v. M/s Patnaik Steel & Alloys Ltd & Ors*).

[57] National Green Tribunal judgment dated 19 January 2014 in Appeal No. 68 of 2012 (*State Pollution Control Board, Odisha v. M/s Swastik Ispat Pvt. Ltd. & Ors*) and Appeal No. 69 of 2012 (*State Pollution Control Board, Odisha & Ors v. M/s Patnaik Steel & Alloys Ltd & Ors*).

[58] A judicial writ seeking the Supreme Court to command or give directions to an inferior court or a person to perform a public or statutory duty.

to ensure, that no industry which requires 'consent to operate' from the concerned Pollution Control Board, is permitted to function, unless it has a functional effluent treatment plant, which is capable to meet the prescribed norms for removing the pollutants from the effluent, before it is discharged.[59]

In this case, neither the petitioner nor the respondents disputed the requirement and efficiency of individual industrial units setting up ETPs as a pollution control measure.

The case mainly focused on two questions:

- Why industries requiring consents from PCBs were functioning without ETPs/STPs/CETPs when these are an integral part of the project?

- Why had the PCBs not issued notices or directions to the defaulting industries?

The judgment emphasised that ETPs would be a prerequisite for the grant of consent to operate. The apex court directed that SPCBs should advertise widely that the concerned industries should set up primary ETPs within three months. The SPCBs were required to carry out inspections and take action in case the Court's directions were not complied with.

The Court also recorded that industries need to set up CETPs which was delayed because of both financial reasons and unfinished process of land acquisition. However, the judgment ordered that CETPs should be set up on an urgent basis within three months of the date of the judgment. Where CETPs were dysfunctional, they had to be operationalised.

The apex court entrusted the NGT to monitor the implementation of the Court's orders and receive complaints against non-compliance. After monitoring the matter for more than a year,[60] the NGT concluded that the matter is in need of regular follow-up by statutory authorities entrusted with the task. In its final order dated 3 August 2018, the NGT circled back to the CPCB to prepare an action plan, publish the compliance reports on their website and take penal action for failure. The NGT also reiterated the

---

[59] Supreme Court judgment dated 22 February 2017 in Writ Petition (Civil) No. 375 of 2012 (*Paryavaran Suraksha Samiti v. Union of India*).

[60] National Green Tribunal order dated 3 August 2018 (revised on 31 August 2018) in Original Application No. 593 of 2017 (*Paryavaran Suraksha Samiti & Anr. v. Union of India & Ors*).

Supreme Court's directions for setting up pending ETPs/STPs/CETPs and making operational the ones that were non-functional.

HIGH-LEVEL ASSESSMENTS OF POLLUTION CONTROL BOARD FUNCTIONING

> Without casting any aspersions on the statutory bodies, it is an acknowledged fact that the Pollution Control Boards have not been able to take adequate steps for keeping the standards of water within the prescribed limits. They have not been able to stop dumping of wastes, discharge of municipal or industrial effluents in rivers and water bodies.

This was the NGT's observation in its order dated 20 September 2018 in Original Application No. 673/2018.[61] This was a *suo moto* case taken up by the NGT based on a news item titled 'More River Stretches Are Now Critically Polluted: CPCB' published in *The Hindu* newspaper on 17 September 2018. While this news report highlighted that a CPCB study has concluded that 351 river stretches have been recorded as polluted by the CPCB, there is a shortfall in implementing several directions and judicial pronouncements. The order once again emphasised SPCBs' inability to enforce compliance. The judgment set in motion a detailed set of actions to assist SPCBs including by setting up special task forces.

The PCBs have been the subject of scholarly research, judicial scrutiny and government assessments. The incapacity of PCBs to regulate large number of industrial and infrastructure facilities was recognised years ago but still remains unresolved. The Public Accounts Committee of the Lok Sabha Secretariat submitted a report on the functioning of the CPCB in 1994. It identified that the composition of PCBs remains incomplete, and there are also shortcomings in the setting up and functioning of monitoring stations for both the Air and the Water Acts. The report also pointed out that there have been financial, administrative and other irregularities in the CPCB functioning.[62]

---

[61] National Green Tribunal order dated 20 September 2018 in Original Application No. 673 of 2018 (*News Item Published in 'The Hindu' Authored by Shri Jacob Koshy: 'More River Stretches Are Now Critically Polluted: CPCB'*).

[62] Public Accounts Committee, *Central Pollution Control Board: Audit Review*.

The Planning Commission initiated an extensive review of 25 SPCBs between 1992–93 and 1997–98 coinciding with the eighth five-year plan cycle of the commission. This programme evaluation study was carried out with the objectives of studying SPCBs organisational set-up, staffing pattern, finances and training requirements and examining their functioning with reference to the Pollution Acts. It pointed out that PCBs have been reduced to merely 'Industrial Pollution Control Boards'. Adequate attention has not been given to non-industrial sources of pollution. In practice, implementation of the regulations related to the vehicular pollution, a significant contributor to the ambient air pollution, does not fall within the jurisdiction of SPCBs. The report also recorded, 'SPCBs are, sometimes, not able to exercise the powers to force compliance by stopping electricity supply or water because of interference by powerful pressure groups'.[63]

These reviews identified the constraints in their functioning and suggested remedial measures. Their recommendations suggested how the implementation of the pollution control laws could be improved. The 192nd Report on Functioning of Central Pollution Control Board by the Department-Related Parliamentary standing Committee on Science and Technology and Environment and Forests suggested that CPCB should be restructured as an autonomous statutory authority.[64] This report also said that it should have the mandate not only to develop regulations and fixing standards but also to ensure enforcement and compliance.

IV ACTION PLANS

The environment ministry and PCBs have also devised action plans to guide how pollution prevention or remedial action could be taken up.

---

[63] Planning Commission of India, *Evaluation Study on Functioning of State Pollution Control Boards (PEO Study 180)* (New Delhi: Planning Commission of India, n.d.), 5, available athttps://niti.gov.in/planningcommission.gov.in/docs/reports/peoreport/cmpdmpeo/volume1/180.pdf, accessed on 26 July 2019.

[64] Department-Related Parliamentary Standing Committee on Science and Technology and Environment and Forests, *One Hundred and Ninety Second Report on Functioning of Central Pollution Control Board* (New Delhi: Rajya Sabha Secretariat, 2008).

COMPREHENSIVE ENVIRONMENTAL POLLUTION INDEX (CEPI) AND POLLUTION MITIGATION ACTION PLANS

In 2009, the CPCB developed a CEPI with the objective of identifying critically polluted areas (CPAs) and severely polluted areas (SPAs) across the country. The exercise was carried out in collaboration with Indian Institute of Technology, Delhi. The index is a number between 0 and 100, arrived at by weighing several environmental parameters of a given location. The index is assigned to the given location to indicate its environmental quality.[65] Environmental assessments were carried out in 88 industrial clusters. Among the industrial clusters, 43 received a score above 70 and were declared CPAs and 32 as SPAs. Subsequently, the number of areas assessed increased from 88 to 100. Once the areas were ranked based on the CEPI, PCBs were to initiate steps in preparing comprehensive remedial action plans.

The CEPI was revised in 2016 by changing and leaving out critical parameters on the ground that they were subjective and, therefore, not measurable. A 2016 CPCB document says,

> it has been experienced that some factors are difficult to measure objectivity like potentially affected population and assessment of health impacts etc.

A CPCB document dated 15 August 2015[66] states that the CEPI requires reliable health impact studies on humans, flora and fauna. This was one of the criteria removed from the revised index. The 2015 document justified 'eliminating the subjective factors' by stating that the studies related to health are time consuming. Further, data is not readily available and the exercises to generate fresh data require substantial funding support. Another justification was that the 'existing criteria also lacks clarity with respect to potentially affected population'. This too was removed as a parameter to calculate the CEPI.

---

[65] Description of CEPI on the CPCB website, available at http://cpcb.nic.in/comprehensive-environmental-pollution-index-cepi/, accessed on 29 July 2019.

[66] CPCB letter no. B2902/ESS/CPA/2015-16 dated 15 August 2015 inviting comments on draft document on Revised CEPI Version–2015.

When the CEPI parameters were altered, areas like the Vapi industrial cluster in Gujarat were no longer listed as critically polluted.[67] Previously, the environment ministry had imposed a moratorium on new projects in the area based on the CEPI. After the revised index, the moratorium was lifted and new industries were allowed to be set up in Vapi[68] as long as SPCB ensured compliance with the Vapi Action Plan. However, pollution severity and impacts have remained at critical levels.[69] The case of the Korba Action Plan in Chhattisgarh is similar as most of the steps listed in the plan remain unenforced.[70]

In December 2018, the NGT took *suo moto* cognizance of this issue based on a news report and directed the CPCB to develop time-bound action plans for the critically polluted sectors with the objective of bringing down pollution levels. The status of pollution in all the polluted industrial corridors was to be uploaded on the website of the SPCB/PCC. It also asked all chief secretaries to regularly review environmental issues, including pollution, in their states and submit a report before the tribunal. The time frame set for the same was three months.[71] The NGT continued to monitor the case with detailed observations even as reports were being filed by SPCBs/PCCs.

In its order of 14 November 2019, the NGT commented on the report that mitigation or management measures cannot replace enforcing the rule of law. The order observed,

---

[67] Hardik Shah, 'Vapi Most Polluted Industrial Cluster Again', *Times of India*, 20 May 2012, available at https://timesofindia.indiatimes.com/city/surat/Vapi-most-polluted-industrial-cluster-again/articleshow/13320549.cms, accessed on 10 June 2020.

[68] MoEFCC Office Memorandum No. J1103/5/2010/-I.A-II (I) dated 25 November 2016.

[69] Krithika Dinesh and Bharat Patel, 'Why Does Vapi in Gujarat Continue to Be Critically Polluted?' *The Wire*, 5 January 2017, available at https://thewire.in/environment/vapi-polluted-gpcb-cepi, accessed on 6 May 2019.

[70] Rashmi Drolia, 'Raipur, Korba Lack Action Plan to Tackle Air Pollution in Chhattisgarh', *Times of India*, 22 September 2018, available at https://timesofindia.indiatimes.com/city/raipur/raipur-korba-lack-action-plan-to-tackle-air-pollution-in-chhattisgarh/articleshow/65910704.cms, accessed on 11 September 2020.

[71] National Green Tribunal order dated 13 December 2018 in Original Application No. 1038 of 2018 (*News Item Published in 'The Asian Age' Authored by Sanjay Kaw Titled: 'CPCB to Rank Industrial Units on Pollution Levels'*).

[W]hile every mitigation measures must be taken, this cannot be ground not to take any legal action for violation of law.

The CPCB and SPCBs were directed to submit details of

action taken report furnished showing the number of identified polluters in polluted industrial areas mentioned above, the extent of closure of polluting activities, the extent of environmental compensation recovered, the cost of restoration of the damage to the environment of the said areas, otherwise there will be no meaningful environmental governance.

The case was pending final orders or judgment as of January 2020.

National Clean Air Programme

The visible rise of air pollution in India and citizens' campaigns led the environment ministry to launch the NCAP in January 2019. The ministry recognised air pollution as one of the biggest environmental challenges after years of government denial. It proposed a five-year programme starting in 2019. The ministry also indicated the possibility of its extension into a long-term plan following a mid-term review. The NCAP is designed to be a national level inter-ministerial and inter-state collaborative response to address the problem of air pollution.

The priorities of the plan are threefold.

First, the NCAP acknowledges that there are multiple sources of air pollution where action needs to be taken. This includes power, transport, industry, construction and agriculture. The NCAP gives recommendations on how the current contribution of these sectors can be reduced.

Second, the programme emphasises all the existing measures that have been undertaken at the national level to monitor and regulate air pollution, which need to be supported and strengthened. This includes putting in place National Ambient Air Quality Standards (NAAQS) and the National Air Quality Index (AQI). In addition, there are specific measures for Delhi and the NCR. It also includes the activities of the EPCA[72] and notifying a Graded

---

[72] This authority was notified in 1998 under Section 3(3) of EPA, 1986, in pursuance orders of the Supreme Court in Writ Petition (Civil) No. 13029 of 1985 (*M. C. Mehta v. UoI & Ors*) in relation to pollution in Delhi-NCR. It is headed by Shri Bhure Lal, former Secretary, Central Vigilance Commission, since the time the authority was first constituted.

Response Action Plan for the Delhi-NCR. This is in addition to other specific directions issued by the CPCB which are clubbed together as 'Forty-Two Action Points'.

Third, the programme focusses on 102 'non-attainment cities' at the national level. These have been identified by the CPCB as cities where the NAAQS have been violated. The NCAP seeks to develop city-specific action plans and develop mitigation measures to address the problem of air pollution in these cities. Forty-three of these, which fall within the government's 'Smart Cities' framework, will be prioritised.

The NCAP has been critiqued for being a non-prioritised wish list with a short-term focus. It has also been called out for its narrow approach,[73] for being city specific rather than taking a regional or air shed-level approach.[74] The NCAP is not legally binding which raises questions of enforcement.[75] Some reviews have pointed to the non-viability of the action plan as it is without adequate financial support.

In 2018, the principal bench of the NGT took *suo moto* cognizance of a news report in the *Times of India*. The article stated that the environment ministry would unveil the NCAP on 15 August 2018.[76] Based on this article, the NGT ordered all the states and UTs with non–attainment cities to

> prepare action plans to bring the air quality up to the prescribed norms. The action plans were to be forwarded by 31 December 2018 to the CPCB to be evaluated by a Committee constituted by the Tribunal.

---

[73] Anumita Roychowdhury, 'National Clean Air Programme: Good Idea but Weak Mandate', *Down to Earth*, 11 January 2019, available at https://www.downtoearth. org.in/blog/air/national-clean-air-programme-good-idea-but-weak-mandate-62785, accessed on 11 September 2020.

[74] Santosh Harish, Shibani Ghosh and Navroz K. Dubash, 'Clearing Our Air of Pollution: A Road Map for the Next Five Years', in Centre for Policy Research, *Policy Challenges (2019–2024): Charting a New Course for India and Navigating Policy Challenges in the 21st Century* (New Delhi: Centre for Policy Research, 2019), 39–43.

[75] Jayashree Nandi, 'Air Quality Plan Not Legally Binding', *Hindustan Times*, 27 November 2018, available at https://www.hindustantimes.com/india-news/air-quality-plan-not-legally-binding/story-h9hAsRDBJq0YnBiGrqygqN html, accessed on 11 September 2020.

[76] National Green Tribunal order dated 15 March 2019 in Original Application No. 681 of 2018 (*News Item Published in 'The Times of India' Authored by Shri Vishwa Mohan Titled 'NCAP with Multiple Timelines to Clean Air in 102 Cities to Be Released around August 15'*).

The environment ministry released the NCAP on 10 January 2019, but it was not made effective even a year after its announcement.[77] During this time, the NGT continued to monitor the preparation of the action plans. The tribunal recorded that non-compliance of its orders is a criminal offence.[78] On 16 January 2019, the NGT summoned 'Chief Secretaries of all the States to appear in person and furnish compliance of various orders' of the tribunal. On 15 March 2019, the NGT asked the governments of six states, namely, Assam, Jharkhand, Maharashtra, Punjab, Uttarakhand and Nagaland to furnish their action plans by 30 April 2019. These plans were to be implemented within six months and needed clear financial allocations. Non-compliance of the tribunal's order would make state governments liable to pay an environment compensation of ₹10 million. For the states where action plans were made and where deficiencies were not removed, they would be asked to pay ₹2.5 million each. The NGT continues to monitor this case.

V   APPEALS BEFORE APPELLATES AND TRIBUNAL

This section deals with two appeal mechanisms for pollution-related matters:

*   Appeals before the pollution appellate
*   Appeals before the NGT

Both these appellate mechanisms do not foreclose the right of citizens to approach the district and state courts and the Supreme Court for relief. The role of courts in environmental matters has been discussed in Chapter 2.

APPEALS BEFORE POLLUTION CONTROL BOARD APPELLATE

Appeals against orders of the PCB or UT PCCs are discussed in Section 31 of the Air Act and Section 28 of the Water Act. These sections allow any person aggrieved by an order made by the state board to appeal to the appellate authority, which is constituted through the Act. Respective state governments and the UT administration appoint the appellate. Appeals against an order need to be filed within 30 days of the issuance of the order.

---

[77] Mayank Agarwal, 'Is India's National Clean Air Plan on Track?' *Mongabay–India*, 15 January 2020, available at https://india.mongabay.com/2020/01/is-indias-national-clean-air-plan-on-track/, accessed on 10 February 2020.

[78] Section 26 of the National Green Tribunal Act, 2010.

Even though these mechanisms have been in existence since the time the law has been in place, they have remained grossly underutilised. Researchers argue that the pollution control appellate pre-dates environmental courts and 'green benches' in the higher judiciary.[79] They also have the potential of delivering effective decisions and are more accessible than courts. The paper highlights that appellate authorities are not set up in four states in India.[80]

The Air Act or the Water Act does not specify the definition of 'aggrieved person'. The study by Ghosh, Lele and Heble review cases filed before 22 authorities. This analysis shows that most cases are filed by industries as parties aggrieved by the orders of PCBs. Of the 15 judgments uploaded on the website of the Odisha Pollution Appellate up to April 2019, none were filed by aggrieved citizens or communities. There could be multiple reasons for this, including lack of knowledge, tedious procedure for filing, fear of backlash or lack of legal representation.

The interpretation by the Karnataka Pollution Control Board that only project proponents could be aggrieved parties for the purposes of these clauses was challenged in an appeal before the NGT.[81] The judgment dated 11 January 2013 clarified aggrieved parties are not just project proponents as 'the statutory provisions are always subservient to the mandate of the Constitution'. Therefore, the concerned sections of the Air Act and the Water Act has to be read along with 'every citizen' in Article 51A(g) of the Constitution of India. Article 51A relates to fundamental duties, which includes the duty of every citizen '(g) to protect and improve the natural environment including forests, lakes, rivers and wildlife, and to have compassion for living creatures'.

This interpretation substantially increases the scope of appeal for citizens and communities affected by pollution. The appellate mechanism also does not necessitate the need for a lawyer. However, this interpretation has not been exercised before specific appellate cases.

---

[79] Shibani Ghosh, Sharachchandra Lele and Nakul Heble, 'Appellate Authorities under Pollution Control Laws in India: Powers, Problems and Potential', *Law, Environment and Development Journal* 14, no. 1 (2018): 47–58, available at http://www.lead-journal.org/content/18045.pdf, accessed on 10 June 2020.

[80] These are Assam, Chhattisgarh, Meghalaya and Nagaland.

[81] National Green Tribunal Appeal No. 56 of 2012 South Zone (*Janajagrithi Samithi [Regd] v. Karnataka State Pollution Control Board*).

A case[82] related to non-compliance of consent conditions at a port in Gujarat helps to understand the difficulties in invoking the jurisdiction and expedient action from the pollution appellate.

The case related to the violation of the Air Act in a port on the Gujarat coast was argued before the Gujarat Pollution Appellate and the NGT. When a case was first filed in the pollution appellate in 2013, the Gujarat Pollution Control Board (GPCB) and Gujarat Maritime Board (GMB) argued that the appeal was time barred, and therefore should be rejected. The appellants argued that they had approached the NGT for relief and were directed to approach the appellate. The delay was condoned by the appellate as 'sufficient reasons' were provided for the same. The respondents, GPCB and GMB, then questioned the locus of the appellant arguing that he lives 235 kilometres from the port and hence cannot be considered an aggrieved person. The appellant relied on the *Janajagrithi Samithi (Regd) v. Karnataka State Pollution Control Board*[83] and another judgment of the Madras High Court, and the appeal was admitted.

The appellant also simultaneously pursued the NGT for relief.[84] The petitioner alleged that the port had been operating without complying with the necessary conditions listed in the consent and, therefore, should not be allowed to carry out any further handling of coal at the terminal. It was argued that this non-compliance was occurring despite two notices from the GPCB. In July 2014, the NGT issued detailed directions and asked the GPCB to monitor the same. In February 2015, additional time was granted to the GMB to ensure compliance with the NGT's orders and conditions of the consent.

Meanwhile, the appellant's earlier appeal before the pollution appellate was pending. In May 2015, the appellate finally rejected this appeal, observing that the NGT had already passed judgment with directions.

---

[82] National Green Tribunal order dated 23 May 2015 in Appeal No. 2 of 2014 before the Gujarat Pollution Appellate Authority (*Rajendrasinh Mansinh Kashtrya v. Gujarat Pollution Control Board & Ors*).

[83] National Green Tribunal judgment dated 11 January 2013 in Appeal No. 56 of 2012 (*Janajagrithi Samithi (redg) & Ors v. Karnataka State Pollution Control Board & Anr*).

[84] National Green Tribunal judgment dated 17 July 2014 in Application No. 41 of 2013 West Zone (*Rajendrasinh Mansinh Kashtrya v. Gujarat Pollution Control Board & Ors*).

APPEALS BEFORE NATIONAL GREEN TRIBUNAL

Section 33B of the Water Act and 31B of the Air Act provide for appeals before the NGT. Any person aggrieved by an order or decision of the appellate authority, an order passed by the state government or directions issued by pollution control boards can appeal to the tribunal. There are corresponding provisions in the NGT Act, 2010 for appeals and applications seeking remedial measures against pollution related impacts. This includes raising substantial questions related to the environment as well as seeking compensations against damages. The jurisdiction of NGT is discussed in Chapter 2.

OVERLAPPING JURISDICTION

A case in the Supreme Court[85] clarifying the legal relationship between the PCB, the pollution appellate, the NGT and the high court is discussed here. In 2018, a copper smelter in Thoothukudi, one of India's largest copper production plants in operation since 1997, initiated expansion activity allegedly without seeking environment clearance.[86] The local communities living in the project area had been regularly complaining against the pollution and other illegalities by the company. Based on these complaints, the Tamil Nadu Pollution Control Board (TNPCB) had issued several notices, against which appeals had been heard by the Supreme Court in 2013. The relief in these cases had allowed the plant to continue operations, subject to compliance of conditions.

This case illustrates a regulatory conflict that lasted a decade. The TNPCB had filed several notices related to non-compliance which had been challenged before the NGT and Supreme Court. Three orders were passed by the TNPCB in April and May 2018. On 9 April 2018, the TNPCB refused renewal of consent to operate allegedly due to repeated non-compliance of conditions. On 24 April 2018, the TNPCB passed an order restraining the company to

---

[85] Supreme Court judgment dated 18 February 2019 in Civil Appeal Nos. 4763–4764 of 2013 (*Tamil Nadu Pollution Control Board v. Sterlite Industries India Ltd*) with Civil Appeal Nos. 8773–8774 of 2013, Civil Appeal Nos. 9542–9543 of 2013, Civil Appeal No. 5782 of 2014, Civil Appeal Nos. 1552–1554 of 2019, Civil Appeal No. 23 of 2019, and Civil Appeal No. 1582 of 2019.

[86] Times News Network, 'Environment Ministry Team Checks Sterlite Expansion Site', *Times of India*, 25 April 2019, available at https://timesofindia.indiatimes.com/city/madurai/environment-min-team-checks-sterlite-expansion-site/articleshow/69033138.cms, accessed on 11 September 2020.

continue production without seeking consent. Since March 2018, the local communities had staged a peaceful protest against the project. The company claimed that it had an impeccable environmental record and that the protest was instigated and motivated. On May 22, the 100th day of the protest, police firing resulted in the death of 13 people. On 28 May 2018, the board issued closure orders and also directed the district collector to seal the premises.[87] In June 2018, the Tamil Nadu government announced the permanent closure of the copper smelter plant.

The company challenged these orders before the NGT.[88] The NGT directed the CPCB to submit a report in this matter. Based on the recommendations of this report, the NGT concluded that the TNPCB orders could not be sustained, as they do not uphold the principles of natural justice. This is because the company had not been provided an opportunity for hearing or served a notice before ordering closure. The judgment says,

> [T]he impugned orders cannot be sustained as it is against the principles of natural justice. No notice or opportunity of hearing was given to the appellant. The grounds mentioned in the impugned orders are not that grievous to justify permanent closure of the factory.

A civil appeal was filed by the TNPCB in the Supreme Court challenging the jurisdiction of the NGT's decision to revoke its closure orders to the plant. In a judgment dated 19 February 2019, the apex court concluded that the NGT did not have jurisdiction to adjudicate the matter. The Supreme Court concluded that a case should have first been filed before the pollution appellate, after which the tribunal could be approached.

This conclusion foregrounded the role of the pollution appellate. The Supreme Court did not comment on the merits of the TNPCB's enforcement directions. However, the apex court's judgment provided observations on how the company could seek relief. The judgment states,

---

[87] Press Trust of India, 'Tamil Nadu Orders Permanent Closure of Sterlite Plant', *BloombergQuint*, 29 May 2018, available at https://www.bloombergquint.com/business/tamil-nadu-orders-permanent-closure-of-sterlite-plant, accessed on 10 February 2020.

[88] National Green Tribunal judgment dated 28 November 2018 in Appeal No. 87 of 2018 (*Vedanta Ltd v. State of Tamil Nadu & Ors*).

Given the fact that we are setting aside the NGT judgments involved in these appeals on the ground of maintainability, we state that it will be open for the respondents to file a writ petition in the High Court against all the aforesaid orders. If such writ petition is filed, it will be open for the respondent to apply for interim reliefs considering that their plant has been shut down since 09.04.2018.

The company subsequently approached the Madras High Court, where the court upheld the orders of the TNPCB and dismissed the petitions filed by the company. [89]

---

[89] High Court of Madras judgment in 18 August 2020 Writ Petition Nos. 5756, 5764, 5771, 5772, 5773, 5774, 5776, 5792, 5793, 5801 and 21547 of 2019 (*Vedanta Limited, Unit: Sterlite Copper v. State of Tamil Nadu & Ors.*) and W.M.P. Nos. 6575, 6590, 6596, 6598, 6602, 6628, 6630, 6703, 6634 of 2019 and W.M.P.SR No. 102459 of 2019.

5

Environmental Protection

INTRODUCTION

The legal mandate for environmental protection is outlined through specific articles of the Constitution of India and the division of responsibilities between the centre and state governments as listed in its Seventh Schedule. Article 48A of the Constitution holds both the national and state governments accountable to 'protect and improve the environment and to safeguard the forests and wildlife of the country'. The protection of the environment is also the fundamental duty of every citizen of the country as enshrined in Article 51A (g) of the constitution.[1] Article 21 of the constitution, which is the right to life and personal liberty, has also been interpreted to include the right to clean and healthy environment through several court judgments. This is discussed in detail in Chapters 1 and 7.

The protection of the environment is the stated overarching objective of the central environment ministry. As discussed in Chapter 2, the environment ministry was carved out of the Department of Agriculture in 1984 primarily to realise the objective of environmental protection. The specific legal framework to achieve this objective is provided by the Environment Protection Act (EPA), 1986. Using the jurisdiction provided in this law, the environment ministry has introduced many regulations for its purpose. These regulations are the largest group of rules, notifications and guidelines that have dominated India's enviro-legal landscape. They have also led to the creation of an elaborate and multilayered institutional framework that is under the control of the central government.

---

[1] Article 51A (g) states, 'It shall be the duty of every citizen of India to protect and improve the natural environment including forests, lakes, rivers and wildlife and to have compassion for living creatures.'

The Indian parliament has exercised oversight in the enactment of rules and notifications only in specific instances where its intervention has been sought. But more generally, there has been little monitoring or oversight. Courts and tribunals have created a wide jurisprudence on a range of themes that are regulated by the EPA such as impact assessments, coastal areas regulation and environmental damages. The enactment of the National Green Tribunal (NGT) Act, 2010, institutionalised the role of this quasi-judicial body to respond to grievances and also influence environmental policymaking.

So far, there has only been one National Environment Policy (NEP), 2006,[2] that laid down the contours of how environmental protection is to be prioritised in the country. The policy attempts to achieve environmental protection while continuing to advance economic ambition. The NEP was strongly criticised by both environmental groups and industries. Their views reflected the practical difficulties of achieving environmental protection or development using the notion of 'balance' that is legitimised by the concept of sustainable development.

This chapter includes the following sections:

I.   *EPA's Legal Framework*
II. *Subordinate Legislations under the EPA*
III. *Proposal for a New Environment Regulator*

I   EPA'S LEGAL FRAMEWORK

The EPA, 1986, was enacted by the central government using Entry 13 of List I (Union List) of the Seventh Schedule[3] of the Constitution of India. Entry 13 is 'Participation in international conferences, associations and other bodies and implementing of decisions made thereat'. The international conference referred to in the EPA's preamble is the United Nations Conference on the Human Environment held at Stockholm in June 1972, where participating countries, including India, committed to take appropriate steps for the protection and improvement of the human environment.

---

[2]   Ministry of Environment and Forests (MoEF), *National Environment Policy* (New Delhi: Government of India, 2006).

[3]   This schedule specifies allocation and distribution of powers and functions between union territories and states. It contains three lists—List 1: Union List, List 2: State List and List 3: Concurrent List, specifying matters where both the centre and state have joint jurisdiction.

In a recent judgment in the case of an airport in the western Indian state of Goa, the Supreme Court reiterated these reasons.[4] The judgment says,

> Following the decisions taken at the United Nations Conference on the Human Environment held at Stockholm in June 1972 in which India participated, Parliament enacted the Environment Protection Act 1986 to protect and improve the environment and prevent hazards to human beings, other living creatures, plants and property.

This law for environmental protection was enacted 14 years after the Stockholm Conference. It can be argued that although the law was introduced as an obligation to the international convention, the Act was a response to the growing domestic environmental challenges. Scholars have pointed that the industrial accident of methyl isocyanate gas leak in the Union Carbide factory in the city of Bhopal was the trigger for several actions on environment protection, including the enactment of the EPA, 1986.[5] The accident, which took place on the night of 2 December 1984, killed 5,300 people[6] and exposed over 600,000 workers and residents to the toxic gas. The impacts of this disaster are unfolding as an intergenerational health catastrophe as children are born with genetic abnormalities to affected residents. The toxic dump is yet to be cleaned up as the question of liability remains undecided legally.[7]

The EPA is often referred to as 'the umbrella law' as it has several legal measures to protect the environment, restore environmental damage and charge penalties in instances of violations. The EPA needs to be read with

---

[4] Supreme Court judgment dated 29 March 2019 in Civil Appeal 12251 of 2018 (SC) (*Hanuman Laxman Aroskar v. Union of India*).

[5] C. M. Abraham and Sushila Abraham, 'The Bhopal Case and the Development of Environmental Law in India', *International and Comparative Law Quarterly* 40 (2): 334–365; Edward Broughton, 'The Bhopal Disaster and Its Aftermath: A Review', *Environmental Health* 4, no. 6 (2005), available at https://ehjournal.biomedcentral.com/articles/10.1186/1476-069X-4-6, accessed on 11 September 2020.

[6] Adam Withnall, 'Bhopal Gas Leak: 30 Years Later and After Nearly 600,000 Were Poisoned, Victims Still Wait for Justice', *Independent*, 14 February 2019, available at https://www.independent.co.uk/news/world/asia/bhopal-gas-leak-anniversary-poison-deaths-compensation-union-carbide-dow-chemical-a8780126.html, accessed on 26 February 2020.

[7] Alan Taylor, 'Bhopal: The World's Worst Industrial Disaster, 30 Years Later', *The Atlantic*, 2 December 2014.

the Environment (Protection) Rules (EPR), 1986, as well as several other subordinate legislations that have been discussed further in this chapter.

The EPA presents the definition of the environment as

> 'environment' includes water, air and land and the inter-relationship which exists among and between water, air and land, and human beings, other living creatures, plants, micro-organism and property.

The short five-page framework of the law envisages environmental protection through three main functions.

- *Preventive*: To put in place specific regulations, coordination mechanisms and planning nation-wide programmes for environmental protection.
- *Mitigative*: To put in place specific standards for environmental pollutants and emissions from industries as in the case of thermal power plants.
- *Punitive*: To carry out enforcement action and issue-specific penalties in case of violations.

Section 3 of the EPA, 1986, gives power to the central government, that is, the environment ministry, to take all measures that it feels is necessary to protect and improve the quality of the environment and to prevent and control environmental pollution.

Section 3(2)v gives powers to the central government to restrict areas in which any industries, operations or processes or class of industries, operations or processes shall not be carried out or shall be carried out subject to certain safeguards. This section is to be read with Section 5(1) of the EPR, 1986, which states that the central government may consider standards for environmental quality while prohibiting or restricting industrial processes or activities. Several flagship regulations discussed here such as the Environment Impact Assessment (EIA) and the Coastal Regulation Zone (CRZ) notifications have been issued using these clauses.

Section 3(3) of the Act allows the central government to constitute an authority or authorities to perform functions and exercise powers to take measures to protect the environment or enforce short-term or long-term regulation. This also includes enforcement action. Several authorities like the coastal zone management authorities (CZMAs), Central Empowered Committee on forest matters and ecologically sensitive area (ESA) committees discussed in this book and further in this chapter were set up using this section.

Section 15 to 17 deals with fines and punishments for offences and violations of the Act by private parties and public entities.

II   SUBORDINATE LEGISLATIONS UNDER THE EPA

The EPA has an elaborate set of legal instruments for environment protection such as rules, notifications, circulars and guidelines. These are called subordinate legislations and are designed by the executive. The Rajya Sabha Committee on Subordinate Legislations has defined these legislations as follows:[8]

> Subordinate legislation is the legislation made by an authority subordinate to the legislature. According to Sir John Salmond, 'Subordinate legislation is that which proceeds from any authority other than the sovereign power and is, therefore, dependent for its continued existence and validity on some superior or supreme authority'. Most of the enactments provide for the powers for making rules, regulations, byelaws or other statutory instruments, which are exercised by the specified subordinate authorities. Such legislation is to be made within the framework of the powers so delegated by the legislature and is, therefore, known as delegated or subordinate legislation.

This section provides information about some prominent subordinate legislations of the EPA. These include the EIA notification, the CRZ notification, the rules and notifications for the management of waste, and notifications identifying ESAs.

ENVIRONMENT IMPACT ASSESSMENT NOTIFICATION

The EIA Notification, 1994, of the central government established a procedure through which certain industrial, infrastructure, mining or energy generation projects would be appraised for their environmental impacts. This appraisal is done prior to the setting up of projects and it determines if the project should be allowed. The process under this notification is called 'environmental clearance' by the environment ministry.

Until January 1994, when the EIA notification was issued, obtaining environmental clearance for developmental projects from the environment ministry was only a requirement for a few projects that involved public

---

[8]   Available at https://rajyasabha.nic.in/rsnew/practice_procedure/book13.asp, accessed on 25 June 2019.

investments such as large dams. This notification, issued under the EPA, 1986, made it legally mandatory for 29 industrial and developmental activities (increased to 32 by subsequent amendments) to obtain environmental clearance from the central government before being set up. The projects listed in the notification are required to follow a stepwise procedure which includes the preparation of a detailed EIA report and organising a public hearing.

In the first 11 years of this notification until 2005, the notification went through 12 amendments,[9] some of which included exemptions from the environment clearance process or from public hearings (discussed later in this section). For instance, in December 2001, the notification was amended to exempt defence-related road construction in border areas from the purview of it.[10]

The draft of the subsequent amendment to the EIA Notification, 1994, was made public in 2005. The amendment was referred to as 'reforms' for 'streamlining the environment clearance process'. The basis for the amendment was in a note of the environment ministry titled 'Reforms in grant of Environmental Clearances'.[11] This note stated,

> [I]n order to further improve the EC process and to make it more effective and time bound the MoEF had undertaken a comprehensive review of the existing EC process as a sub component of the World Bank Assisted Environmental Management Capacity Building (EMCB).[12] This project which had started in 2001 has been completed on 30th June 2004.

---

[9] S.O. 60(E) dated 27 January 1994 was amended through S.O. 356(E) dated 4 May 1994, S.O. 318(E) dated 10 April 1997, S.O. 319 dated 10 April 1997, S.O. 73(E) dated 27 January 2000, S.O. 1119(E) dated 13 December 2000, S.O. 737(E) dated 1 August 2001, S.O. 1148(E) dated 21 November 2001, S.O. 632(E) dated 13 June 2002, S.O. 248(E) dated 28 February 2003, S.O. 506(E) dated 7 May 2003, S.O. 1087(E) dated 22 September 2003, S.O. 891(E) dated 4 August 2003 and S.O. 801(E) dated 7 July 2004.

[10] K. Kohli and M. Menon, *Eleven Years of the Environment Impact Assessment Notification, 1994* (New Delhi: Kalpavriksh, Just Environment Trust, Environment Justice Initiative [HRLN], 2005), 94.

[11] Ibid.

[12] M. Menon and K. Kohli, 'The World Bank and Environmental Policy Reform', in *The World Bank in India: Undermining Sovereignty, Distorting Development*, ed. Michelle Kelly and Deepika Dsouza (New Delhi: Orient Blackswan Pvt Ltd, 2010), 431–434.

The amendment also claimed to follow the recommendations of the Govindarajan Committee (discussed in Chapter 9) set up by the Cabinet Secretariat in November 2002. The committee's report titled 'Reforming Investment Approval and Implementation Procedures' observed that the environment clearance process was delaying economic investments.[13]

The draft notification was strongly opposed by citizens, non-governmental organisations (NGOs) and political parties. Their critiques claimed that the environment clearance process was being diluted in favour of economic investments.[14] The draft notification issued on 15 September 2005 was held back for a year and was issued a day before it expired. The notification currently in operation was issued on 14 September 2006.

The present EIA notification structures the process for environmental approval as a four-step process.

- *Screening* is the first step in the process to identify if the project comes under the purview of the notification. The notification splits the list of projects in Appendix I of the notification as Category A (to be approved by the central government) or Category B (to be approved by the State Environment Impact Assessment Authority [SEIAA]). Category B projects are further divided into B1 and B2 with only the former requiring to carry out an EIA study. It was only in December 2013 that the environment ministry issued guidelines on how to determine whether a project falls under Category B1 or B2.[15] Until then, it was decided by the discretion of the SEIAA. All Category B projects are treated as Category A if the project is to be located within 5 kilometres of a national park, sanctuary, ESAs or a state boundary.

- *Scoping* is when project proponents are issued terms of reference (ToRs) for carrying out an EIA. Until recently, the practice was that proponents suggest a ToR which was modified by the thematic Expert Appraisal Committees (EACs) before whom an application for environment clearance was listed.

---

[13] GoI, *Report on Reforming Investment Approval and Implementation Procedure*, Part II (New Delhi: Government of India, 2002), ii.

[14] Leo F. Saldanha, Abhayraj Naik, Arpita Joshi and Subramanya Sastry, *Green Tapism: A Review of the Environmental Impact Assessment Notification: 2006* (Bengaluru: Environment Support Group, 2006), 1.

[15] Office Memorandum No. J-13012/12/2013-IA-1I (I) dated 24 December 2013 titled 'Guidelines for Consideration of Proposals for Grant of Environmental Clearance Environmental Impact Assessment (EIA) Notification, 2006 and Its Amendments regarding Categorization of Category "B" Projects/Activities into Category "B1" & "B2".

Since 2015,[16] the environment ministry has instituted the process of standardising these ToRs based on specific project types: mining, thermal power, industry, infrastructure or new construction projects and industrial estates. The ToR lists the items that need to be studied to predict the environmental and social impacts of projects. Legally, there is no restriction on the EAC to modify a standard or proponent provided ToR, for which they can also carry out a site visit. The EIA is then commissioned to an accredited EIA consultant who is contracted by the project proponent for a fee. The completion of a draft EIA marks the end of this step.

- *Public consultation* is the third step in the environment clearance process. The objective is to ascertain the concerns of 'locally affected people' and those 'with plausible stake in the environment'. It includes two components: a public hearing and written submissions. A minimum 30-day notice needs to be given for a public hearing along with the details of the venue and where documents such as the draft EIA report and other project documents would be available to the public. The EIA notification has a detailed Appendix IV on the 'Procedure to Conduct a Public Hearing' in a free, fair and transparent manner, ensuring maximum public participation. The proceedings of the public hearing have to be officially documented in writing and as a video recording.

- The fourth process in the pre-approval phase is *appraisal*, where the EAC that issued a ToR to the project proponent carries out a detailed scrutiny of all the documents generated by the earlier steps including submissions made at the time of the public hearing. They can also call upon additional expertise and information to make a decision in favour of or against an application for environment clearance. They also suggest the conditions that should be imposed on the project if an approval is being recommended. These conditions are for the mitigation of environmental and social impacts of the project.

The EIA notification dedicates a small section to the *monitoring and compliance* of environment clearance conditions.[17] These safeguards range from general ones about following standards and stipulations prescribed by environment laws to more specific ones based on the nature of the project and the region where it is to be set up. For instance, clearance conditions for hydroelectric projects include conditions regarding the dumping of debris that is generated during construction and controlled blasting.

---

[16] S.O. 996 (E), amendment dated 10 April 2015.

[17] Section 10 of the EIA notification.

Each project proponent with an environment clearance needs to submit a six-monthly compliance report to either the regional office of the Ministry of Environment, Forests and Climate Change (MoEFCC) or the SEIAA depending on whether they are Category A or Category B projects. There are 10 regional offices of the MoEFCC who have the responsibility of monitoring compliance in specific states of their jurisdiction. The regional offices need to generate six-monthly monitoring reports. These offices are located at Chandigarh, Lucknow, Shillong, Bhubaneshwar, Nagpur, Ranchi, Dehradun, Bengaluru, Bhopal and Chennai.[18]

Every environment clearance is issued with the condition that non-compliance of environment safeguards can make the environment clearance granted liable for revocation or suspension. The details related to the compliance of clearance conditions is also required to be considered at the time of expansion or modernisation of any project. The concerned EAC needs to make a request for compliance reports from the regional offices at the time of the issuance of a ToR.[19]

The implementation of the EIA notification has been researched and litigated upon extensively. The following are understood to be the unresolved legal and enforcement challenges of the EIA notification.

- *Independence of impact assessments* has been a question as the project proponent hires and funds EIA consultants for the project. Independent research and reports of the erstwhile Planning Commission[20] have recommended delinking the preparation of the EIA report from the project finances.[21]

---

[18] Prior to 2011, the number of regional offices was 6 and was increased to 10 pursuant to the Supreme Court of India judgment dated 6 July 2011 in Interlocutory Application (I.A.) Nos. 1868, 2091, 2225–2227, 2380, 2568 and 2937 (*Lafarge Umiam Mining Pvt. Ltd*) in Writ Petition (Civil) No. 202 of 1995 (*T.N. Godavarman Thirumulkpad v. Union of India & Ors*).

[19] Circular dated 7 September 2017 titled 'Environmental Clearance to the Expansion Projects/Activities under the EIA Notification, 2006—Certified Compliance Report Regarding'.

[20] Planning Commission of India, *Reports of the Task Forces on Governance, Transparency, Participation & Environmental Impact Assessment and Urban Environmental Issues: In the Environment and Forests Sector for the Eleventh Five Year Plan (2007–2012)* (New Delhi: Government of India, 2007), 11.

[21] Vaibhav Chaturvedi, Vaibhav Gupta, Nirmalya Choudhary, Sonali Mittra, Arunabha Ghosh and Rudresh Sugam, *State of Environmental Clearances in India:*

- *Quality of EIA* reports remains an unaddressed issue.[22] Even though the environment ministry has formalised a process for accrediting EIA consultants, this has not substantially altered the quality of the EIAs. Complaints have been filed in courts as well as before the accreditation agency and that has only partially addressed the problem.

- *Restrictions in public participation* has been a challenge in the environment clearance process and is hampered by practical difficulties such as lack of information, incorrect venue of the public hearing, poor recording of minutes and the presence of official security personnel hindering free and fair participation.[23] Despite a detailed procedure listed in the EIA notification and court judgments,[24] most public hearings have remained incomplete or inadequate as per the law.

- *Limitations in the appraisal by expert committees* has been highlighted in media reports and is a subject matter of several appeals filed against the environmental clearances. The principles of detailed *scrutiny* and *application of mind* by an expert body has been discussed in several court and tribunal judgments.[25] For example, the recent judgment related to an airport in Goa has observed,[26]

> The EAC is an expert body. It must speak in the manner of an expert. Its remit is to apply itself to every relevant aspect of the project bearing upon

---

Procedures, Timelines and Delays Across Sectors and States, report (New Delhi: CEEW [Council on Energy, Environment and Water], 2014).

[22] Manju Menon and Vidya Vishwanathan, 'How Not to Do an Environmental Assessment', *The Hindu*, 29 August 2018, available at https://www.thehindu.com/opinion/op-ed/how-not-to-do-an-environmental-assessment/article24813642.ece, accessed on 11 September 2020.

[23] M. P. Ram Mohan and Himanshu Pabreja, 'Public Hearings in Environmental Clearance Process: Review of Judicial Interventions', *Economic and Political Weekly* 51, no. 5 (2016): 68–75.

[24] Kanchi Kohli, 'Cleared, Denied Cleared', *India Together*, 18 May 2010, available at http://www.indiatogether.org/athena-environment, accessed on 12 September 2020. It is related to the public hearing of a thermal power plant by M/s. Athena Chhattisgarh Power Ltd in Chhattisgarh, challenged in Appeal No. 26 of 2009 before the National Environment Appellate Authority by residents of the Singhitarai village.

[25] Writ Petition (Civil) No. 9340 of 2009 and CM Appl Nos. 7127 of 2009, 1249 of 2009 (*Utkarsh Mandal vs Union of India*) dated 26 November 2009.

[26] Supreme Court judgment dated 29 March 2019 in Civil Appeal 12251 of 2018 (SC) (*Hanuman Laxman Aroskar v. Union of India*).

the environment. It is not bound by the analysis, which is conducted in the EIA report. It is duty bound to analyse the EIA report. Where it finds it deficient it can adopt such modalities which, in its expert decision-making capacity, are required.

• *Instability with repeated amendments involving exemptions* to the EIA notification has continued to remain a problem with the 2006 version. Frequent amendments have arbitrarily exempted particular projects from public hearings or from the requirement of environment clearance completely. This has been the case with sectors such as coal mining, real estate and highways.[27] In April 2019, media reports informed that the government was proposing substantial reforms to the EIA notification. This version proposed exemptions to a number of project sectors from public hearings.[28] In March 2020, the environment ministry disclosed a draft notification proposing several controversial amendments to the EIA process.[29] There have been widespread objections to this proposal,[30] including litigation in various High Courts.[31]

• Post facto *approvals* relate to instances where either the environment ministry or courts have allowed a single or a set of projects to secure *post facto* approvals. Affected parties have brought several instances before the regulatory authority or courts when construction activities have been undertaken without prior approval. This issue was raised in a comprehensive Public Interest Litigation before the Supreme Court in 2004, which is pending a final disposal.[32] In 2015, the environment ministry introduced

---

[27] MoEFCC (Ministry of Environment, Forests and Climate Change), *Compendium of Gazette Notifications, Office Memoranda under Environment Impact Assessment Notification, 2006* (New Delhi: Government of India, 2014).

[28] Mayank Aggarwal, 'Amid Elections, Government Proposes Overhaul of Environment Clearance Rules', *Mongabay-India*, 22 May 2019, available at https://india.mongabay.com/2019/05/amid-elections-government-proposes-overhaul-of-environment-clearance-rules/, accessed on 12 September 2020.

[29] MoEFCC (Ministry of Environment, Forests and Climate Change) Notification S.O. 1199(E) dated 23 March 2020.

[30] Soutik Biswas, 'The Environment Law That Mobilised Two Million Indians', BBC News, 25 August 2020, available at https://www.bbc.com/news/world-asia-india-53879052, accessed on 24 September 2020.

[31] High Court of Karnataka Writ Petition 8632 of 2020 (*United Conservation Movement Charitable and Welfare Trust (UCM) v. Union of India*).

[32] Supreme Court Writ Petition (Civil) No. 460/2004 (*Goa Foundation v. Union of India*) challenging the grant of *post facto* environment clearances.

a one-time amnesty scheme for all projects that admit to having violated the EIA notification and seek post facto approvals. A special 'Violations' expert committee is set up within the ministry, which has received over 3,000 project applications.[33]

• *Compliance of environmental safeguards* is one of the weakest aspects of the environment clearance process. In a 2009 study, the non-compliance of these conditions was recorded as more than 90 per cent.[34] Since then, several cases have been filed in courts and complaints have been made to the environment ministry. The environment ministry is considering introducing a third-party audit mechanism by expert institutions to carry out monitoring in addition to ministry's regional offices. This was yet to be finalised as of June 2019.[35] Currently, the environment ministry has three kinds of responses to the challenges of non-compliance: self-regulation by project proponents and penalties such as fines, bank guarantees and one-time amnesty schemes.[36]

The following are two case examples to illustrate the issues with the EIA notification.

The first case is related to an allegedly plagiarised EIA and public hearing related to a proposed refinery in Ratnagiri. In this case, affected people and concerned citizens also complained to the accreditation agency against poor quality reports by the EIA consultants.

In March 2017, a public hearing for the construction of a captive jetty was held for the second time. The first time, it was postponed as affected people and concerned citizens had complained that the basic documents required for an informed hearing were not made available in the local language as required by the law.[37]

---

[33] Krithika Dinesh and Kanchi Kohli, 'From Prior to Post: Legalising Environmental Violations?' Working Paper, Centre for Policy Research, New Delhi, 2017.

[34] K. Kohli and M. Menon, *Calling the Bluff: Revealing the State of Monitoring and Compliance of Environmental Clearance Conditions* (New Delhi: Kalpavriksh, 2009).

[35] Ministry of Environment Forest and Climate Change draft notification S.O. 4271 (E) dated 10 September 2018.

[36] Manju Menon and Kanchi Kohli, 'Regulatory Reforms to Address Environmental Non-Compliance', in Centre for Policy Research, *Policy Challenges (2019–2024): Navigating Policy Challenges and Charting a New Course for India in the 21st Century* (New Delhi: Centre for Policy Research, 2019), 44–48

[37] Priyanka Sahoo, 'Activists Oppose Captive Jetty in Ratnagiri', *Indian Express*, 15 March 2017.

The site of the jetty is Nate village, Rajapur *taluka* of Ratnagiri district in Maharashtra. When the documents were made available during the second public hearing, it was revealed that the EIA was highly inadequate. More importantly, the section on impacts to the biodiversity and livelihood of the ecologically fragile Ambolgad Nate area had text which had been copied from a report by the Bombay Natural History Society (BNHS). In a letter dated 7 March 2017, the director of BNHS wrote to the Maharashtra Pollution Control Board to say 'The biodiversity section of the current EIA is fully plagiarized and copied from the BNHS report and warrants to be rejected outright, making public hearing untenable'.[38]

The EIA consultant was an officially accredited one to carry out such studies. While the public hearing was cancelled, the issue of the EIA quality was not recorded. The Konkan Vinashkari Prakalp Virodhi Samiti, a community-based organisation, also complained about this to the National Accreditation Board for Education and Training (NABET) (a constituent board of Quality Council of India) which is the agency responsible for accreditation of EIA consultants. A complaint was also registered before the environment ministry. While there was no response from the ministry, the accreditation agency issued a warning to the consultant.[39] The EIA report was not revised. The final decision on this project by the ministry is pending.

The second case relates to the NGT's interpretation leading to the inclusion of roads and bridges within the purview of the EIA notification.

On 12 January 2015, the NGT issued an important judgment[40] interpreting the applicability of the EIA notification on a Category B project. As recorded in the judgment, the question before the tribunal was:

---

[38] Letter from Director, BNHS, to Maharashtra PCB, 7 March 2017, 'Public Hearing for Proposed Captive Jetty of 4.5 MTPA near Nate Village, Rajapur Taluka, Ratnagiri District, Maharashtra and BNHS Objections towards the Project', Ref 374/2017. This letter was reported in Priyanka Sahoo, 'Activists Oppose Captive Jetty in Ratnagiri', *Indian Express*, 15 March 2017, available at https://indianexpress.com/article/cities/mumbai/activists-oppose-captive-jetty-in-ratnagiri-4569383/, accessed on 13 September 2020.

[39] Letter from NABET to Bhagwati Ana Labs, 27 September 2017, Reference No. QCI/NABET /ENV/AC0/17/0416.

[40] National Green Tribunal judgment dated 12 February 2015 in Original Application No. 137 of 2014 (*Vikrant Kumar Tongad v. Delhi Tourism and Transportation Corporation & Ors*).

Whether, constructing a 'bridge' across Yamuna is a 'project' or 'activity' that shall require prior Environmental Clearance from the Regulatory Authority.

The project in question here is a bridge of a total area of 155,260 square metres that was being constructed by the Delhi Tourism and Transport Development Corporation (DTTDC) allegedly without prior approval under the EIA notification. The DTTDC submitted that while they applied to the environment ministry for approval, the ministry wrote back to them to say that 'Bridges' are not covered under the EIA Notification, 2006.

The NGT judgment concluded that bridges should be treated as 'Area Development Projects' as covered under Section 8(b) of Appendix 1 of the notification. Therefore, this bridge or any other such construction with an area of 150,000 square metres and above or 50 hectares and above would be treated as Category B project requiring an approval from the concerned SEIAA.

COASTAL REGULATION ZONE NOTIFICATION

India's 8,000 kilometre-long coastline along the western and eastern borders of the Indian land mass and its island complex is considered one of the most diverse and unique ecosystems in the world. This coastline is covered by several official legislations that regulate activities including construction, industrial activity and coastal infrastructure. The laws related to forests, wildlife, pollution control, water extraction and environment protection are all applicable to coastal spaces.

A specific regulation issued under the EPA, 1986, is the CRZ notification. It was first issued in 1991[41] and has gone through several amendments and two major revisions in 2011 and 2019. This notification identifies the coastline of the country as ecologically sensitive, where industrial and infrastructure activities are to be regulated. According to the notification, 500 metres from the high tide line (HTL) is the regulated coastal zone. The CRZ law classifies the coast into different categories based on levels of sensitivity and population density and imposes a range of permissions and prohibitions to regulate activities in these areas.

The notification requires every state to prepare a coastal zone management plan (CZMP) (discussed in detail later in this section). Research studies, statements issued by fisher and coastal communities, and judicial

---

[41] Notification no. S.O.114 (E) dated 19 February 1991.

pronouncements have highlighted the limitations and ineffectiveness of the CRZ, 1991.[42]

There are three following factors that influenced the multiple amendments to the 1991 notification and its 2011 version:

- First, the World Bank through its Environment Management Capacity Building (EMCB) project[43] engaged in a long-term process of revising and re-engineering environment regulations including the CRZ notification since the late 1990s. One proposal under the EMCB project was to develop an appropriate environmental regime for Integrated Coastal Marine Area Management.[44] The World Bank aimed at providing expert assistance for the same.[45]

- Second, a committee was set up under the chairmanship of M. S. Swaminathan, a senior agricultural scientist with a key role in the India's First and Second Green Revolution and member of parliament (Rajya Sabha). This committee was to review the CRZ notification and suggest an appropriate framework of coastal management.[46]

- Third, the 2004 Indian Ocean Tsunami and its impacts on India's southern coastal regions influenced the changes to the notification. The recommendations of the Swaminathan Committee were under finalisation when the disaster struck. The committee submitted its report to the environment ministry in February 2005.[47] Their report recommended a

---

[42] M. Menon, M. Kapoor, P. Venkatram, K. Kohli and S. Kaur, *CZMAs and Coastal Environments: Two Decades of Regulating Land Use Change on India's Coastline* (New Delhi: Centre for Policy Research and Namati, 2015), 12–14.

[43] This technical assistance project between the World Bank's International Development Agency (IDA) and the Ministry of Environment and Forests (MoEF) was approved in 1998.

[44] Website of the Ministry of Earth Sciences, https://www.moes.gov.in/programmes/coastal-research, accessed on 12 September 2020).

[45] Menon and Kohli, 'The World Bank and Environmental Policy Reform'.

[46] Ministry of Environment and Forests Order No. 15 (8)/2004-IA-III dated 19 July 2004.

[47] A detailed assessment of this report is available in A. Sridhar, R. Arthur, D. Goenka, B. Jairaj, T. Mohan, S. Rodriguez and K. Shanker, *Review of the Swaminathan Committee Report on the CRZ Notification* (New Delhi: UNDP, 2006), available at https://www.dakshin.org/wp-content/uploads/2017/08/Review-of-the-Swaminathan-Report.pdf, accessed on 3 July 2006.

new coastal management strategy and the outline of a proposed regulatory regime for the coasts.

On 1 May 2008, the environment ministry issued a draft Coastal Management Zone (CMZ) notification that can be traced back to the controversial annexure inserted in the Swaminathan Committee Report of the Ministry of Environment and Forests (MoEF).[48] This draft CMZ notification was issued amidst much controversy. It was reissued in July 2008 with the ministry providing more time for the public to give their feedback.

Fisher unions and campaign networks like National Fishworkers' Forum and the National Coastal Protection Campaign publicly opposed the CMZ. This pushed the environment ministry to conduct several public hearings to determine the next steps. The then Environment Minister Jairam Ramesh who took charge of the environment ministry in May 2009 directed this initiative.[49] The CRZ, 2011, which was in operation until recently, was an outcome of negotiations between the concerned ministries, scientists, industry groups, fishing unions, and coastal and environmental campaign networks.

Since its issuance, the CRZ, 2011, has also seen over ten amendments.[50] Most of these were after 2014, when the government at the centre changed. These amendments were also issued by leaving aside any process to receive public comments. The government justified that it was in public interest to do so. While the environment ministry issued these amendments, it also undertook another national-level review process through a special committee set up in June 2014.[51]

---

[48] A. Sridhar, M. Menon, S. Rodriguez and S. Shenoy, *Coastal Management Zone Notification '08: The Last Nail in the Coffin* (Bangalore: ATREE [Ashoka Trust for Research in Ecology and the Environment], 2008), 2.

[49] MoEF, *Report of the Public Consultation with Fisherfolks and Community to Strengthen Coastal Regulation Zone (CRZ) Notification, 1991* (New Delhi: Centre for Environment Education, Ahmedabad, and MoEF, 2010), 1.

[50] Meenakshi Kapoor, Dinesh Krithika, 'Tourism Scores Over Ecology, Livelihoods as India's Coastal Law Is Amended Yet Again', *Scroll.in*, 9 April 2018, available at https://scroll.in/article/874451/tourism-scores-over-ecology-livelihoods-as-indias-coastal-law-is-amended-yet-again, accessed on 3 June 2019.

[51] This committee was set up in June 2014 'to review the issues relating to the Coastal Regulation Zone Notification, 2011' (F.No.19/112/2013/IA/3).

A committee headed by Shailesh Nayak, the then secretary, Ministry of Earth Sciences, was constituted to address the issues raised by various coastal states/union territories (UTs) regarding the CRZ Notification, 2011. The committee submitted its report in the early 2015. The report was never made public by the ministry, even though the Central Information Commission (CIC) set up under the Right to Information Act, 2005, ordered for the same.[52] The recommendations of this report influenced the CRZ, 2019.

The present environmental legal regime for the coast is realised through two notifications:

- Coastal Regulation Zone Notification, 2019:[53] This notification regulates coastal areas of 500 metres from the HTL[54] and 50 metres or width of the creek, whichever is less, on the landward side along tidal-influenced waters, such as estuaries and creeks. It also regulates the territorial waters of the country. The notification classifies the CRZ into four zones: CRZs I, II, III and IV, which are further divided into subzones.

- Island Protection Zone (IPZ) Notification, 2019:[55] This notification regulates the Andaman and Nicobar group of islands and Lakshwadeep. These islands are divided into Group I and Group II islands with the CRZ areas up to 200 metres from HTL for the former and 100 metres for the latter. Here too, the Island Coastal Regulation Zone (ICRZ) is split into four zones: ICRZs I, II, III and IV; these are further divided into subzones. For tidal-influenced waterbodies, ICRZ is up to 20 metres from the HTL or the width of the creek, whichever is lesser.

This legal structure prescribed in both these notifications can be understood through the following functions that the notifications seek to perform:

- *Zoning*: Both the CRZ and the IPZ notifications divide the area under their jurisdiction into four zones (the CRZ and the ICRZ), the classification of which is similar for both notifications (Table 5.1).

---

[52] Central Information Commission judgment dated 13 May 2016 in CIC/SA/A/2016/000209 (*Kanchi Kohli v. PIO, M/o Environment & Forest*).

[53] Ministry of Environment, Forests and Climate Change Notification No. 37 (E) dated 18 January 2019.

[54] The law defines high tide line (HTL): 'The HTL means the line on the land up to which the highest water line reaches during the spring tide, as demarcated by the National Centre for Sustainable Coastal Management (NCSCM) in accordance with the laid down procedures and made available to various coastal States and Union territories.'

[55] Ministry of Environment, Forests and Climate Change Notification No. 35 (E) dated 17 January 2019, amended as S.O. 1242(E) on 8 March 2019.

Table 5.1 Regulatory aspects of CRZ and ICRZ

Coastal Regulation Zone (CRZ)	Island Coastal Regulation Zone (ICRZ)
CRZ I is split into two subzones	The same criteria as CRZ I and its subzones are applicable to ICRZ I
CRZ IA is the most ecologically fragile part of the coast including areas with mangroves of more than 1,000 square metres, coral reefs, sand dunes, mudflats, salt marshes, archeological sites, and nesting grounds of several species such as birds, turtles and seagrass	
CRZ IB is the inter-tidal area between HTL and low tide line (LTL), which can differ from place to place	
CRZ II is a developed area within municipal limits or other legally designated urban areas. These are substantially built up with drainage, approach road and sewage facilities	The criteria are the same as for CRZ II. However, land areas along creeks or tidal-influenced[i] waterbodies in municipal areas are also earmarked as ICRZ II. This is limited to the land area of 20 metres from HTL or the width of the creek
CRZ III are 'relatively undisturbed' rural areas. This is divided into two subzones based on their population density	There is no subzoning of ICRZ III. It comprises 'relatively undisturbed' rural areas
CRZ IIIA are densely populated rural areas with a population density of more than 2,161/square kilometre as per the 2011 population census	The NDZ will be up to 100 metres for Group I islands and up to 50 metres for Group II islands
CRZ IIIB are all other CRZ III areas not falling under CRZ IIIA	For tidal-influenced waterbodies, the NDZ would be 20 metres from HTL or the width of the creek
CRZ IIIA has a 'No Development Zone' (NDZ) of 200 metres from the HTL until coastal zone management plans (CZMPs) are not finalised, which is when the NDZ will be 50 metres	
For tidal-influenced waterbodies in the CRZ, the NDZ would be 20 metres from HTL or the width of the creek	

(*Contd*)

(*Contd*)

Coastal Regulation Zone (CRZ)	Island Coastal Regulation Zone (ICRZ)
CRZ IV is for territorial waters up to 12 nautical miles. This area is also divided into two subzones	ICRZ IV with two subzones is the same as prescribed for CRZ IV
CRZ IVA is the water area and the seabed between LTL and 12 nautical miles	
CRZ IVB relates to tidal-influenced waterbodies on the seaward side and is defined as the area from LTL at the bank of such a waterbody to the LTL of the opposite bank	

Source: Authors.
Note: i. Tidal influence is understood as salinity of 5 parts per thousand during the driest season of the year.

- *Regulation*: The notifications lay out the kinds of activities that are prohibited or restricted in the CRZ or ICRZ and those that are exempted. The CRZ Notification, 2019, lists 11 activities which would be prohibited within the CRZ in general, with exceptions stated in the notification. The prohibited activities include expansion of new industries, ports and harbours, effluent discharge, land reclamation and altering sand dunes. But land reclamation and setting up of ports are permissible activities in CRZ IA. Similarly, sewage treatment plants can be set up in CRZ III areas, including in the NDZ, with prior permission from the state pollution control boards (SPCBs) (see Chapter 4 for details on this institution).

In instances where prior approvals need to be sought under the CRZ or the IPZ notifications, the following procedure is applicable. An application for the clearance needs to be made before a requisite State Coastal Zone Management Authority (SCZMA).

- If the activity is listed in Appendix I of the EIA Notification, 2006 (as discussed earlier in this chapter), then the SCZMA would forward the application and its recommendations to the environment ministry. These will be considered for 'a composite Environmental and CRZ clearance under EIA Notification'.

- All activities listed in CRZ/ICRZ I and CRZ/ICRZ IV should go to the environment ministry for approval along with the SCZMA's recommendation.

- All activities listed in CRZ/ICRZ II and CRZ/ICRZ III and not listed in the EIA Notification, 2006, would need to get an approval from the SCZMA.

Both notifications provide a detailed procedure for the prerequisite studies and impact assessment requirements mandatory for seeking approvals.

- *Management Plans*: The requirement of CZMPs was mandated from the time the first CRZ notification was issued in 1991. At the time, a one-year time frame was given to all state governments to finalise these plans. Several states did prepare these plans, which were approved by the environment ministry or were nearly approved. CRZ Notification, 2011, required these plans to be updated within two years of the notification being issued, and the date for the same was repeatedly extended.[56] Since most of the CZMPs could not be finalised at the time the CRZ Notification, 2019, was issued, the task of completing CZMPs is listed in the notification.

- *Conservation*: The preambles of all three versions of the coastal regulation listed conservation as one of their objectives. The CRZ notification seeks to enforce conservation measures by the identification of ESAs in CRZ I areas and the declaration of Critically Vulnerable Coastal Areas (CVCAs). In the 2011 notification, there were no limits to the number of CVCAs that could be declared, and the notification had provided an indicative list. However, the 2019 notification has limited CVCAs to Sundarbans region of West Bengal; Gulf of Khambat and Gulf of Kutchh in Gujarat; Malvan and Achra-Ratnagiri in Maharashtra; Karwar and Coondapur in Karnataka; Vembanad in Kerala; Gulf of Mannar in Tamil Nadu; Bhitarkanika in Odisha; and Coringa, East Godavari, and Krishna in Andhra Pradesh. These areas are to be managed with the involvement of coastal communities including fisherfolk.

The implementation of the CRZ notification has been a challenge from the outset. Between 1991 and 2015, there were as many as nine committees that were set up to review the regulation and its implementation challenges.[57] A variety of recommendations emerged from these committees, but there are

---

[56] Menon, Kapoor, Venkatram, Kohli, and Kaur, *CZMAs and Coastal Environments*.

[57] These were the B. B. Vohra Committee (1992), Prof. N. Balakrishnan Nair Committee (1996), Fr Saldanha Committee I (1996), Dr Arcot Ramachandran Committee (1996), Fr Saldanha Committee II (1997), D. M. Suthankar Committee I (2000), D. M. Suthankar Committee II (2000), Swaminathan Committee (2005) (Menon et al, 2015) and Shailesh Nayak Committee (2014).

the following five sets of issues that have historically constrained the CRZ's implementation:

- *Dynamic interpretation of coastal zones*: The interpretation of the CRZ areas by the SCZMAs have been a critical issue in the CRZ and IPZ notifications' implementation. A study of the functioning of SCZMAs revealed arbitrariness in interpretation by these authorities. One of the reasons identified for this was the lack of official records of HTLs in specific coastal areas. This had resulted in CRZ I areas being regarded as CRZ III, thereby allowing for lower restrictions on land use change.

- *Finalisation* of CZMPs has been lagging behind the schedule since the mid-1990s. Some of the CZMPs approved by the environment ministry were also questioned in courts and during public hearings. After 2011, the procedures related to CZMP preparation has been contested both in courts and during public hearings held by state governments to discuss draft CZMPs.

- *Coastal zone management authorities* were set up in every state following the judgment of the Supreme Court in 1996.[58] Until then, the responsibility of coastal regulation was with the pollution control boards (PCBs). In 2011, the new CRZ notification added another layer of decision-making through the District-Level Coastal Committees in every district that has a coastal stretch and where the CRZ notification is applicable. The functioning of both these bodies has suffered due to delays in suitable appointments and lack of financial resources, especially from state governments.

- *Multiple amendments and project approvals* have also made the CRZ an unstable law just like the EIA notification. Lenient approvals have made coastal areas available to land use change and led to reclamation of the seas for the promotion of tourism, trade and commerce.[59]

- *Action against violations and post-clearance compliance* is an area of concern with the CRZ and the IPZ notifications. Several violations have been recorded through complaints and others have been litigated upon. In

---

[58] Supreme Court judgment dated 18 April 1996 in Writ Petition 664 of 1993 (*Indian Council for Enviro-Legal Action v. Union of India*).

[59] Jayashree Nandi, 'Luxury Tourism Projects on Andamans May Get Nod Soon', *Hindustan Times*, 19 February 2019, available at https://www.hindustantimes.com/india-news/andamans-luxury-tourism-projects-may-get-nod-soon/story-Cl9GWkXhEOUFdKLBNLlUFL.html, accessed on 12 September 2020.

March 2018,[60] the environment ministry issued a notification allowing for projects or activities that have not complied with the regulatory requirements to report to the ministry so that they can be considered for *post facto* approval. This notification acknowledged that several projects were operating on the coasts without the requisite approvals.

Waste Generation and Management

Various subordinate legislations issued under EPA have sought to address the generation and management of industrial and domestic waste to minimise environmental damage. These rules and guidelines, issued at different times, pointed to the problem of excessive waste generation but limited themselves to regulating the storage, collection and management of wastes and the creation of waste management facilities. Their aim was to reduce the occurrence of accidents, contamination and pollution that may occur during the management of wastes.

Until 2016, there were legal instruments dealing with municipal solid waste, biomedical waste, plastics, hazardous wastes including manufacture, storage and use of genetically engineered (GE) organisms. In 2016, the environment ministry substantially amended all of these rules except for the GE Rules. The government also enacted new laws for wastes generated from activities such as construction and demolition and electronic waste. All these rules have been issued under the EPA, 1986.

A 2008 audit report of the Comptroller and Auditor General (CAG) [61] assessed if there had been an environment and health risk assessment of waste and whether policies had given priority to waste reduction and minimisation instead of waste disposal. The audit also looked into whether waste legislations had penal provisions and whether the existing laws and monitoring mechanisms were followed. The institutional roles and performance were also audited. One of the recommendations of this audit was:

> MoEF should consider framing laws/rules for the management of all major kinds of waste like construction & demolition waste, end of life vehicles, packaging waste, mining waste, agriculture waste and e-waste being generated in the country.

---

[60] Ministry of Environment, Forests and Climate Change Notification No. S.O. 1002 (E) issued on 6 March 2018.

[61] Comptroller and Auditor General (CAG), *Performance Audit on 'Management of Waste in India': Report No. PA 14 of 2008* (New Delhi: Government of India, 2008), 2.

Immediately after the audit, the environment ministry set up a committee 'to evolve a roadmap for management of wastes in the country'.[62] This committee worked with two working groups. One working group looked at municipal solid wastes, plastic waste, packaging waste, and construction and demolition waste and the other at biomedical waste, hazardous waste and e-waste. The final report of this committee was submitted in March 2010.[63] It gave 'recommendations on the legal, administrative and technological interventions required for managing each type of waste as categorized in the CAG performance audit report'. Following this, the laws for different waste types were revised or newly enacted first in 2011 and then in 2016.

This section deals with laws on municipal solid waste, hazardous waste, plastic waste and construction and demolition waste. It does not deal with biomedical waste[64] and electronic waste[65] as these rules follow a similar legal framework as the 2016 Plastic Waste and Construction and Demolition (C&D) Waste Management Rules elaborated in detail.

Municipal Solid Waste

One of the longest-standing issues before the judiciary and the executive was the management of solid waste generated in domestic, commercial or institutional (for example, government offices and schools) facilities in municipal areas. It does not include industrial or hazardous waste. A case that pushed the government to put a regulatory mechanism in place was filed in the Supreme Court in 1996.[66] During its pendency, the apex court passed various directions and orders to regulate garbage heaps and dump sites. It

---

[62] The committee was appointed with a letter number F. No. 12-17/2008-HSMD dated 3 September 2010 under the chairmanship of R. H. Khwaja, the then Additional Secretary, MoEF. It also had a representative of the CAG as a member.

[63] MoEF, *Report of the Committee to Evolve Road Map on Management of Wastes in India* (New Delhi: Government of India, 2010).

[64] Biomedical waste is regulated through Notification G.S.R. 343(E) issued on 28 March 2016 as the Bio-Medical Waste (Management and Handling) Rules, 2016, replacing the earlier 1996 version.

[65] Electronic waste is regulated through Notification G.S.R. 3338(E) issued on 23 March 2016 as the E-waste (Management and Handling) Rules, 2016, replacing the earlier 2011 version.

[66] Supreme Court Writ Petition (Civil) No. 888 of 1996 (*Almitra H. Patel & Anr v. Union of India & Ors*).

also led to the setting up of high-level committees to study and recommend management measures. This petition was transferred to the NGT in 2014.[67] The judgement in this case informed the legal framework that is currently operational for the management of solid waste in urban areas.

The current regulations applicable are the Solid Waste Management Rules, 2016,[68] (under the EPA, 1986) which defines 'solid waste' as

> solid or semi-solid domestic waste, sanitary waste, commercial waste, institutional waste, catering and market waste and other non-residential wastes, street sweepings, silt removed or collected from the surface drains, horticulture waste, agriculture and dairy waste, treated bio-medical waste excluding industrial waste, bio-medical waste and e-waste, battery waste, radio-active waste generated in the area under the local authorities and other entities mentioned in rule 2.

These rules primarily look at five kinds of actions:

- Duties of waste generators: It includes segregation of waste into different categories along with specifying storage and disposal mechanism. This waste has to be handed over to authorised waste pickers and not burnt or disposed on the streets.

- Responsibility of the local bodies: Urban local bodies[69] are to set up waste disposal facilities like landfills and segregation centers. They can also partner with Resident Welfare Associations (RWAs) for this purpose. Each landfill has to strictly follow the safeguards listed in the 2016 Rules.

- Identification of landfill sites: This is the responsibility of the district collectors and to be done in close coordination with the secretary of the Urban Development Department of the state government. These rules do not mandate the involvement of rural locals and bodies, for example, panchayats, while identifying the landfill sites. However, 'no-objection' from these bodies needs to be sought under the relevant Panchayat Acts.

---

[67] National Green Tribunal judgment dated 22 December 2016 in Original Application No. 199 of 2014 (*Almitra H. Patel & Ors v. Union of India & Ors*).

[68] These rules were issued vide Notification S.O. 1357(E) dated 8 April 2016 and replaced the Municipal Solid Waste (Management and Handling) Rules, 2000.

[69] Urban local bodies are defined as 'municipal corporation, nagar nigam, municipal council, nagarpalika, nagar palika parishad, municipal board, nagar panchayat and town panchayat, census towns, notified areas and notified industrial townships with whatever name they are called in different States and union territories in India'.

- Pollution control measures: These are to be adopted by the Central Pollution Control Boards (CPCB) and SPCBs. The state-level boards have the responsibility to authorise the setting up of new waste management sites and ensuring the compliance of stipulated standards on and off these waste management sites. They can also cancel such authorisations in instances of non-compliance. The CPCB puts in place standards and coordinates the activities of the pollution control boards.

- Monitoring by the environment ministry: This is done through the Central Monitoring Committee, which is to meet once a year to review the implementation of these rules. The committee's composition includes representatives of ministries of urban development, chemical and fertilisers, rural development and agriculture. It also includes two expert members and representatives of urban local bodies. It also has the CPCB and three state or UT pollution control bodies through rotation.

The rules also list specific responsibilities of ministries that regulate activities that generate waste. This includes urban development, agriculture, chemicals and fertilisers, renewable energy, and so on. For instance, the ministry of chemicals and fertilisers has to ensure market development and promotion of city composts with chemical fertilisers. This is by engaging contributions from fertiliser companies.

The rules have a section dedicated to waste-to-energy (WtE) plants, which have been promoted as the solution for management of large amounts of waste that is presently a critical urban governance challenge. The technological effectiveness, siting and environmental compliance concerns of these plants can be understood through the experience of the Okhla WtE plant in New Delhi. Both the present plant and its expansion have been challenged before the environment ministry,[70] the NGT[71] and the Supreme Court[72] by the plant's neighbouring RWAs. The NGT in O.A.199/2016 has clearly directed the need to segregate waste before it is incinerated in any WtE plant, which allegedly remains to be complied in the Okhla plant.

Accidents at any of the solid waste processing, treatment facilities and landfill sites need to be reported to the relevant local body. It is the responsibility

---

[70] Office Order No. J-13012/09/2018-IA.I(T) dated 20 November 2011.

[71] National Green Tribunal Original Application No. 22 (THC) of 2013 (*Sukhdev Vihar Residents Welfare Association & Ors v. the State of NCT of Delhi & Ors*).

[72] Supreme Court Civil Appeal No. 13130 of 2017 (*Ravinder Chnana & Ors v. Delhi Secretariat & Ors*).

of the local body to 'review and issue instructions if any, to the in-charge of the facility'. No time frame within which action is to be taken has been specified in the rules.

Hazardous Waste

The EPA regulates   hazardous waste through the Hazardous and Other Wastes (Management and Transboundary Movement) Rules, 2016. These rules were notified on 4 April 2016 as G.S.R No. 395 (E). This replaced the Hazardous Wastes (Management, Handling and Transboundary Movement) Rules, 2008. The 2016 Rules defines hazardous waste as

> any waste which by reason of characteristics such as physical, chemical, biological, reactive, toxic, flammable, explosive or corrosive, causes danger or is likely to cause danger to health or environment, whether alone or in contact with other wastes or substances.

The rules list 38 industrial processes, which are hazardous. These include plastics, petrochemicals, paints, fertilisers, asbestos, steel or pharmaceuticals. There is also a list of 79 waste types, which are regulated under these rules whether or not they are part of the 38 processes listed in the rules. If these wastes such as arsenic, cyanide, ammonia, nitrate, endosulfan, fluoride and zinc are used or found to be used to the extent prescribed in the rules, they would be considered as hazardous waste.

Under these rules, the following are prescribed:

- *Responsibility of waste generator:* A hazardous waste generator needs to follow specific norms for managing waste within the premises and while transferring the same to specific waste facilities. It also binds the waste generator to provide safety equipment and information to all the workers.

- *Prior approval from SPCB*: Any occupier of a facility engaged in 'handling, generation, collection, storage, packaging, transportation, use, treatment, processing, recycling, recovery, pre-processing, co-processing, utilisation, offering for sale, transfer or disposal of the hazardous waste' needs to take prior permission from PCBs. This permission can be cancelled in case of violation of any safeguard conditions.

- *No storage beyond 90 days*: No hazardous waste can be stored in a storage facility beyond 90 days and has to be transferred to a processing site, for example, a cement plant where it is utilised, managed or exported.

- *Establishing treatment, storage and disposal facility (TSDF) sites*: Sites for TSDF for hazardous and other wastes can be set up with prior permissions from SPCBs and with strict management protocol.

- *No import only for disposal purposes*: No hazardous waste can be imported into India from any other country only for the purposes of disposal. It can be imported with prior permission and conditions for the 'recycling, recovery, reuse and utilisation including co-processing'. However, this is not applicable to any waste listed in the Fifth Schedule of the notification such as for waste related to copper, zinc, lead acid batteries, e-waste, paint and sludge residue and oil waste.

In 2014, the farmers of Kalvad village in Valsad district of Gujarat filed an application before the NGT seeking compensation against damage caused from improper handling and spillage from a common hazardous waste treatment, storage and disposal facility site in Vapi. They claimed that their agricultural fields, a stream, Bil Khadi, and groundwater around the site were contaminated due to a breach in one of the walls at the TSDF site.[73]

The judgment upheld the contention of the petitioners, which was also affirmed by reports of the PCB. The bench applied the 'polluter pays' principle as specified in Section 20 of the NGT Act, 2010.[74] The judgment directed the district collector to determine the compensation due to the farmers based on the actual losses already calculated by the collector in his/her earlier order, probable future loss and non-pecuniary damages such as those that cannot be quantified. The same amount was to be transferred by owners of the facility to the office of the collector within one month of the said order. In addition, TSDF operators were asked to deposit ₹1,000,000 for environmental damage and another ₹500,000 towards the expenditure of monitoring, sampling/ analysis, investigations and supervision conducted by the Gujarat Pollution Control Board and the CPCB.

Pursuant to the directions in this case, the CPCB also issued 'Guidelines on Implementing Liabilities for Environmental Damages due to Handling & Disposal of Hazardous Waste and Penalty' in January 2016 which are in force.

---

[73] National Green Tribunal judgment dated 18 February 2014 in Application No. 87 of 2013 West Zone (*Ramubhai Kariyabhai Patel v. Union of India & Ors*), 6.

[74] Section 20 of the NGT Act, 2010, states, 'The Tribunal shall, while passing any order or decision or award, apply the principles of sustainable development, the precautionary principle and the polluter pays principle.'

Plastic Waste

The problem of managing plastic waste was in discussion since the mid-2000s. While some state governments responded to the concern by introducing a ban on the use of single use plastic,[75] the management of plastic waste was regulated under the law related to managing municipal solid wastes. The environmental impacts of plastics and their management have also been challenged before high courts, and these matters have been transferred to the NGT for adjudication.[76]

In 2016, the government enacted the Plastic Waste Management Rules, 2016, under the EPA, 1986.[77] The rules define plastic as

> material which contains as an essential ingredient a high polymer such as polyethylene terephthalate, high density polyethylene, Vinyl, low density polyethylene, polypropylene, polystyrene resins, multi-materials like acrylonitrile butadiene styrene, polyphenylene oxide, polycarbonate, polybutylene terephthalate.

Plastic waste is defined as

> any plastic discarded after use or after their intended use is over.

The rules also defines 'compostable plastic', 'virgin plastic', plastic sheets and multi-layered packaging, all of which are regulated. Any manufacturer or importer who is involved with stocking, distribution, sale or use of plastic bags, sheets and covers made of plastic sheets and multi-layered packaging can do

---

[75] Neha Lalchandani, '3-Year-Old Plastic Bag Ban Fails, Govt Drafts New Law', *Times of India*, 10 May 2012, available at https://timesofindia.indiatimes.com/city/delhi/3-year-old-plastic-bag-ban-fails-govt-drafts-new-law/articleshow/13072442.cms, accessed on 13 September 2020.

[76] National Green Tribunal order dated 5 August 2019 in Original Application No. 56 (THC) of 2013 (*Satish Kumar v. Union of India*) and related matters; National Green Tribunal judgment dated 8 August 2013 in Application No. 26 of 2013 (THC) (*Goodwill Plastic Industries & Anr v. Union Territory of Chandigarh & Ors*).

[77] ET Bureau, 'Government Notifies Plastic Waste Management Rules', *The Economic Times*, 18 March 2016, available at https://economictimes.indiatimes.com/news/economy/policy/government-notifies-plastic-waste-management-rules-2016/articleshow/51459885.cms?utm_source=contentofinterest&utm_medium=text&utm_campaign=cppst, accessed on 13 September 2020.

so only if the material in use adheres to certain standards. For instance, 'Carry bags and plastic packaging shall either be in natural shade which is without any added pigments or made using only those pigments and colourants which are in conformity with Indian Standard: IS 9833:1981....' Similar standards are prescribed for other kinds of plastic products such as carry bags, plastic sheets and sachets.

There are responsibilities assigned to different agencies involved in waste management, producers, distributers and users of plastics. The primary responsibility of enforcement of these rules rests with the state and UT PCBs. Plastic waste recycling or processing units need to take consents from PCBs for establishing and operating these facilities (see Chapter 4 on these consents).

- *Urban* (municipalities or municipal corporations) *and rural local bodies* (gram panchayats) need to develop infrastructure for 'segregation, collection, storage, transportation, processing and disposal of the plastic waste'. This can be done by themselves or by engaging other agencies or producers. This process also includes channeling waste to recycling facilities. The rules specify that there should be no damage caused to the environment in this process of waste management by local bodies.

- *Waste generators*, whether individual or institutional, need to minimise the generation of waste, segregate it at source and hand it over to an urban or a rural local body without any littering. This handover can be done through "registered waste pickers', registered recyclers or waste collection agencies'. Waste generators also need to pay a fee or charge as prescribed in the bye-laws of the concerned local body. Events held in open spaces also need to segregate and manage waste.

- *Retailers or street vendors* are not to allowed 'to sell or provide commodities to consumer in carry bags or plastic sheet or multilayered packaging' that are not manufactured or labelled or marked as per the requirements of the 2016 Rules.

- *Producers, importers and brand owners* of plastic had to establish a system of waste management based on 'Extended Producer Responsibility' within six months of the enactment of the rules along with state urban development departments. Project proponents who introduce products such as sachets and pouches into the market 'need to establish a system for collecting back the plastic waste generated due to their products'. This plan of collection is to be submitted to the SPCBs while applying for consent to establish or operate or renewal.

The NGT also monitors the compliance with these rules and has asked state governments to submit action plans for waste management before it.[78] In March 2019,

> the Tribunal directed the States/UTs to ensure compliance of the PWM Rules requiring furnishing of reports as well as taking other steps. In default, the defaulting States were to be required to pay compensation @ Rs. 1 crore per month after 01.05.2019. CPCB was to furnish status of compliance.

Since then, state governments have submitted their action plans to the NGT[79] and the hearings are underway.

Construction and Demolition Waste

The working sub-group on C&D waste constituted by the environment ministry discussed earlier in this section proposed that an institutional mechanism is required for the management of C&D waste which could be done through an amendment to the Municipal Solid Waste Management Rules, 2000. At that point, the only document focusing on C&D waste was the 'Manual on Municipal Solid Waste Management' prepared by the Ministry of Urban Development in 2000. This only gave basic guidelines for the handling of such waste.[80]

This issue was the core focus of discussions at a December 2013 conference organised by the Environment Pollution (Prevention and Control) Authority

---

[78] Press Trust of India, '25 States Miss Deadline for Plan on Plastic Disposal, Face Rs 1-cr Penalty', *Business Standard*, 26 May 2019, available at https://www.business-standard.com/article/pti-stories/over-25-states-face-rs-1cr-ec-for-no-action-plan-on-plastic-waste-disposal-119052600168_1.html, accessed on 13 September 2020.

[79] National Green Tribunal Execution Application No. 13/2019 in Original Application No. 247/2017 (*Central Pollution Control Board v/s State of Andaman & Nicobar & Ors*).

[80] Centre for Science and Environment, *Waste to Resource: Briefing Note CSE Warns Cities Are Choking on and Because of Their Own Construction and Demolition (C&D) Waste with Serious Environmental Consequences* (New Delhi: Centre for Science and Environment, 2013), available at https://www.cseindia.org/waste-to-resource-briefing-note-cse-warns-cities-are-choking-on-and-because-of-their-own-construction-and-demolition-cd-waste-with-serious-environmental-consequences-5248, accessed on 2 August 2019.

(EPCA) and the Centre for Science and Environment. The conference titled 'Waste to Resource' took into account that '70 percent of buildings that will stand in India in 2030 are yet to be built'. This construction boom needs to take steps to 'recycle and reuse construction waste and turn it into a resource'. The participants discussed that mature technologies are available to help with this process.[81]

The C&D waste rules were enacted in 2016 under the EPA, 1986.[82] The March 2016 press release of the environment ministry records that C&D waste in India amount to 530 million tonnes annually, and that it should not be treated as a waste but as a resource.[83]

These rules define C&D waste as

> the waste comprising of building materials, debris and rubble resulting from construction, re-modeling, repair and demolition of any civil structure.

The rules require every 'waste generator', be it an individual, institution, public or commercial establishment generating more than 20 tonnes of C&D waste in a day or 300 tonnes in a month, to segregate the waste into four streams: concrete, soil, steel, wood and plastics, bricks and mortar. Environment Minister Prakash Javadekar while releasing these rules said, 'Permission for construction will be given only when the complete construction and demolition waste management plan is presented.'[84]

The rules lay out the legal framework for proper in-site and off-site management of C&D waste, including transportation, storage and utilisation. Following these rules, specific state governments have issued directions and clarifications for the implementation of C&D waste rules. For instance, the Karnataka state government issued a circular in January 2019 flagging the

---

[81] Ibid.

[82] Ministry of Environment, Forests and Climate Change. G.S.R. 317(E) dated 29 March 2016.

[83] Press Trust of India, 'Govt Frames Rules to Manage Construction, Demolition Waste', *Times of India*, 29 March 2016, available at http://timesofindia. indiatimes.com/articleshow/51599447.cms?utm_source=contentofinterest&utm_ medium=text&utm_campaign=cppst, accessed on 13 September 2020.

[84] Ibid.

responsibilities of waste generators and the role of the SPCBs to monitor the compliance with these rules.[85]

GENETIC ENGINEERING

Since 1989, the EPA regulates the manufacture, use, import, export and storage of hazardous microorganisms and GE organisms and cells.[86]

These rules were a response to the growing understanding that 'indiscriminate use of modern biotechnology in healthcare, agriculture, environment and process industries' poses potential risks that need to be regulated. They came nine years after the setting up of the Department of Biotechnology (DBT) created under the Ministry of Science and Technology to advance India's research and development infrastructure in modern biotechnology including genetic engineering.[87]

Since then, regulating genetic engineering has been a contested domain between the environment and science and technology ministries. Since 2008,[88] several attempts were made to create a biotechnology regulator within the DBT and shift the regulatory oversight functions away from the environment ministry. Farmer's organisations, environmentalists, agricultural scientists and NGOs have strongly opposed this on various grounds, including that of conflict of interest.[89]

The 1989 rules that deal with regulating the use (including cultivation, production, sale and storage) of the GE material are in operation today. Genetic engineering is defined as

---

[85] Circular of the Karnataka PCB No. PCB/031/C&D/2016 dated 30 January 2016 to all regional offices of SPCBs.

[86] These rules are notified as G.S.R. 1037(E) on 5 December 1989 (amended in 1993, 2005, 2006, 2007 and 2009 through specific notifications by the environment ministry).

[87] CEE (Centre for Environment Education) and MoEF, *National Consultations on Bt Brinjal: A Primer on Concerns, Issues and Prospects* (New Delhi: Government of India, 2010).

[88] Prashant Reddy, 'SpicyIP Tidbit: The National Biotechnology Regulatory Authority Bill', *SpicyIP*, 6 July 2008, available at https://spicyip.com/2008/07/spicyip-tidbit-national-biotechnology.html, accessed on 29 August 2019.

[89] Shaheen Contractor, 'No BRAI Please', *Hindu Business Line*, 23 August 2013, available at https://www.thehindubusinessline.com/opinion/No-BRAI-Bill-please/article20651786.ece#!, accessed on 13 September 2020.

the technique by which heritable material, which does not usually occur or will not occur naturally in the organism or cell concerned, generated outside the organism or the cell is inserted into said cell or organism. It shall also mean the formation of new combinations of genetic material by incorporation of a cell into a host cell, where they occur naturally (self cloning) as well as modification of an organism or in a cell by deletion and removal of parts of the heritable material.

The 1989 Rules require prior approval from the Genetic Engineering Appraisal Committee (GEAC)[90] for any activity involving large-scale use of hazardous microorganisms and recombinants. This can be in both research and industrial production. The GEAC is also responsible for approving or rejecting the release of GE organisms and products into the environment including experimental field trials of GE crops.

Any person who has received an approval from the GEAC can use GE and other hazardous microoganisms only for the specified purposes. It is also their responsibility to inform the district-level committee or the State Biotechnology Coordination Committee set up under these rules in case of any accidents or interruption in operations that would lead to discharge of GE organisms or cells which are harmful to environment, health or can cause danger. These accidents do not absolve the person or facility that has received prior approval to prevent such accidents or interruptions.

There have been several controversies around the use of GE materials, especially in agriculture, due to their risks or due to lack of due procedure in grant of approvals (as in the case of Bt [*Bacillus thuringiensis*] cotton) given under these rules or demanded restriction of such approvals (as in the case of Bt brinjal or genetically modified [GM] maize).[91]

There is an ongoing case in the Supreme Court[92] since 2012 that has raised several biosafety concerns related to genetically modified organisms (GMOs) and has asked the apex court to direct that those organisms should

[90] On 16 July 2010, the Genetic Engineering Approval Committee was renamed as the Genetic Engineering Appraisal Committee (GEAC).

[91] Jyotika Sood, 'Monsanto Conducted Trials of GM Maize without Approval: Reveals RTI Reply', *Down to Earth*, 4 July 2015, available at https://www.downtoearth.org.in/news/monsanto-conducted-trials-of-gm-maize-without-approval-reveals-rti-reply-35916, accessed on 13 September 2020.

[92] Supreme Court Writ Petition (Civil) No. 260 of 2005 (*Aruna Rodrigues & Ors v. Union of India & Ors*).

not be released in the environment 'without proper scientific examination of bio safety concerns and affecting both the environment and human health'. The public interest litigation was filed with 'the intent and substance' to 'put in place a protocol that shall maintain scientific examination of all relevant aspects of bio safety before such release, if release were to be at all permissible'.

During the course of the hearing, the petitioner gave several instances of GM crops being grown illegally in different parts of the country and sought a stay on the commercial release of crops such as GM mustard.[93] The Supreme Court appointed a Technical Expert Committee, which in 2013, raised serious concerns and instances of conflict of interest in the manner in which GMOs are regulated in India.[94] The hearings in this case are still underway.

ECOLOGICALLY SENSITIVE AREAS

The concept and practice of demarcating ESAs or ecologically sensitive zones (ESZs) can be traced back to the 1980s, when Section 3(2) (v) of the EPA was used to secure areas that were affected by large-scale, rapid land use change. Specific notifications were introduced under the EPA to impose, prohibit or regulate activities or processes that were impacting or likely to threaten the ecological sensitivity or regions identified as ESAs. These have included mining, tourism, real estate, industry or power generation. The implementation of these regulations is supervised and monitored by specialised authorities or committees as stated in the notifications.[95] These bodies have varied powers as defined in the notifications. They have specific tenures; some of which were

---

[93] Press Trust of India, 'SC Extends Stay on Commercial Release of GM Mustard Crop', 24 October 2016), available at http://ptinews.com/news/7979223_SC-extends-stay-on-commercial-release-of-GM-Mustard-crop, accessed on 11 February 2020.

[94] Colin Todhunter, 'Challenging the Flawed Premise behind Pushing GMOs into Indian Agriculture', *CounterPunch*, 22 January 2020, available at https://www.counterpunch.org/2020/01/22/challenging-the-flawed-premise-behind-pushing-gmos-into-indian-agriculture/, accessed on 13 September 2020.

[95] These committees include the Dahanu Taluka Environment Protection Authority (DTEPA) to monitor the compliance of the Dahanu ESA (Ecologically Sensitive Area) Notification, 1996; the state-level expert and monitoring committees to grant permissions and oversee the mining and other permitted developmental activities in the Aravalli region of Rajasthan and Haryana (1999); the high-level monitoring committee for Mahabaleshwar-Panchgani ESA (2000) and the monitoring committee for the Matheran ESA (2003).

not renewed after their first term had expired.[96] However, the notifications for regulating activities in ESAs have no time limitations. In the 1990s, this concept was also understood for its potential to realise landscape-level conservation in the country.

These legal measures were prompted by environmental groups, court judgments or through *suo moto* actions by the environment ministry to design environment protection and land use planning frameworks and plans for specific regions. However, the declaration and enforcement of ESA notifications has faced hurdles at the following two levels:

- The first is related to the acceptance and implementation of these centralised decisions by the state and local governments.

- The second is in obtaining the co-operation or compliance of local communities and citizens of these areas who were not involved in these decisions.

As a result of these issues, the implementation of the regulatory provisions of the ESA notifications has remained patchy.

Until May 2009, the environment ministry issued 8 ESA notifications. The notified ESAs included Murud-Janjira, Dahanu, Mahabaleshwar-Panchgani and Matheran in Maharashtra; Doon Valley in Uttarakhand; Aravalli Hills in Haryana and Rajasthan; 15 kilometre around the Numaligarh refinery in Assam and the Taj Trapezium Zone around Taj Mahal in Agra, Uttar Pradesh. The draft notifications for regulating land use in Pachmarhi (Madhya Pradesh) and in the entire Himalayan region lapsed[97] before they could be finalised. Other notifications that have also remained pending include the proposal for Sahyadri Ecological Sensitive Area (SESA, covering contiguous forested stretches of Karnataka, Maharashtra and Goa),[98] Rishi Valley and Araku Valley (Andhra Pradesh), Kasauli (Himachal Pradesh) and Kodachadri ESA (covering Kodachari Hills in Karnataka).

---

[96] Meenakshi Kapoor, Kanchi Kohli and Manju Menon, *India's Notified Ecologically Sensitive Areas (ESAs): Story So Far* (New Delhi: Kalpavriksh and WWF-India, 2009), 2–3.

[97] Two notifications were issued for the proposed Pachmarhi ESA. The first was in 17 September 1998 and the second was in 18 September 2000. Final drafts could not be issued before 365 days which is when such notifications were due to lapse according to the provisions of the EP Rules, 1986.

[98] This is discussed further in this chapter in the section titled 'Western Ghats as Ecologically Sensitive Area'.

These notifications can be understood as the first phase of ESA zoning in India. During this phase, the environment ministry also attempted to systematise the identification and declaration of ESAs by introducing criteria for identification. In 1999, a committee was set up under the chairmanship of Pronab Sen, the then adviser to the Planning Commission on Perspective Planning and Statistics and Surveys, with an explicit mandate to frame parameters for designation of areas as ESAs. This committee submitted its report to the MoEF in 2000 in which it defined 'ecological sensitivity' and presented a list of primary and auxiliary criteria for designating ecological sensitivity and accordingly identifying ESAs.[99]

This report was the first institutional commitment in formalising the process of setting up ESAs in the country. This prompted the setting up of another committee under the chairmanship of H. Y. Mohan Ram, a retired professor of botany, University of Delhi, to review and appraise specific ESA proposals. The Mohan Ram committee was constituted through a notification dated 1 January 2001.[100] This expert body of six members functioned until July 2006, during which ESAs such as Mahabaleshwar-Panchgani, Matheran and Mount Abu (draft) ESA notifications were formalised through gazette notifications. In July 2006, the term of the above committee expired. Since then, no new committee was constituted.

Ecologically Sensitive Areas around Protected Areas

The environment ministry engaged in another process of declaring ESAs around protected areas (PAs), following the recommendations of the 2002 National Wildlife Action Plan. Protected areas are demarcated under the Wild Life Protection Act,1972 discussed in detail in Chapter 6 of this book. The declaration of ESAs as buffer zones around officially declared national parks and sanctuaries was introduced into wildlife conservation policies since the early 2000s. The National Wildlife Action Plan, 2002–2016, said,

> All identified areas around Protected Areas and Wildlife Corridors to be declared as Ecologically Fragile under the Environment (Protection) Act, 1986.

---

[99] Ministry of Environment and Forests, *Report of the Committee on Identifying Parameters for Designating ESAs in India* (New Delhi: MoEF, 2000).

[100] Kapoor, Kohli, and Menon, *India's Notified Ecologically Sensitive Areas (ESAs)*.

Section XI(5.2) of the plan stated that the environment ministry should complete the aforementioned task by 2004. Section 9 of the Wildlife Conservation Strategy, 2002, substantiated this and specified that

> lands falling within 10 km of the boundaries of national parks and sanctuaries should be notified as eco-fragile zones under section 3(2) (v) of the Environment (Protection) Act and Rule 5 Sub-rule 5 (viii) & (x) of the Environment (Protection) Rules.

These policy prescriptions were formalised after the intervention of the Supreme Court in Writ Petition No. 460/2004 filed by Goa Foundation on 17 December 2004. In addition to seeking the closure of defaulting units operating in violation of environment laws, the petitioner, Goa Foundation, sought the Court's attention on the issue of declaring ESAs around PAs for which the environment ministry had already written to the state governments. In this case, the Supreme Court passed an order on 4 December 2006 asking the environment ministry to write again to the states and UTs and give them a final opportunity to respond. The letter would also direct the state governments to send appropriate proposals, failing which the Court may consider passing orders for notifying 10 kilometre areas around all PAs as ESAs. The Court also observed that the environment ministry should refer all cases, where environment clearance was given to projects within 10 kilometres of PAs, to the National Board for Wild Life (as discussed in Chapter 6).[101]

This can be understood as the second phase of ESAs where state governments directly negotiated with the environment ministry on ESA boundaries for PAs within their states. On 9 February 2011, the ministry also issued detailed guidelines for declaration of eco-sensitive zones around national parks and wildlife sanctuaries.

Even before the decision was finalised, several state governments expressed reservations at being able to comply with the 10 kilometre buffer requirement. For instance, the Himachal Pradesh state called the policy a 'sweeping notification' and feared that if the aerial distance were to be used, the state would have no space left for development. The Government of Goa

---

[101] These details are drawn from the record of proceedings, Writ Petition No. 460 of 2004 (*Goa Foundation v. Union of India*), Supreme Court of India, 12 December 2006; correspondence between the MoEFCC and state governments, from February 2002 to December 2008.

communicated to the then Prime Minister Manmohan Singh that if the policy was formalised, two-thirds of the available land in Goa would fall under the protected area network. This would have implications for the socio-economic development of the state.[102]

The reduction of boundaries for ESAs around PAs was legally challenged in specific states. One such case before the NGT was filed by Tseten Lepcha, former honorary Wild Life Warden and member of the Sikkim State Wildlife Advisory Board. This petition[103] challenged the reduction of the ESA buffer area highlighting 'that the Eco-sensitive Zones which usually serve as shock absorbers, have been notified keeping a mere 25–200 metres'. The petition challenged eight such notifications by the environment minister. The judgment states that

> it becomes clear that adequate exercise was undertaken both by the state and the Union Government to follow the 2011 Guidelines to the extent possible and there has been no violation of the order of the Hon'ble Apex Court nor any procedural lapses and violation of the existing Acts and Rules. The comments of the people on draft notifications were also considered....

The NGT dismissed the petition with the conclusion that there was no illegality in the notification of the ESA boundaries.

Western Ghats as Ecologically Sensitive Area

The Western Ghats of India is one the world's richest biodiversity hotspots. On 4 March 2010, the environment ministry announced the constitution of the Western Ghats Ecology Expert Panel (WGEEP) to explore the possibility of a landscape-level conservation strategy for this ecoregion.[104] This arose out of a proposal that was pending before the environment ministry to declare parts of the Western Ghats covering Maharashtra, Karnataka and Goa as the

---

[102] For more details, refer to M. Kapoor and K. Kohli, *The Second Act: Centre–State Conversations on Ecologically Sensitive Areas (2009–2012)* (New Delhi and Anand: Kalpavriksh and Duleep Matthai Nature Conservation Trust, 2012), 9.

[103] National Green Tribunal judgment dated 21 August 2017 in Original Application No. 15 of 2015 Eastern Zone (*Tseten Lepcha v. Union of India & Ors*).

[104] MoEF Office Order No. l/1/2010- RE (ESZ) dated 10 March 2010: Constitution of Western Ghats Ecology Expert Panel.

SESA.[105] This proposal for SESA was made by researchers and NGOs from this region.

The 14-member WGEEP comprised ecologists, representatives of NGOs, scientific institutions and representatives of the environment ministry. It was set up with a term of one year to assess the ecology of the Western Ghats and suggest the demarcation of an ESZ under the EPA, 1986, and also the modalities of the establishment of a Western Ghats Ecology Authority.

The WGEEP finalised its report over 12 meetings between March 2010 and May 2011 and submitted it to the environment ministry in October 2011. The report was sent out to the state governments for their comments. The recommendations in the report faced backlash from the state governments. Due to this rejection by the state governments, the report was not made public by the environment ministry until a decision by the Central Information Commission (CIC) under the Right to Information Act, 2005, directed the ministry to do so.[106]

As the reactions from the state governments grew, the environment ministry constituted a high-level working group under the chairmanship of K. Kasturirangan, Member (Science), Planning Commission[107], to review the findings of the WGEEP. This committee popularly known as the Kasturirangan Committee finalised its report in 2013[108] after reviewing all submissions from the state governments and independent submissions. This committee suggested another strategy to declare ESAs that was less restrictive

---

[105] Madhav Gadgil, 'Western Ghats Ecology Expert Panel: A Play in Five Acts', *Economic and Political Weekly* 49, no. 18 (2014): 38–50.

[106] An RTI application and two appeals by G. Krishnan, a Kerala-based activist, sought the summary of the WGEEP report. However, the ministry denied the information with the reasoning that this would have implications on the economic interests of the states. Overruling the environment ministry's decision, the CIC directed the MoEF on 9 April 2012 to make the report public. See Krishnaraj Rao, 'Nothing Secret about Panel Reports', *India Together*, 14 April 2012, available at http://www.indiatogether. org/cicruling-rti, accessed on 12 September 2020.

[107] MoEF Office Order dated 17 August 2012, 'Constitution of High Level Working Group to study the preservation of the ecology, environmental integrity and holistic development of the Western Ghats in view of their rich and unique biodiversity' (No.1/1/2010- RE (ESZ) Pt.).

[108] The MoEFCC Office Memorandum No. 1-4/2012-RE (Pt.) dated 19 October 2013 indicating acceptance of the high-level committee report.

of permissible land use change in the region. This was not received well by the members of the WGEEP as they felt that the intent of conserving the Western Ghats was diluted by the Kasturirangan Committee.[109]

A draft notification[110] was issued on 27 February 2017 seeking comments from the state governments and the citizens. This notification was based on the recommendations of the Kasturirangan Committee, which 'identified approximately thirty-seven percent the Western Ghats ecologically sensitive which covers an area of 59,940 square kilometre'. The final notification was not issued. Another notification was issued on 3 October, 2018 and is yet to be finalized.[111]

III PROPOSAL FOR A NEW ENVIRONMENT REGULATOR

In its 25th year, the environment ministry made public a discussion paper which presented its limited institutional capacity to deal with the growing number of cases for which environment clearances are sought and in which pollution monitoring is required. The discussion paper dated 17 September 2009 stated that the environment institutions are unable to enforce environmental laws and ensure that environmental safeguards are effectively monitored.[112] This note acknowledged the observations in both the Planning Commission documents and independent NGO studies on the failure of decision-making structures to achieve environment protection over the years.

In response to these challenges, the environment ministry proposed setting up of a National Environment Protection Authority (NEPA) as an independent regulator for the environment. This announcement coincided with the Prime Minister Manmohan Singh's visit to the United States of America in November 2009 to discuss the United States (US)–India partnership. The government's intent to set up a new regulator was part of the agenda for discussions during his visit. A November 2009 press release of the US Senate and Prime Minister's Office stated that the United Nations Environment Protection Authority would help India to set up a NEPA.

---

[109] Gadgil, 'Western Ghats Ecology Expert Panel'.

[110] MoEFCC S.O. 667(E) dated 27 February 2017.

[111] MoEFCC S.O. 5135(E) dated 3 October 2018 (F. No. 25/03/2010-ESZ).

[112] MoEF, 'Towards Effective Environmental Governance: Proposal for a National Environment Protection Authority', Discussion Paper, Government of India, New Delhi, 2009.

The environment ministry subsequently revised the discussion paper on the proposed NEPA prior to a public consultation held in New Delhi on 25 May 2010.[113] The revised proposal was discussed at this meeting. By November 2010, the design of the institutional framework for a new environment regulator was re-announced as the National Environment Assessment and Monitoring Authority (NEAMA).[114] This body was proposed as an autonomous institution with scientific and professional rigour to appraise projects for environmental approvals as well as to monitor their implementation. The environment ministry proposed to retain the law and policymaking functions within its ambit.

The four points on which the proposal for a new environment regulator was assessed are as follows:

- Autonomy: The NEAMA was proposed as an independent authority under the EPA, 1986. However, in its design, it continued to be tied to the environment ministry. The environment ministry would hold the functions of law-making and amending regulations. The ministry would also be responsible for constituting the authority and disbursing funds for its functioning. These functions of the ministry would limit the independence of the authority.

- Conflict of interest: The NEAMA note mentioned that it is the dual role of the government, in both appraisal as well as approval of projects, which results in a perception of conflict of interest. However, the design of the new regulator did not replace the expert committees, which is where most of the conflict of interest was reported. It primarily reformed the EACs that were in use in the environment clearance processes to Thematic Appraisal Committees located in the NEAMA. There was no indication that the compositions would be any different than what they were.

- Poor compliance: The note emphasised that conditions levied at the time of environment clearance should be objective, measurable, fair and consistent as well as should not impose inordinate financial or time costs

---

[113] MoEFCC, 'Workshop on Reforms in Environmental Regulation: With Specific Reference to Establishment of National Environment Protection Authority', Discussion Paper, Government of India, New Delhi, 2010.

[114] MoEF, *Reforms in Environmental Governance with Special Reference to Establishment of National Environment Assessment and Monitoring Authority (NEAMA)* (New Delhi: Government of India, 2010).

on the proponents. However, the problem of poor compliance record of approved projects was not addressed.

- Capacity shortfall: While the discussion papers highlighted the capacity of the ministry to handle many and varied tasks, the structure of the NEAMA did not address this issue. The tasks of the environment ministry's impact assessment division, monitoring tasks of the MoEF's regional offices, and mapping and management functions of CZMAs were grouped into one authority, called the NEAMA. The discussion notes did not highlight any studies that assured that the capacity issues would not emerge for the new regulator as well.

The need for an environment regulator was also emphasised in the Supreme Court in its judgment in the 2011 Lafarge Mining case.[115] The Court stated,

> Thus, we are of the view that under Section 3(3) of the Environment (Protection) Act, 1986, the Central Government should appoint a National Regulator for appraising projects, enforcing environmental conditions for approvals and to impose penalties on polluters.

In a follow-up order dated 6 January 2014, the Supreme Court asked the environment ministry to set up the regulator by 31 March 2014. With the change in the central government in May 2014, this proposal has been on the back-burner. In January 2021, the Supreme Court issued notices in a petition asking for directions to set up an environment regulator in compliance with the 2011 Lafarge Mining case.[116]

---

[115] Supreme Court judgment dated 6 July 2011 in I.A. Nos. 1868, 2091, 2225–2227, 2380, 2568 and 2937 in No. 202 of 1995 (*Lafarge Umiam Mining Pvt. Ltd. v. Union of India & Ors*).

[116] Livelaw News Network, 'Supreme Court Issues Notice on a Petition Seeking Appointment of a National Regulator to Oversee the Environmental Violations in the Country', 29 January 2021, available at https://www.livelaw.in/news-updates/supreme-court-appointment-national-regulator-environmental-violation-169068, accessed on 27 February 2021.

6

Wildlife and Biodiversity Conservation

INTRODUCTION

Wildlife and biodiversity conservation can be understood as the proactive protection of species and habitats, both wild and cultivated. The central focus of the Wild Life Protection Act (WLPA) introduced in India in 1972 was to provide protection to wildlife and wildlife habitats. This law largely focuses on reducing human use and developmental pressures on areas important for wildlife species. But this law did not aim to conserve biodiversity, a much wider concept than wildlife. India enacted a law to protect biodiversity only in 2002. This was also the year when legal frameworks for wildlife conservation carved out space for greater local community participation. These two changes reflect, to some extent, the evolution of international and domestic discourses on wildlife and biodiversity conservation.

This community-based approach to wildlife conservation soon came under pressure after the enactment of the Scheduled Tribes and Other Traditional Forest Dwellers (Recognition of Forest Rights) Act, 2006. After the passing of this law, exclusionary forms of wildlife conservation regained support within the government system and among some organisations working on wildlife conservation. One illustration of this is in the push to create inviolate areas like critical tiger habitats (CTHs) under the WLPA.

Both wildlife and biodiversity conservation laws are land centric in their approach. Marine conservationists have argued that this approach is unfit for the conservation of marine ecosystems that require distinct management strategies. The reliance on WLPA to provide a 'one size fits all' model for conservation of different habitats has failed to bring attention to the conservation needs of marine ecosystems. The creation of Marine Protected Areas (MPAs) has not gained from the lessons of managing terrestrial protected areas (PAs).[1]

In this chapter, we discuss two significant laws related to wildlife and biodiversity conservation in India: the WLPA, 1972, and the Biological Diversity Act (BDA), 2002. Both these laws were enacted at very different points in India's modern environmental history and are designed to respond to different problems. The BDA comes with the stated purpose of conservation but is not limited to wildlife. It seeks to cover all areas that are important for biological diversity. The BDA defines biodiversity as 'the variability among living organisms from all sources and the ecological complexes of which they are part, and includes diversity within species or between species and of ecosystems'. Therefore, the jurisdiction of this law is across all administrative categories and ecosystem types.

India enacted the BDA in response to its commitment to the Convention on Biological Diversity (CBD). The BDA served two functions—responding to the global concern over the depletion of biological diversity and providing regulated legal access to bioresource-based industries dealing with pharmaceuticals, seeds and biofuels that wanted access to international germplasm without being accused of biopiracy.[2] In recent years, the contentious issue of 'benefit sharing' has dominated the BDA's implementation landscape. As required by international agreements and domestic law, clear contractual agreements need to be created as soon as access to biological material or related people's knowledge is permitted for research or commercial use.

Both these laws operate through different federal arrangements and institutional frameworks. However, their overlapping jurisdiction and specific enforcement actions help to understand how these laws are related. This

---

[1] Sridhar Aarthi and Shanker Kartik, 'Lessons from Marine Paradigms', *Seminar*, no. 577 (2007), available at http://www.india-seminar.com/2007/577/577_aarthi_sridhar,_shanker.htm, accessed on 28 August 2019.

[2] Biopiracy has not been defined in either the Convention on Biological Diversity (CBD) or India's the Biological Diversity Act (BDA). However, it has been defined by the ETC Group engaged with the international discussion on the issue as 'appropriation of the knowledge and genetic resources of farming and indigenous communities by individuals or institutions that seek exclusive monopoly control (patents or intellectual property) over these resources and knowledge', available at https://www.etcgroup.org/issues/patents-biopiracy, accessed on 21 June 2019.

chapter presents details of both these laws and their current frameworks and administrative priorities. It also highlights some overlaps in institutional coordination for enforcement such as in the case of hunting and biopiracy and in the designing of conservation management areas such as heritage sites in places which are protected for wildlife.

This chapter includes the following sections:

I.   *Legal Framework for Wildlife Protection*
II.   *Legal Framework for Biodiversity Conservation*
III. *Overlaps between Wild Life Protection Act and Biological Diversity Act*
IV.   *Issues Arising out of Legal Frameworks for Conservation*
V.   *Wildlife Action Plans*

I   LEGAL FRAMEWORK FOR WILDLIFE PROTECTION

This section deals with the provisions of the WLPA, 1972, that is specifically designed to protect and prevent damage to wildlife and wildlife supporting areas. The scope of this law rests on how the term 'wildlife' is defined in the law and elaborated through the Act's annexures. The jurisdiction of the law extended to the whole of India 'except the State of Jammu and Kashmir'. However, this status changed with the reorganisation of Jammu and Kashmir and Ladakh as Union Territories.[3]

The WLPA defines wild life as 'any animal, aquatic or land vegetation which forms part of any habitat'. It also defines habitat as 'land, water, or vegetation which is the natural home of any wild animal'. The scope of these definitions needs to be read with the five schedules, which list the species of flora and fauna upon which the provisions of the law are differentially applicable.

The WLPA essentially uses two strategies to protect wild life. These are the creation of PAs and prevention of activities that are harmful to wildlife, found within or outside officially demarcated PAs. It also has an array of enforcement actions that can be taken in case there is a contravention or abetment through actions such as poaching, hunting, illegal entry, damage to wildlife and animal trade.

---

[3]   Ministry of Law and Justice (Legislative Department), The Jammu and Kashmir Reorganisation Act, 2019, 9 August 2019

CREATION OF PROTECTED AREAS

When the WLPA was first enacted in 1972, it envisaged the creation of two kinds of PAs—national parks (NPs) and wildlife sanctuaries (WLS)—each having different levels of protection and restrictions on human presence. These can be declared in any area that the state government considers important for its ecological, faunal, floral or geomorphological characteristics. In 2002, through an amendment in the law, two more categories were added to the types of PAs: community reserves and conservation reserves. These areas are presented in the law as areas adjoining NPs and sanctuaries that were not necessarily under government 'ownership' but could strengthen the existing PA network. For the declaration of conservation and community reserves, the law prescribes mandatory consultations and discussions with local communities living in and using these areas. In the case of NPs and sanctuaries, the local community is only given the opportunity to prove their rightful use and inhabitation of the area, following an initial notification.

According to the WLPA:

- Protected areas are to be notified by the state government following an approval of the state legislature. The central government can also declare areas as PAs if the state government transfers or leases an area to the centre.

- They can be land based or in territorial waters of the country. The areas under territorial waters are defined separately as MPAs.

- Protected areas require clear delineation of boundaries and details of existing use and habitation rights that will be impacted.

- The final notification of a PA is complete only after rights have been recorded and settled. This process is facilitated through the district collector and Chief Wild Life Warden (CWLW) of the area.

- Where rights have been extinguished, the necessary relief, rehabilitation and relocation plans are to be prepared by the district collector along with the CWLW.

- Protected Areas cannot be denotified or modified except by the state legislature through which it is first approved. On 13 November 2000, the Supreme Court also directed that no forest, NP or sanctuary could be de-reserved without the approval of the Supreme Court.[4]

---

[4] Supreme Court order in Writ Petition (Civil) 337 of 1995 (*The Centre for Environmental Law [CEL], WWF v. Union of India and Ors.*).

All four categories of PAs have distinct objectives laid out in the law. They also determine the extent to which existing rights of communities living or dependent on these areas can continue or be extinguished.

- National parks can be declared in any area of the country, and the law does not specify any administrative restrictions. In such areas, all rights are recorded, extinguished and subsequently vested with the state government. After the notification of any area as an NP, all activities are governed through the CWLW. No seasonal grazing activity can also take place in an NP.

- Wildlife sanctuaries can be declared in any area, irrespective of administrative jurisdiction. In these areas, rights of communities can be extinguished, acquired[5] or allowed to continue following approval of the CWLW of the area. After the notification, all activities are governed through the CWLW.

- Conservation reserves are declared on government land particularly in areas that are adjacent to PAs or link one PA to another. These reserves help to connect landscapes and seascapes and secure larger wildlife habitats with fewer restrictions. These areas attract the same process of settlement of rights as in the case of WLSs. The state government appoints a Conservation Reserve Management Committee to advise the CWLW. The committee needs to include representatives of all village panchayats of the reserved area. There is no extinguishment of rights in these areas.

- Community reserves are declared on any private or community land where the community or an individual has volunteered to conserve wildlife and its habitat. These are not set up in areas that are already declared National Park, sanctuary or conservation reserve. It can be declared as a community reserve for protecting the flora and fauna and also for 'traditional or cultural conservation values and practices'. The authority responsible for conserving, maintaining and managing the community reserve is a management committee comprising five representatives nominated by the village panchayat (council). There is no extinguishment of rights in these areas.

As of May 2019, 165,088.36 square kilometres were under 869 PAs in India. This includes 104 NPs, 551 WLSs, 87 conservation reserves and 127

---

[5] This acquisition was to be done under the Land Acquisition Act, 1894, which is now replaced by the Right to Fair Compensation and Transparency in Land Acquisition, Rehabilitation and Resettlement Act, 2013.

community reserves. This covers 5.06 per cent of the area of the country.[6] The area under WLSs is 72.5 per cent of this coverage.

CREATING SPECIAL WILDLIFE RESERVES

Wildlife conservation has historically had a special focus on flagship species such as tiger, elephant, rhinoceros, lion or leopard. In all these, tiger conservation has seen several high-level political and executive interventions since the 1970s. In 1972, the then Prime Minister Indira Gandhi set up a special task force on tiger conservation, and the report of the task force submitted in August that year formed the blueprint of 'Project Tiger'.[7] 'Project Tiger' was launched in 1973 and has undertaken conservation initiatives for the protection of tigers in India. One such initiative was the declaration of all tiger habitats including existing PAs and their adjoining areas as tiger reserves.

In 2005, there was a national concern on the disappearance of tigers in Sariska Tiger Reserve.[8] This prompted the government to set up a Tiger Task Force (TTF) to address the specific problem of Sariska and recommend other measures for tiger conservation in the country.[9] The report submitted by the TTF states that the suggestion to set up such a task force was made in March 2003 during the second meeting of the National Board for Wildlife (NBWL) constituted under the WLPA.[10] On the recommendation of this task force, a separate chapter on tiger conservation was introduced to the WLPA through

---

[6] The data on the protected areas (PAs) of India (May, 2019) is accessed from WII-ENVIS Centre on Wildlife & Protected Areas, available at http://www.wiienvis. nic.in/Database/Protected_Area_854.aspx, accessed on 18 June 2019. This portal is hosted by Wildlife Institute of India and sponsored by the Ministry of Environment, Forests and Climate Change (MoEFCC).

[7] Project Tiger, *Joining the Dots: Report of the Tiger Task Force* (New Delhi: Ministry of Environment and Forests (MoEF), Government of India, 2005), vi.

[8] Bisheshwar Mishra, 'No Tigers Left in Sariska: CBI', *Times of India*, 7 May 2005, available at https://timesofindia.indiatimes.com/india/No-tigers-left-in-Sariska-CBI/articleshow/1102282.cms#:~:text=Visit%20TOI%20daily%20%26%20earn%20TimesPoints!&text=NEW%20DELHI%3A%20A%20CBI%20probe,out%2C%20thanks%20to%20greedy%20poachers, accessed on 14 September 2020.

[9] The MoEF (Project Tiger) set up the task force vide notification no. 6 (4)/2005-PT dated 19 April 2005.

[10] Project Tiger, *Joining the Dots*.

an amendment in 2006 where CTHs were to be declared as a targeted protection for tigers. As of 2019, there are 50 tiger reserves covering an area of 71,027.10 square kilometres.[11]

The environment ministry has also taken special policy measures for the protection of elephants by setting up an Elephant Task Force (ETF) in 1992 and subsequently in 2010.[12] The 1992 task force was part of 'Project Elephant', a centrally sponsored scheme providing for finances to protect elephants and their habitats. It was also to take measures to reduce human–elephant conflict. At present, there are 30 elephant reserves in India.[13] However, these are located within existing PAs. The implementation of the recommendations of the ETF is through the provisions of the WLPA. Unlike for tiger reserves, there is no special legal category for elephant reserves that has been included in the national law. In August 2019, the Government of Chhattisgarh proposed 1,995.48 hectares as Lemru Elephant Reserve to conserve elephant habitat and minimise human–elephant conflict.[14] The surrounding forests are disturbed due to the operations of some of Asia's largest coal mines. The final notification for Lemru is yet to be issued.[15]

---

[11] Available at the WII-ENVIS Centre on Wildlife & Protected Areas database http://wiienvis.nic.in/Database/trd_8222.aspx, accessed on 13 July 2020, supported by the MoEFCC.

[12] MoEF, *Gajah: Securing the Future for Elephants in India; the Report of the Elephant Task Force, Ministry of Environment and Forests* (New Delhi: MoEF, Government of India, 2010), i–ii.

[13] *Northeast Now*, 'Nagaland Govt Declares Singphan Wildlife Sanctuary as Singphan Elephant Reserve', 31 August 2018, available at https://nenow.in/north-east-news/singphan-wildlife-sanctuary-as-singphan-elephant-reserve.html, accessed on 14 September 2020.

[14] Amitabh Choubey, 'Lemru Elephant Reserve Announced by Chhattisgarh Cabinet, Natural Habitat for Wildlife for Protection and Promotion of Elephants', *Divyadoot*, 2019, available at http://www.divyadoot.com/home/newsdetail/573, accessed on 28 August 2019.

[15] Ritesh Mishra, 'Chhattisgarh Govt Considering Extending Area of Proposed Lemru Elephant Reserve', *Hindustan Times*, 2020, available at https://www.hindustantimes.com/india-news/chhattisgarh-govt-considering-extending-area-of-proposed-lemru-elephant-reserve/story-TDYYe1FgSFfANsjHUu8MYN.html, accessed on 24 September 2020.

PREVENTION OF ACTIVITIES HARMFUL TO WILDLIFE

The legal framework of the WLPA aims to prevent activities harmful to wildlife by limiting human presence and activities within PAs as well as specifying other measures outside the PA system. Specialised institutions set up under this law such as the national and state advisory boards for wildlife are given the responsibility to monitor the adherence with the legal requirements and to suggest special policy measures for the same. The NBWL and its standing committee is constituted under Subsection (1) of Section 5A of the Wild Life (Protection) Act, 1972 (53 of 1972), read with Sub-rule (4) of Rule 3 of the National Board for Wild Life Rules, 2003.[16] The State Board for Wild Life, chaired by the state minister in-charge of forests, is constituted by each state government drawing upon Section 6 of the WLPA, 1972. These institutions are explained in Chapter 2.

We can understand the legal measures for the prevention of harmful activities or damage to wildlife through the following aspects.

- Prevention of hunting: This law defines hunting as any attempt of 'killing or poisoning' or 'capturing, coursing, snaring, trapping, driving or baiting' of any 'wild or captive animal'. It also includes any attempt of 'injuring or destroying or taking any part of the body of any such animal'. In case of wild birds or reptiles, damaging or disturbing their nests or eggs is also considered hunting. Chapter III of the law highlights that no person can hunt any flora and fauna species listed in the First, Second, Third and Fourth Schedules of the Act.

  There are instances where hunting can be allowed with prior permit or where it will not be considered as an offence.[17] If the CWLW is satisfied that any wild animal in the First Schedule has become dangerous to humans or is beyond recovery from disability or disease, he/she can permit its hunting. This includes flagship species such as tigers, leopards, rhinos, elephants, Indian bison and rare ones such as the Himalayan ibex or dugongs. It is also allowed in cases where prior permission is sought for any activity covered by the definition of hunting for the purposes of education, scientific research or scientific management. The law also makes an exception for the killing or wounding of a wild animal in self-defence or to protect another person. But this exception will not apply if this act of

---

[16] These rules were enacted through notification no. S.O. 1092(E) dated 26 September 2003.

[17] Sections 11 and 12 of Chapter III of the Wild Life Protection Act (WLPA), 1972.

defence is in contravention of any other provision of the Act, such as if it involves illegal entry into a PA.

In 2018, a public controversy erupted in a case related to the killing of tigress Avni in Yavatmal district of Maharashtra. The permit to hunt was granted by the CWLW on the grounds that the tigress had become a man-eater and, therefore, a threat. There was public outrage on the killing of the tigress, and it generated debates on the difference between animal rights and wildlife conservation and the clauses of the law that allow government to permit hunting of animals. The case also highlighted that the tigress was killed in a night-time operation and that the standard operating procedures were flouted. Conservationists also raised broader challenges of conserving wildlife in India in areas that are not legally protected, as the tigress was not within a PA.[18]

• Restriction on entry and removal of any wildlife: There are several ways in which the law restricts entry into areas protected for wildlife and the removal of any plant, animal, insect and other species listed in schedules appended with the law.

There are clear restrictions of entry and exit in NPs and sanctuaries. Entry into an NP is allowed only with prior permission of the park management authorities. In sanctuaries, entry is permitted with prior permission of the CWLW. A public servant on duty can enter the sanctuary without such permission. Further, whoever has a right over immovable property within a sanctuary, including local communities whose other rights may have been extinguished or acquired, would have no restrictions. Passing through a sanctuary on a public highway is also allowed.

Habitation within the sanctuary and related activities are regulated. People living within such areas are also expected by law to assist the government to protect the wildlife as well as report any wildlife killings or damage.

In both these types of PAs, there are restrictions on the removal of or destruction of wildlife and damage of the wildlife habitat. The CWLW can grant permissions for the removal or damage if it is 'necessary for the improvement and better management of wildlife'. This legal requirement has come into serious questioning after large parts of PAs have been diverted for construction of highways or other energy and infrastructure

---

[18] Neha Sinha, 'To Kill a Tigress', *Economic and Political Weekly* 53, no. 48 (2018): 12–15.

projects. This is discussed in detail in the section on current conservation challenges later in this chapter.

• Regulating wildlife trade: The list of wild animals is specified in the first to fifth schedules of the Act. The law restricts the possession, custody and control of wild animals or sale, gift or any other form of transfer. Any person can access or retain custody of wild animals only with the permission of the CWLW. The WLPA considers wild animals hunted or traded as a property of the state government.[19] If any individual had any wild animals in their possession prior to the enactment of this law, they have to be declared before the CWLW.[20] Any trade in animal, 'trophy and animal articles' without a licence, is prohibited.

There are international regimes that regulate wildlife trade across borders. One of these is the Convention on International Trade in Endangered Species (CITES, discussed in Chapter 1) of Wild Fauna and Flora. It was adopted in 1963 at a meeting of members of the International Union for Conservation of Nature (IUCN) (the World Conservation Union).[21] The WLPA provides for restricting and regulating trade in wild animals within India. The definition of 'wild animals' for the purposes of trade includes animal articles, trophies or meat. This definition also includes vehicles, vessels, weapons, traps or tools that have been used in illegal trade, which can be seized in instances of an offence or illegality. The WLPA is not as specific about trade in wild plants. Its access for trade is regulated only through the legal provisions that prevent damage to wildlife or restrict access without prior permissions.

Wildlife trade has been an acute problem in India and internationally. The website of the World Wide Fund for Nature-India states that much of the trade in materials such as rhino horn, tiger and leopard claws, turtle shells, bear bile and medicinal plants is for markets outside India.[22] In 2016, the Wildlife Protection Society of India recorded over 20,000 wildlife crime cases

---

[19] Chapters III and V of the WLPA, 1972.

[20] A detailed discussion on enforcement procedures related to wildlife trade can be found in Samir Sinha, *Handbook on Wildlife Law Enforcement in India* (New Delhi: TRAFFIC-India, 2010).

[21] Details are available at https://www.cites.org, accessed on 1 July 2020.

[22] Available at https://www.wwfindia.org/about_wwf/enablers/traffic/illegal_wildlife_ trade_in_india/, accessed on 19 June 2019.

related to poaching and illegal trade in the country. Their report also revealed 781 kilograms of ivory and 69 tusks seized from illegal traders and poachers.[23]

II   LEGAL FRAMEWORK FOR BIODIVERSITY CONSERVATION

The BDA, 2002, was enacted under India's commitment to the United Nations' CBD with three objectives: conservation of biological resources, sustainable use of these resources and ensuring equitable sharing of benefits arising out of this use. The discussions at the CBD concluded that biodiversity conservation is a 'common concern of humankind', and more importantly states had sovereign rights over biological resources. This law granted ownership of biological material to national governments and required them to put in place mechanisms to regulate its access and use.

DEFINING BIODIVERSITY AND ITS USE

There are three definitions that are important to navigate the framework of the BDA. Biological resources have already been defined in the introductory section of this chapter. The other two definitions are about the kinds of access that the BDA regulates. These are biosurvey, bioultilisation and commercial utilisation.

'Biosurvey and bioutilization' means survey or collection of species, subspecies, genes, components and extracts of biological resource for any purpose and includes characterization, inventorisation and bioassay.

'commercial utilization' means end uses of biological resources for commercial utilization such as drugs, industrial enzymes, food flavours, fragrance, cosmetics, emulsifiers, oleoresins, colours, extracts and genes used for improving crops and livestock through genetic intervention, but does not include conventional breeding or traditional practices in use in any agriculture, horticulture, poultry, dairy farming, animal husbandry or bee keeping.

Any access to material resources or knowledge without prior permissions is considered theft or biopiracy and, therefore, punishable under law. There are four kinds of access that are exempt from seeking prior approval under the BDA:

---

[23] Pandey Kiran, 'Over 20,000 Wildlife Crimes Recorded in India', *Down to Earth*, 3 March 2016.

- All local people and communities of the area, including growers and cultivators of biodiversity: It includes indigenous medicine practitioners such as *vaid*s and *hakim*s.

- Access for bioresources exported as commodities: The list of species was notified in 2009[24] and is regularly updated by the National Biodiversity Authority (NBA). The 2009 list included 190 species including vanilla, tamarind, lemon and eggplant. The NBA has clarified that these species will be exempt from prior permissions only if they are being traded as commodities. If these species or their germplasm are being accessed for research or commercial purposes, that would require approval under the BDA.

- Collaborative research projects including those involving an Indian and a foreign entity: This legislation comes into play only if Intellectual Property Rights (IPR) are involved in such collaborative research projects.[25]

- Any access being made under the Protection of Plant Variety and Farmers Right Act, 2001, for Plant Breeder Rights.

THREE-TIER FRAMEWORK FOR PERMISSIONS

The first versions of the BDA primarily intended to put in place a legal regime to check biopiracy of the country's genetic wealth and associated knowledge. While critics saw it as a mechanism to facilitate access to bioresources, the drafters of the law created a mechanism where both domestic and foreign access to biological resources and related knowledge are allowed only after due permissions.

The law prescribes a procedure for any individual or institution seeking to access biological resources, whether wild, domesticated or in government or scientific gene collections. The procedures are further elaborated through the 2004 Rules. For any kind of access, there is a prescribed form and fee that needs to be deposited. The law also has clauses related to conservation and knowledge protection. The environment ministry is the nodal ministry for the law.

There is a three-tier institutional hierarchy through which this regulatory regime is implemented. These are the NBA, state biodiversity boards (SBBs) at every state and biodiversity management committees (BMCs) at panchayat/ municipality levels. Table 6.1 presents the functions and roles of the three bodies under this law.

---

[24] MoEF notification of 26 October 2009 (F. No. 28-13/2008-CS-III) and clarification on 16 February 2010 by Jairam Ramesh, Minister, MoEF.

[25] MoEF notification dated 8 November 2006 (F. No. 26/4/2006-CSC).

Table 6.1 Institutional Framework of the Biological Diversity Act

Institution	Powers and responsibilities			
	Regulation of access	Conservation	Documentation	Benefit sharing
NBA	Permission to be sought by foreign nationals before accessing biodiversity or people's knowledge or seeking Intellectual Property Rights (IPR) on the same. Indian nationals to seek permission for IPR	Advise the central government on matters relating to the conservation of biodiversity and sustainable use and protection of people's knowledge Advise state governments in selection of areas of biodiversity importance as biodiversity heritage sites	NBA's expert committee has designed the format based on which documentation is to be carried out by BMCs	Ensure and determine equitable sharing of benefits arising out of the use of access between the persons applying for approval, the local bodies concerned and the 'benefit claimers'
SBB	To be intimated by Indian nationals before accessing biodiversity and/or traditional knowledge, can suggest prohibitions and conditions and charge access fee	Powers to restrict activities in the state, which are likely to be detrimental to biodiversity	Guide the BMCs to document information related to biodiversity and traditional knowledge in people's biodiversity registers (PBRs), with the help of a technical support group	Ensure benefit sharing following consultation with BMC

(Contd)

(*Contd*)

| Institution | Powers and responsibilities | | | |
	Regulation of access	Conservation	Documentation	Benefit sharing
BMC	To be consulted by the NBA and SBB prior to grant of permission or intimation	Broad conservation role in the Act, not further defined	Document resources and knowledge with the help of thematic support group and the guidance of SBB using the format prepared by NBA expert committee	Members of the BMC may be benefit claimers, though no specific role prescribed in determining benefit sharing

Source: K. Kohli, M. Fareedi and S. Bhutani, *Six Years of the Biological Diversity Act* (New Delhi: Kalpavriksh and GRAIN, 2009), 9.

PEOPLE'S BIODIVERSITY REGISTERS

The BDA and corresponding 2004 Rules require all BMCs to prepare people's biodiversity registers (PBRs). The history of this process goes back to several community-based efforts to document all the resources they access and the knowledge that has been developed over the years. This is necessary to understand the biological wealth around villages and towns and make specific efforts to conserve them either through existing practices or by devising new ones.

The BDA uses PBRs not only to prioritise local conservation but also to prove as 'prior art' (a previous claim) against any IPR claim that may be made on an existing resource or associated knowledge. The World Intellectual Property Organisation recognises 'publication in any document or used in India or elsewhere in the world before the filing date (priority date)'[26] as a proof of claims that exist prior to a patent application being filed. Therefore, PBRs are seen necessary if national governments are to

---

[26] WIPO (World Intellectual Property Organisation), *Certain Aspects of National/ Regional Patent Laws: Status as on March 2019* (Geneva: WIPO, 2019), available at https://www.wipo.int/export/sites/www/scp/en/national_laws/prior_art.pdf, accessed on 3 August 2019.

secure their ownership of biodiversity from patents and other IPRs. The controversies around the patenting of basmati rice, turmeric or neem[27] reflect the importance of prior art.

The preparation of PBRs is essentially an exercise in documentation. Scientists appointed by the NBA provide a database format which is filled by BMCs or with the assistance of colleges or local civil society organisations. This has raised at least three following problems since the time the requirement was incorporated in the law:

- The documentation exercise for PBRs does not legally require prior informed consent of the knowledge holders.

- The BDA does not provide mechanisms to legally protect the knowledge collected through PBRs. Security and leakage of documented knowledge could encourage biopiracy.

- The quality of databases is varied given the diverse methodologies adopted in documentation and no legal requirement for validation. So, it is unclear if they can serve as evidence to defend against IPRs.

Today, printed PBRs and online digitised databases like PeBInfo and Indian Biodiversity Information System that are maintained by the NBA and the SBBs are used to put in place benefit sharing agreements once the access is approved by either of these bodies.

Legal Challenges

The BDA has also attracted several legal challenges. In this section, we look at two such cases in detail.

Is Coal a Bioresource?

On 11 January 2013, the Madhya Pradesh State Biodiversity Board (MPSBB) served notices to the three subsidiaries of Coal India Limited—South Eastern Coalfields Limited, Western Coalfields Limited (WCL) and Northern Coalfields Limited (NCL) suggesting that coal is a 'bioresource' and, therefore, its extraction requires prior intimation to the board and the payment of an

---

[27] Emily Marden, 'The Neem Tree Patent: International Conflict over the Commodification of Life', *Boston College International & Comparative Law Review* 22, no. 2 (1999): 279, available at http://lawdigitalcommons.bc.edu/iclr/vol22/iss2/3, accessed on 22 June 2019.

access fee as per the BDA. Two months later, NCL responded to the notice by stating that definition of bioresources does not include coal and, therefore, there is no liability to pay. The MPSBB's position was that coal is a fossil fuel and, therefore, a plant byproduct, attracting the provisions of the BDA.[28]

While this resolution was pending, the local BMC of the village Eklahara in Madhya Pradesh filed a case before the Bhopal Bench of the National Green Tribunal (NGT)[29] as a 'benefit-claimer'. They stated that WCL did not adhere to the BDA, and that the coal mining company had denied the villagers their share of benefits. The MPSBB supported the contention of the BMC; the environment ministry and the NBA disagreed. They took the position that neither the CBD nor the BDA had ever been conceived to regulate fossil fuels and that the law was for access to genetic material and people's knowledge on bioresources. This case created a debate on whether coal should be treated as a bioresource or not. Mainstream newspapers and researchers also discussed whether the levying of an access fee on coal would be the same as a 'tax'.[30]

The MPSBB argued,

> Coal is a fossil fuel because it is formed from the remains of vegetation that grew as far as back 400 millions ago and that it is often referred to as 'buried sun shine' because the plant which formed coal captured energy from the sun through photosynthesis created the compounds that makes up plant tissues. By that logic, coal being of plant origin has to be treated as a biological resource.

---

[28] Shalini Bhutani and Kanchi Kohli, *Litigating India's Biological Diversity Act: A Study of Legal Cases* (New Delhi: Kalpavriksh and Foundation for Ecological Security, 2016), 19–20.

[29] National Green Tribunal judgment dated 16 October 2015 in Original Application No. 28 of 2013(CZ) (*Bio Diversity Management Committee v. Western Coalfields Ltd. & Ors*) and Original Application No. 17 of 2014(CZ) (*Bio Diversity Management Committee v. Union of India & Ors*).

[30] Times News Network, 'Ministry of Environment and Forests Says Coal Is Mineral, Green Tribunal Slaps Notice', *Times of India*, 12 August 2014, available at https://timesofindia.indiatimes.com/city/bhopal/Ministry-of-environment-and-forests-says-coal-is-mineral-green-tribunal-slaps-notice/articleshow/40079739. cms, accessed on 14 September 2020; K. Kohli and S. Bhutani, 'The Legal Meaning of Biodiversity', *Economic and Political Weekly* 48, no. 33 (2013): 15–17; Shashikant Trivedi, 'Is Coal a Mineral or Bio-resource?' *Business Standard*, 12 September 2013, available at https://www.business-standard.com/article/economy-policy/is-coal-a-mineral-or-bio-resource-113091100946_1.html, accessed on 14 September 2020.

The environment ministry and the NBA responded that coal is not covered in the definition of biological resources. Moreover,

> coal is combustible, sedimentary and organic rock and, therefore, cannot be compared to a living organism. It takes approximately 300 million years to form coal and thus it is fossil and by no stretch of imagination coal can be categorised as biological resource as defined in the Act.

The NGT's judgment engages with both these contentions in detail and concluded in favour of the environment ministry and the NBA. The judgment relied on a letter from the ministry to the MPSBB, which said, 'It is hereby clarified that on the issue of whether coal is a "biological resource" or not, the NBA and the MoEF&CC have unequivocally concluded that coal is not "biological resource" under Section 2(b) of the Biodiversity Rules, 2004.'

However, the discussion on the definition on bioresources is still wide open for interpretation, as the present case decided only against the inclusion of coal into this law.

Who Can Regulate Access and Benefit Sharing?

The access question has been a bone of contention between the environment ministry, the NBA and SBBs, especially after the CBD's Nagoya Protocol on Access and Benefit Sharing (ABS) was formalised[31] (more in Section IV). The protocol requires countries to put in place a robust legal regime for regulating ABS. Several SBBs sought legal clarity on their role in determining access fees and benefit sharing safeguards for Indian companies. It was only through litigation in the NGT[32] that the NBA put in place ABS Guidelines in 2014.

The power of the SBBs with respect to determining the terms of ABS has also been contested. While SBBs argue that they are legally empowered to set the terms for access by Indian entities, user/accessor Indian companies

---

[31] The Nagoya Protocol on access to genetic resources and the fair and equitable sharing of benefits arising from their utilisation to the CBD is an international agreement. It was adopted by the Conference of the Parties to the Convention on Biological Diversity at its 10th meeting on 29 October 2010 in Nagoya, Japan.

[32] National Green Tribunal order dated 1 August 2014 in Original Application No. 62 of 2013 Central Zone (*Som Distilleries Pvt. Ltd. V. MP State Biodiversity Board & Ors*) and another 13 related original application numbers before the Bhopal Bench of the NGT.

have challenged the same saying that the BDA only requires them to inform or intimate the SBBs.

This aspect was clarified in a 2018 Uttarakhand High Court judgment,[33] which related to notices sent to an Indian company for accessing bioresources in violation of the BDA. The company is very popular in India for its ayurvedic products made from bioresources sourced by its units based in the state of Uttarakhand. It had been issued notices by the Uttarakhand Biodiversity Board (UBB) to pay fee to access biological resources. In turn, the company argued that the payment should not be charged to them as they were an Indian company:

> UBB cannot raise a demand under the Head of 'Fair and Equitable Benefit Sharing' (FEBS) as the Board neither has powers nor the jurisdiction to do that and, secondly, the petitioner in any case is not liable to pay any amount or make any kind of contribution under the head of 'FEBS'

The UBB's response was as follows:

> in the concept of FEBS, no distinction is made between a foreign entity and an the Indian entity, and the only distinction which the Act makes within Indian entities is in proviso to Section 7 of the Act of 2002 where an exception has been created for local people and communities in that area, including growers and cultivators of biodiversity, and vaids and hakims, who have been practicing indigenous medicine.

The judgment concluded that the SBB's core function is to regulate the use of biological resources:

> In view of the above, this Court is of the opinion that SBB has got powers to demand Fair and Equitable Benefit Sharing from the petitioner, in view of its statutory function given under Section 7 read with Section 23 of the Act and the NBA has got powers to frame necessary regulations in view of Section 21 of the Act. The challenge of the petitioner to the validity of the Regulations fails.

---

[33] High Court of Uttarakhand judgment dated 21 December 2018 in Writ Petition (M/S) No. 3437 of 2016 (*Divya Pharmacy v. Union of India*).

This function includes asking for benefit sharing and determining the terms and conditions to be imposed on the user/accessor for granting access to genetic resources and knowledge. The court did not give any observations on the amount that needs to be paid and if any additional mechanisms need to be determined for the same.

Delays in Constituting BMCs and Preparing PBRs

In 2016, a case was filed in the NGT[34] seeking the tribunal's intervention to expedite the setting up of BMCs and preparing PBRs. The petitioner argued that there had been non-compliance with specific provisions of the BDA several years after it was enacted.

In its order dated 12 April 2019, the NGT directed immediate compliance with these provisions and report back by the state governments within three months, with clear explanations. As the matter continued to be heard, the tribunal observed,

> Having regard to the laudable objective to meet the necessity of conservation of biological diversity, delay of more than 16 years in complying with the mandate of law is a matter of serious concern.

In its order dated 9 August 2019, the NGT gave all state governments time until 31 January 2020 to comply with these orders, barring which

> [t]he States will be accountable for the defaults and required to deposit a sum of Rs. 10 lakhs per month each from 01 February 2020 with the CPCB to be utilized for restoration of the environment. The States will be at liberty to recover the said amount from the persons committing the default.

The final judgment in this case was pending as of March 2020.

III   OVERLAPS BETWEEN WILD LIFE PROTECTION ACT AND BIOLOGICAL DIVERSITY ACT

There are several instances where the WLPA and BDA jurisdictions overlap. These include:

---

[34] National Green Tribunal Original Application No. 347 of 2016 (*Chandra Bhal Singh v. Union of India & Ors*).

- Instances when approvals are required for hunting and access to bioresources in PAs.

- Enforcement coordination in cases of theft from PAs or any other activity attracting the definition of biopiracy.

- In designating conservation management areas, while WLPA lists different kinds of PAs, the BDA also provides for the declaration of biodiversity heritage sites in areas considered important for biodiversity. These areas could overlap with existing PAs or core zones of CTHs.

- Accessing biological resources from zoos registered with the Central Zoo Authority[35] constituted under the WLPA. In several cases, the NBA or SBBs receive applications for accessing biological material from animals in zoos. These species are also 'wild animals' protected under the WLPA.[36]

These overlaps can be understood through a 2008 case that convicted two Czech scientists for illegal entry and collecting rare insects in Singhalila National Park, West Bengal. In July 2008, they were arrested by the State Forest Department officials for 'illegal entry into the park and accessing beetles, butterflies and other insects from the park without permission'.[37] Subsequently, additional grounds related to the violation of Section 3 of BDA, 2002 were added as they were in possession of over 1,500 species of butterflies without the permission of the NBA. The scientists claimed their collection was for the purposes of research; however, the case proceedings revealed evidence of commercial sale. Irrespective of the purpose of access, they were charged to be in violation of both the WLPA and BDA. Both scientists were subsequently fined and sentenced to three years imprisonment.

According to available literature, the Chief Judicial Magistrate in Darjeeling convicted both scientists on 8 September 2008 under the provisions of the WLPA, 1972.[38] While one was fined ₹20,000, the

---

[35] Website of the Central Zoo Authority, available at http://cza.nic.in/page/en/introduction, accessed on 3 August 2019.

[36] Kohli, Fareedi and Bhutani, *Six Years of the Biological Diversity Act*.

[37] C.R. Case 48 of 2008 before the Darjeeling chief judicial magistrate.

[38] Priyadarshini Subhra, 'Entomologists Convicted', *Nature India*, 8 September 2008, available at http://www.natureasia.com/en/nindia/article/10.1038/nindia.2008.277, accessed on 15 July 2016.

other was sentenced to three years of imprisonment and fined ₹60,000.[39] According to a 2009 paper in a journal, *Conservation Biology*, the court took into account the international reputation of one of the scientist for the differential sentencing. The second scientist was granted bail and ordered to stay in India till the matter was heard before the appellate court. He, however, reportedly fled the country.[40]

This case also prompted the issuance of a January 2009 notification by the Environment Ministry where forest officials of the level of Range Officers and above can book offences and take action under Section 61 of the BDA. This also institutionalised the mechanism for collaborative enforcement in cases that involved the violation of both these laws. Range officers are also empowered to take action under the Indian Forest Act, 1927 and specific state forest laws as discussed in Chapter 3.

In addition, there are various institutional overlaps between the WLPA and the BDA. The two laws can also be understood through a comparative lens. Table 6.2 below helps to understand the primary features of these two conservation laws.

Table 6.2 Biological Diversity Act and the Wild Life Protection Act

Biological Diversity Act (BDA)	Wild Life Protection Act (WLPA)
Access to PAs or to wildlife for research	
The BDA requires prior permissions of the NBA or SBBs for accessing biological resources for biosurvey, bioutilisation or commercial utilisation. Research that leads to commercialisation also requires prior permissions	The WLPA requires prior permission of the CWLW for research. No permission is given to foreigners/non-resident Indians unless the project is sanctioned by the Ministry of External Affairs, the Ministry of Home Affairs and the NBA. There are also overall restrictions of entry into PAs for any purpose including research

(*Contd*)

---

[39] Times New Network, 'Forest Officials Elated at Czechs' Conviction', *Times of India*, 12 September 2008, available at https://www.downtoearth.org.in/news/wildlife-and-biodiversity/madhya-pradesh-government-gives-go-ahead-to-ken-betwa-river-link-51218, accessed on 14 September 2020.

[40] David Kothamasi and Toby E. Kiers, 'Emerging Conflicts between Biodiversity Conservation Laws and Scientific Research: The Case of the Czech Entomologists in India', *Conservation Biology* 23, no. 5 (2009): 1228–1230.

(*Contd*)

Biological Diversity Act (BDA)	Wild Life Protection Act (WLPA)
Trade in forest produce	
The BDA requires prior permissions of the NBA or SBBs for accessing biological resources. Local people and communities of the area, including growers and cultivators of biodiversity, and *vaids* and *hakims* are exempt	Rights to access forest produce are regulated under the WLPA even for local people and communities living within and outside PAs. Prior permissions of the CWLW (NPs and sanctuaries) or management committees (community reserves and conservation reserves) are necessary
Conservation areas	
Declaration of biodiversity heritage sites by the state government	Declaration of NPs, sanctuaries, community reserves, conservation reserves and CTHs by the state government or central government

Source: Authors.

IV ISSUES ARISING OUT OF LEGAL FRAMEWORKS FOR CONSERVATION

The implementation of WLPA has faced long and challenging disputes over issues such as the settlement of rights of communities living within habitats that are identified as important for wildlife. The implementation of the law has created a conflict between conservation and local livelihoods in the contexts where one has to be prioritised. The legal obligations to protect wildlife have also been affected by land grabs, 'encroachments' and other project development and infrastructure expansion like mining and road building. Such problems have affected forest-dwelling communities and wildlife populations in profound ways.

The experience with the legal implementation of the BDA is limited and evolving. Conservation objectives are not yet upfront, and the executive push has been towards streamlining access to bioresources and encouraging good practices for benefit sharing. Globally and in India, these discourses have led to executive level negotiations and litigation.

In this section, we discuss four thematic issues that have resulted from the legal frameworks of the WLPA and the BDA. While these have all been exercised in courts, they remain major challenges that need legal and administrative redress.

Relocation

As discussed earlier, the declaration of NPs and sanctuaries triggers the process where existing rights of residents inside these areas may need to be extinguished and acquired under the WLPA. The affected communities need to be compensated and relocated. The administrative practice of relocation has been one of the most contentious issues for wildlife conservation since the time the WLPA was first enacted.[41] The issue has been discussed in special task forces, debated in parliament and continues to be heard by the Supreme Court since 1995. During the ninth five-year plan (1997–2002), the government approved ₹190 million for the purposes of carrying out the relocation process from NPs, sanctuaries and tiger reserves. The 2005 report of the TTF observed that 'between 1997–1998 to 2001–2002, a total of ₹143.9 million was spent to relocate 2,157 families from different protected areas—approximately ₹67,000 per family'. The Task Force report also highlighted that relocated communities have also returned to their villages. It states,

> In Sariska, villagers of Kraska village were offered land by the forest department in a village outside the reserve's core area. They relinquished their landownership certificates and shifted to that village, only to face the wrath of its residents. Selling off the new land they had got at low prices, the villagers went back into the core. Now they live in an atmosphere that is war-like: harassed, forcibly evicted again....

There are two legal processes through which the relocation issue can be understood.

- The *Centre for Environmental Law (CEL), WWF v. Union of India and Ors* (Writ Petition No. 337 of 1995) related to the issue of settlement

---

[41] Mahesh Rangarajan and Ghazala Shahabuddin, 'Displacement and Relocation from Protected Areas: Towards a Biological and Historical Synthesis', *Conservation and Society* 4, no. 3 (2006): 359–378; CFR-LA (Community Forest Rights—Learning and Advocacy), *Promise and Performance: Ten Years of the Forest Rights Act in India*, citizens' report on promise and performance of the Scheduled Tribes and Other Traditional Forest Dwellers (Recognition of Forest Rights) Act, 2006, after 10 years of its enactment, December 2016 (produced as part of CFR-LA, India, 2016), available at http://www.cfrla.org.in/uploads_acrvr/X36BEPromise%20and%20Performance%20National%20Report.pdf, accessed on 14 September 2020.

of Rights in NPs and sanctuaries and other issues under the Wild Life (Protection) Act, 1972.

- Protocol/Guidelines for Voluntary Village Relocation in Notified Core/ CTHs of Tiger Reserves declared as part of the Project Tiger.

Even though the WLPA was enacted in 1972, there were delays in the final declarations of PAs and the settlement of rights processes. This led to a petition being filed in the Supreme Court by World Wildlife Fund India, a leading conservation organisation. The CEL case, also referred to as the 'settlement of rights case', was filed under Article 32 of the Constitution of India seeking directions to expedite these processes. In an interim order of 14 August 1997, the Court directed state governments to complete the process of determining and extinguishment of rights and acquisition of land within a period of one year. This case is still being heard in the Supreme Court on the issue of rights settlement as well as relocation.[42] The experience with relocation that was implemented following the Court's directions has been contentious. This is illustrated through the case of the Great Himalayan National Park, discussed below. The CEL case has also led to several important directions that have shaped conservation practices and the relocation process. For instance, in an order dated 9 May 2002, the apex court directed that no permission for destruction, exploitation or removal of any wildlife (including forest produce) from a sanctuary should be granted without approval of the Standing Committee of Indian Board for Wildlife (notified as NBWL in the 2003 by an amendment to the WLPA).[43]

The Great Himalayan National Park (GHNP) known for its exceptional biodiversity value was considered to have low human presence and, hence, ideal for a PA. The region hosts rare and endangered species such as the Western Tragopan, Himalayan Tahr, Blue Sheep and the musk deer. It was first established in 1984 and finally notified in 1999. The current total area of the NP is about 754.4 square kilometres. Vasant Saberwal and Ashwini Chhatre have highlighted the complex web of socio-ecological interactions in this park. Both settled and seasonal livelihood dependence were directly impacted with the final notification of GHNP. Their paper highlights how the

---

[42] Shyam Divan and Armin Rosencranz, *Environmental Law and Policy in India*, 18th edn (New Delhi: Oxford University Press, 2015), 49.

[43] Wild Life Protection Act, 1972, vide Act 16 of 2003.

'settlement of rights' in GHNP was based on colonial settlement records.[44] This left out several families whose rights of habitation, access and livelihoods were directly affected. The customary use of grazing meadows was dealt with an assurance that alternate areas would be assigned to pastoral communities grazing goat and sheep. The existing rights were only compensated through cash compensations. The park management subsequently introduced programmes such as eco-development, but these did not resolve all the contentious issues.[45]

The WLPA was amended in 2006, and the new section[46] provided for the declaration of Critical Tiger Habitats (CTHs) to promote 'inviolate' or undisturbed areas for tigers. These CTHs would cover core areas of existing Tiger Reserves along with any new areas for tiger conservation. In the same year, the Scheduled Tribes and Other Traditional Forest Dwellers (Recognition of Forest Rights) Act was also enacted. While the WLPA amendment looked to create inviolate areas for tiger conservation, the Forest Rights Act (FRA) was about asserting the rights of tribal and other forest-dwelling communities in forest areas.

Both these changes in the legal landscape once again pit the rights of forest dwellers and the efforts to depopulate conservation enclosures against each other. CTHs are to be scientifically identified and notified by state governments in NPs and sanctuaries and subsequently notified by state governments. However, relocation from these areas is to be voluntary and only with the consent of gram sabhas (village assemblies). The National Tiger Conservation Authority (NTCA) suggested a protocol/guideline in 2008 whereby relocation has to be on 'mutually agreed terms and conditions'. Relocation can take place only after recognition of rights as per the FRA, 2006 is complete.[47]

---

[44] Vasant K. Saberwal and Ashwini Chhatre, 'The Parvati and the Tragopan: Conservation and Development in the Great Himalayan National Park', *Himalaya, the Journal of the Association for Nepal and Himalayan Studies* 21, no. 2 (2001): 79–88, available at http://digitalcommons.macalester.edu/himalaya/vol21/iss2/13, accessed on 23 June 2019.

[45] Times News Network, 'Affected People of Great Himalayan National Park to Discuss Rights Today', *Times of India*, 13 April 2014.

[46] Sections 38V 4 (i)–4 (ii).

[47] Letter of the NTCA dated 8 September 2009 to Chief Wild Life Wardens of all states (File No.9-6/2006-NTCA (Part).

The centrally sponsored scheme on 'Integrated Development of Wildlife Habitats' and 'Project Tiger' offers two 'relocation packages' for communities who agree to voluntarily relocate. Option I is the payment of the entire package amount (₹1 million per family) to the family if they choose, without involving any rehabilitation/relocation by the forest department. Option II is carrying out relocation/rehabilitation of the village from the PA by the forest department.[48] The WLPA amendment also introduces the concept of buffer zone for CTHs. The geographical limits of these buffers would be determined by scientific criteria but not without consultation with the *gram sabha* (village assembly) and an expert committee constituted specifically for this purpose.

An interesting case was transferred from the Bhopal High Court to the NGT in 2014.[49] In this case, the applicants had complained of illegal felling of trees on forest land by the Madhya Pradesh Forest Department. They highlighted that 'in the year 1980, the State Government has spent huge amount of money and raised plantations over an extent of 1400 acres with different species of trees' which it had now felled. In response, the Environment Ministry and the state forest department stated that the felling of trees was not illegal. It was for the relocation and rehabilitation of villagers displaced from Bori WLS part of the Satpuda Tiger Reserve in Hoshangabad district.

According to the Forest Department, the villagers displaced from the forest villages inside the sanctuary were being shifted out of the PA to lands recorded as degraded protected forest after receiving necessary permissions. The petitioners argued that such relocation should have taken place on revenue land, and it was incorrect of the district collector to state that there was no revenue land available for such relocation in Hoshangabad district.

The Environment Ministry submitted,

Village relocation is in the interest of biodiversity conservation and is as per the provisions of the Wildlife (Protection) Act, 1972 and NTCA guidelines.

---

[48] MoEFCC response to Lok Sabha Starred Question No. 4 regarding the 'Relocation of People from Protected Areas' raised by Adv. Joice George and Shri C. Mahendran on 15 December 2017.

[49] National Green Tribunal judgment dated 14 April 2014 in Original Application No. 38 of 2014 Central Zone and Original Application No. 34 of 2014 Central Zone (*Salim Khan & Anr v. Union of India & Ors*).

The NGT concluded in their judgment dated 14 April 2014 that there were no merits in the application and due procedure was being followed for the relocation of the villagers as per the approved plans, guidelines and with permissions.

DIVERSION OF PROTECTED AREAS

The declaration of an area as 'protected' under the WLPA could give the perception that these areas are secure from any activity that could be damaging to wildlife. In recent years, the diversion of land from PAs has been an important issue taken up by conservationists. The regulatory framework of the WLPA does not put any PA out of bounds from economic activities. Each proposal for the use of NPs, sanctuaries and other conservation areas is evaluated on a case-to-case basis on whether it is 'destructive', 'exploitative' or 'damaging' to wildlife. Such activities are not allowed as per Section 35 and Section 29 of WLPA.

There are no legal standards for the extent of acceptable impacts to PAs. The decision-making on what activities to allow or prohibit in a PA are based on the state government's consultation with the NBWL in the case of National Parks, and the SBWL in the case of wildlife sanctuaries. The CWLW is authorised to grant a permit based on these processes. In case of Tiger Reserves, the prior approval of the NBWL on the advice of the NTCA is required. Clause 38 (O) (g) states that NTCA needs to 'ensure' that Tiger Reserves and areas that connect PAs, Tiger Reserves and PAs to Tiger Reserves are 'not diverted for ecologically unsustainable uses, except in public interest....' In recent years, there has been an increased demand for PA land, either in part or full for the purposes of infrastructure projects. Responding to this demand, the Environment Ministry issued a short guidance document[50] for taking up non-forestry activities in wildlife habitats which stated,

[P]roject proponents cannot rely upon the concept of fait accompli, if they have already received any of the clearances. The Environmental, Forest and NBWL clearances will all be processed on their respective merits, and the clearance of one aspect will not confer any right upon the project proponent.

---

[50] Ministry of Environment Forests and Climate Change, *Guidelines for Taking Non-forest Activity in Wildlife Habitats* (F. No. 6-10-2011 WL) dated 19 December 2012.

Between 1998 and 2008, the standing committees of the IBWL/NBWL considered 244 cases for diversion of land from PAs. An analysis of these cases reveals that diversion of 7,949 hectares of land was approved and 2,399 hectares of land was rejected. Also, 4,453 hectares of the approved areas was for de-notification of PAs for unspecified purposes. In all, 2,102.4 hectares of land was allowed for mining. Another 625 hectares of land was for projects such as transmission lines and wind projects. The smaller proportion includes 271 hectares for rehabilitation projects, 237.1 hectares for dams, 170 hectares for road projects and 90 hectares for building and construction projects.[51]

In the last decade, this trend has continued.[52] While there have been several such projects that required the use of wildlife areas, one that received national attention was the Ken–Betwa river interlinking project. The project proposes to transfer surplus water from Ken River in Uttar Pradesh to the Betwa River in Madhya Pradesh. This exercise would submerge 28 per cent of the core tiger habitat of Panna Tiger Reserve, when significant financial and human resources were spent to revive the tiger population to 40 in number after they had 'disappeared in 2009'.[53] While the project impact assessment reports state that 41.4 square kilometres of the Tiger Reserve will be impacted, independent studies have indicated that the submergence area would be 92 square kilometres.[54]

Forest officials, conservationists and rights-based activists have all questioned this project for the impacts it would have on the PA and the surrounding areas. The Madhya Pradesh State Wildlife Advisory Board approved the project in 2015, despite dissenting views from the non-official members.[55] Petitions have been filed before the Central Empowered

---

[51] M. Menon, K. Kohli and V. Samdariya, 'Diversion of Protected Areas: Role of the Wildlife Board', *Economic and Political Weekly* 45, nos. 26–27 (2010): 18–21.

[52] Jayashree Nandi, '680 Projects in Protected Areas Given Wildlife Clearance in 5 Years', *Hindustan Times*, 19 September 2020.

[53] Prerna Singh Bindra, 'India's Fast-Tracked Wildlife Clearances Threaten Last Wild Areas, Water Sources and Hasten Climate Change', *IndiaSpend*, 22 September 2018.

[54] Joanna Van Gruisen, 'Ken-Betwa Link Creating, Not Solving, Water Woes', *Current Conservation* 11, no. 1 (March 2017): 4–10.

[55] Rajeshwari Ganesan, 'Madhya Pradesh Government Gives Go-Ahead to Ken-Betwa River Link', *Down to Earth*, 24 September 2015, available at https://www.downtoearth.org.in/news/wildlife-and-biodiversity/madhya-pradesh-government-gives-go-ahead-to-ken-betwa-river-link-51218, accessed on 14 September 2020.

Committee of the SC and the NGT appealing against the approvals granted to this politically high profile project.[56]

Human–Wildlife Conflict

According to the IUCN World Parks Congress,[57] HWC is defined[58] as a situation when

> needs and requirements of wildlife impact negatively on the goals of humans or when the goals of humans negatively impact the needs of the wildlife.

Such conflict results in the loss of crops, livestock, property and human lives. Conflicts are also recorded when any wildlife is injured, abused or killed 'deliberately' or in self-defence or when wildlife is perceived as a threat to life or property. The fragmentation of habitats due to forest diversions and deforestation is the primary reason for such conflicts. Research indicates that in recent years, linear projects such as highways and commercial railway lines have been particularly damaging to habitats. Forest corridors used by wildlife to travel long distances have also degenerated due to a range of factors resulting in migratory wildlife, such as elephants shifting their routes towards villages and increasingly entering human habitation.[59]

As per a 2017 report of India's Comptroller and Auditor General, nearly 400 people are killed annually in elephant conflict-related incidents.[60] The report states that upto 500,000 families are affected annually by crop damage

---

[56] Singh Bindra, 'India's Fast-Tracked Wildlife Clearances Threaten Last Wild Areas'.

[57] The International Union for Conservation of Nature (IUCN) World Parks Congress (WPC) is a landmark global forum on PAs, available at https://www.iucn.org/theme/protected-areas/about/congresses/world-parks-congress, accessed on 3 August 2019.

[58] The WPC, WPC recommendations, Recommendation 20: Preventing and Mitigating Human–Wildlife Conflicts (8–17 September 2003).

[59] MoEF, *Gajah*.

[60] Comptroller and Auditor General, 'Human Wildlife Conflicts', ch. 4, in *Report 6 of 2017 Compliance Audit* (New Delhi: Comptroller and Auditor General, Government of India, 2017), 18, available at https://cag.gov.in/sites/default/files/audit_report_files/Chapter_4_Human_Wildlife_Conflicts_and_Wildlife_Corridors_of_Karnataka_Report_No_6_of_2017_on_Administration_of_National_Parks_and_Wildlife_Sanctuaries.pdf, accessed on 27 February 2019.

caused by elephants. The report also highlighted that in a bid to remove 'problem' animals, farmers kill elephants in retaliation. Some measures have been taken to tackle human–animal conflicts. The Environment Ministry has issued several circulars on preventive and conflict resolution measures to be taken. One such circular from 2015 suggests state governments to prepare HWC Management strategies for identified conflict zones. This includes training volunteers in management strategies.[61] India's Wildlife Action Plan (2017–2031)[62] suggests the setting up of a centre for excellence for mitigation of conflicts and deploying special conflict mitigation squads. It also looks to set up primary response teams including community representatives in consultation with local bodies.

Another measure taken to deal with this problem is with respect to compensation mechanisms. While each state government has designed its own state-level compensation mechanisms for cases of HWC, the central Environment Ministry also prescribes compensation amounts through centrally sponsored schemes as contribution to the cost incurred by state governments.[63] These also act as guidelines to state governments.

Despite these measures, this has continued to be an area of concern for conservation groups, as conflicts can undermine the goodwill for wildlife conservation among local communities who live in and near areas supporting wildlife populations. Members of Parliament have raised questions seeking explanations from the Environment Ministry on actions taken to address the increased instances of conflicts.[64]

In 2018, a case was filed in the Supreme Court[65] to address the pressing issue of elephant conservation and to formulate and implement measures to effectively deal with human–elephant conflicts. The Supreme Court sought the support of the Additional Solicitor General and specific details from state

---

[61] MoEFCC, *Solution to Hardships of Human Beings and Wild Animals* (New Delhi: Press Information Bureau, 2015), available at http://pib.nic.in/newsite/PrintRelease. aspx?relid=124461, accessed on 23 June 2019.

[62] MoEFCC, *India Wildlife Action Plan (2017–2031)* (New Delhi: Government of India, 2017).

[63] MoEFCC Circular [F.No.14-2/2011 W-I (Pt-II)] issued on 9 February 2018.

[64] The Lok Sabha Starred Question No. *62 (to be answered on 7 February 2017) on human–animal conflicts by Members of Parliament Joice George and P. K. Sreemathi.

[65] Supreme Court Writ Petition (Civil) No. 489 of 2018 (*Prerna Singh Bindra and Ors. v. Union of India and Ors.*).

governments where such conflicts had arisen. In its order dated 4 December 2018, the SC constituted a High-level Committee

> to look into the issues related to monitoring and implementation of various guidelines, decisions, directions and instructions issued by the Ministry of Environment, Forest and Climate Change and compliance of directions given by Hon'ble Courts from time to time.

The SC continued to monitor this issue as of January 2020.

ACCESS TO BIORESOURCES AND BENEFIT SHARING

One challenge that has dominated the legal landscape of the BDA in recent years is that of ABS. Neither the CBD nor India's domestic law refers to the terms access and benefit sharing together. Given the increased demand for bioresources for commercial and intellectual property applications, there was a push to develop protocols that combine objectives two and three of the convention: *sustainable use* and *equitable sharing for benefits arising out of the use*. With differential national systems in place, international pharmaceutical, seed and nano-technology based industries pushed for a legal system based on a common protocol to be formalised. At the same time, domestic companies have resisted from seeking permissions for access and bearing the costs of benefit sharing.[66] This has also been discussed earlier in the section on legal framework where the 2018 judgment of the High Court of Uttarakhand has been discussed in detail.

For the purposes of this section, it is important to understand how the law defines 'benefit claimers', with whom ABS agreements need to be entered into:

> 'benefit claimers' means the conservers of biological resources, their byproducts, creators and holders of knowledge and information relating to the use of such biological resources, innovations and practices associated with such use and application.

A typical example of how ABS is being implemented for domestic companies can be understood through the agreement signed between the UBB and an

---

[66] K. Kohli and S. Bhutani, 'The Legal Meaning of Biodiversity', *Economic and Political Weekly* 48, no. 33 (2013): 15–17.

Indian cosmetics company on 15 April 2015. The company had accessed 39 bioresources for cosmetics in 2013–14 without any intimation to the SBB. They subsequently received notices from the Board in 2015–16 requiring them to comply with the BDA *post facto*. The company claimed that they had bought the bioresources from wholesaler dealers. Therefore, a particular BMC or benefit claimer cannot be identified. Finally, an agreement was signed between the SBB and the company listing eight conditions for access along with a payment of ₹322,991. This figure was arrived at by applying the 2014 ABS guidelines. The company disclosed in the agreement that it made a profit of ₹64,598,179 from one year of access. The conditions emphasised that the agreement was for only one year of access and should not be used for anything else than the stated purpose. The agreement also required that the source of access should not be impacted and the names and photographs of the extractors be shared with the concerned BMC. The monitoring of the access would be carried out by the BMC, and the company would work with the BMC on trainings for sustainable extraction.

This example shows the four areas of conflict applicable to both domestic and international access and requirements of benefit sharing.

- *Establishing ownership and jurisdiction* is one of the most difficult areas in the biodiversity law. A bioresource located in one village may also be found in an adjoining location. In such cases, should the access be with one particular source individual or village from where the material has been collected or from all others who also use the resource, and may be impacted by its access. This becomes even more complex when ownership needs to be attributed to the knowledge associated with the resource. For example, traditional systems of forest produce extraction or knowledge or medicinal properties associated with one plant may lie with more than one group of individuals or villages.

- *Consent* remains a contentious issue especially since the 2010 Nagoya Protocol under the CBD requires countries to put in place systems of *Free Prior and Informed Consent* of custodians and holders of bioresources and associated knowledge. India's BDA only has the requirement of 'consultation' with BMCs, which as discussed in the above example, may not always take place. There are also examples to show that no consultation with BMCs has taken place, prior to approval.[67]

---

[67] K. Kohli and S. Bhutani 'Biodiversity Management Committees: Lost in Numbers', *Economic and Political Weekly* 49, no. 16) (2014): 18–20.

- *Enforcement and compliance* related issues are twofold in the BDA. First, it is difficult for the NBA or the SBB to ensure that the accessed germplasm is being used only for the stated purposes. This is especially difficult when accessed material is taken into a laboratory outside of the country. The second relates to ensuring that all instances of access are reported and benefit-sharing agreements arrived at. As of June 2019, the NBA received only 2,696 applications out of which 904 were approved and agreements signed.[68] There is no record on access made without seeking prior permissions from the NBA or intimation to the SBBs.

V   WILDLIFE ACTION PLANS

Wildlife conservation priorities in India have been driven by long-term and short-term priorities set out in the National Wildlife Action Plans (NWAPs). The Indian Board for Wildlife approved the first such plan for 1983–2001. It drew its emphasis from the World Parks Congress's Bali Action Plan in 1982 and the World Conservation Strategy launched in India in March 1980.[69] Since then, the environment ministry has adopted two more action plans. Table 6.3 presents a comparison of two of India's Wildlife Action Plans (WAPs) on five themes.

Table 6.3 Thematic comparison of two Wildlife Action Plans

	NWAP, 2002–16	NWAP, 2017–31
Landscape-level Approach	No discussion, primary emphasis on increasing the PA system	Introduced in the WAP, calls for identification of critical areas outside PAs and *ex situ* conservation, also suggests setting up of special recovery projects for endemic or endangered species in accordance with IUCN, suggests management and regulation of free-ranging domestic animals in PAs and other sensitive areas

(Contd)

---

[68] Status of applications as on 20 December 2018, available at http://nbaindia.org/content/333/25/1/applicationstatus.html, accessed on 24 June 2019.

[69] Samar Singh, 'India's Action Plan for Wildlife Conservation and Role of Voluntary Bodies', *Environmentalist* 5, no. 1 (1985): 31–37.

	NWAP, 2002–16	NWAP, 2017–31
Mitigation of HWC	Plan calls for mitigation of 'man–animal' conflicts in and around PAs, states that amelioration of these situations is essential for the safety of the animal in question and also prevention of that animal from becoming a subject of illegal trade.	Includes a separate chapter on the subject and lays down policy/schemes that need to be prioritised to address the issue. This includes creation of Centre of Excellence on HWC mitigation, under the aegis of the environment ministry. Proposes Land-Use Practices Assessment and Planning Committee to identify the various land use practices that cause HWC and develop more pragmatic land use practices for various relevant regions.
	Suggests that better coordination between wildlife managers and law and order enforcement agencies is considered essential for speedy and effective amelioration of man–animal conflict situations; approaches such as use of green fences such as bamboo and cactus, alternative cropping patterns, and community watch and ward schemes; and creating insurance schemes.	Emphasises that compensation to the victims should be provided immediately.
Climate Change	No specific discussion on the impact of climatic change on wildlife. Therefore, it does not contain any schemes to deal with such impacts.	Strong emphasis on climate change, plan suggests various mechanisms by which it can be addressed, such as rationalisation of boundaries of PAs based on region-specific projections of climate change impacts and further anticipatory planting of species that will adapt to the changing climatic conditions.
		Also suggests assisting migration of wildlife in highly fragmented landscapes and coastal areas, along with undertaking research on animal responses to climate change.

(*Contd*)

(Contd)

	NWAP, 2002–16	NWAP, 2017–31
Reintroduction of Species	Emphasises responsibility to Central Zoo Authority to use the help of state governments and scientific institutions to develop capabilities for planned breeding and reintroduction of captive bred populations of identified endangered species in accordance with IUCN guidelines.	Emphasises the introduction of the Asiatic cheetah, which is extinct in India, suggests the setting up a special plan for its reintroduction and discusses that the environment ministry has approved cloning of the African cheetah, which will be brought to India from South Africa. (In January 2020, the Supreme Court approved the reintroduction of the African cheetah in India.)*
	Contains extensive policy/schemes on conservation of endangered species.	
Ensuring Sustained Funding for Wildlife	Suggests minimum 2 per cent of the national budget should be allocated to the protection of forests, of which at least 15 per cent should be set aside for wildlife conservation	Suggests the use of extra budgetary resources such as Compensatory Afforestation Management and Planning Authority fund (as discussed in Chapter 3), entry fees of PAs, involvement of corporate sector in special programmes and private sector investment in forestry/tourism to increase funding for wildlife conservation
	Discusses adequate funding allocation for filling of vacancies and providing new posts	
	Provides infrastructure and equipment to forest departments and setting up 'combat forces', legal cells	Presents an economic analysis based on scientific facts to justify funds needed for forest and wildlife conservation
		Suggests a landscape based economic valuation model which relies on the relationship between activities such as agriculture, industry, air quality, water quality and so on with PAs and forests

Source: Authors.

Note: * Interlocutary Application (IA) No.192 of 2017 in Writ Petition (Civil) 337 of 1995 *(Centre for Environment Law WFF v. Union of India & Ors)* filed by National Tiger Conservation Authority.

7

Ground and Surface Water Extraction

INTRODUCTION

According to the Indian Constitution, state governments have exclusive powers to legislate on 'water supplies, irrigation and canals, drainage and embankments, water storage and water power'. Fisheries regulation is also the responsibility of the state.[1] However, these powers interface with the central government's rights to enact and implement laws on the following aspects related to water. These are regulating and development of interstate rivers and river valleys in public interest, declaring rivers as national waterways for the purpose of shipping and navigation, using tidal waters for shipping, and fishing beyond territorial waters.[2]

Individuals have the right to collect and use water on their property. This includes 'all water under the land within his own limits, and all water on its surface that does not pass in a defined channel'. These rights are defined under the Indian Easements Act, 1882.[3] Such easement rights are exercised when an individual seeks to draw groundwater or water from a stream on his or her own property for any purpose, including drinking and irrigation. This right is not absolute, but one regulated by the state or central government. Legal and

---

[1] Seventh Schedule, List 2 (State), Entries 17 and 21.

[2] Seventh Schedule, List 1 (Union), Entries 56, 24, 25 and 57.

[3] Indian Easement Act, 1882, defines easement as 'a right which the owner or occupier of certain land possesses, as such, for the beneficial enjoyment of that land, to do and continue to do something, or to prevent and continue to prevent something being done, in or upon, or in respect of, certain other land not his own'. Section 7 (g) refers to rights to collect and use water on land owned by an individual.

regulatory frameworks on water need to take into account individual easement rights over water resources.

This complicated distribution of rights and powers to govern water resources have led to a web of institutions that cut across the central and state jurisdictions. While the Ministry of Water Resources and Ganga Rejuvenation (water resources ministry)[4] is the central ministry for water governance, state governments have their own departments and authorities that govern the supply, storage and use of surface water. For the regulation of groundwater, 13 state governments and union territories (UTs) have adopted the central government's model groundwater bill.[5] This allows them to set up their own groundwater authorities, while a central authority regulates the groundwater resources in the remaining states. [6] State governments also have their own irrigation laws that regulate, use and supply water available in rivers, dam reservoirs and underground sources. These irrigation laws were not designed to address environmental aspects of water extraction.

Water experts have declared that the institutional frameworks for the management of water in India are outdated. The report of the Mihir Shah Committee of 2016 states that these frameworks were made at a time when the objective of governments was to provide irrigation in the interest of food security.[7] The committee was set up primarily to suggest ways of restructuring the Central Water Commission (CWC) and the Central Ground Water Board, the two main agencies of the central government to provide overarching functions to support water governance in India. Ramaswamy Iyer, a senior retired bureaucrat and scholar, had argued for the need for an overarching

---

[4] In May 2019, the central government merged the water resources and drinking water ministry to form the Ministry of Jal Shakti.

[5] Model Bill for the Conservation, Protection, Regulation and Management of Groundwater, 2016 (draft of 17 May 2016), available at http://mowr.gov.in/sites/default/files/Model_Bill_Groundwater_May_2016_0.pdf, accessed on 27 February 2020.

[6] Notification No. S.O. 6140(E) of the Ministry of Water Resources, River Development and Ganga Rejuvenation, 12 December 2018.

[7] Ministry of Water Resources, River Development and Ganga Rejuvenation, *A 21st Century Institutional Architecture for India's Water Reforms: Report Submitted by the Committee on Restructuring the CWC and CGWB* (New Delhi: Government of India, 2016).

legal framework for water rather than a 'patchwork of laws'[8] that presently regulates water use.

Surface water and groundwater use has been at the heart of serious political and legal debates in India. Controversies around water sharing, groundwater depletion and contamination, excessive extraction of rivers and access to safe drinking water are a few issues that have made water a critical environmental resource to regulate. Policy researchers have made several suggestions for institutional frameworks for the governance of water resources through what is popularly referred to as 'water sector reforms'. They have argued that the present laws for water access and use require several changes based on climate change concerns and the need for water conservation.[9]

There have also been discussions on whether governments should govern water through the principle of eminent domain or public trust.[10] While eminent domain would give powers to the government to appropriate water with due compensations to rights holders, public trust may prevent the government from privatising or commercialising water. Court judgments have upheld both these principles in specific matters. In one such judgment, the Supreme Court recognised the state's eminent domain powers to regulate the appropriation and use of water.[11] In another judgment, the public trust role has been emphasised to bestow the responsibility on the state to protect people's common heritage in streams, lakes, wetlands and tidal areas.[12]

While there is no clarity over the principles that ought to drive water governance, the Supreme Court interpreted Article 21 of the Constitution of India to include 'right of enjoyment of pollution free water and air for

---

[8] Ramaswamy R. Iyer, 'Introduction', in *Water and the Laws in India*, ed. Ramaswamy R. Iyer (New Delhi: Sage Publications, 2009), xi.

[9] Philippe Cullet, 'Governing Water to Foster Equity and Conservation: Need for New Legal Instruments', *Economic and Political Weekly* 51, no. 53 (2016), available at https://www.epw.in/journal/2016/53/web-exclusives/governing-water-foster-equity-and-conservation-need-new-legal, accessed on 16 September 2020.

[10] A. Vaidyanathan and Bharat Jairaj, 'Legal Aspects of Water Resource Management', in *Water and the Laws in India*, ed. Ramaswamy R. Iyer (New Delhi: Sage Publications, 2009), 7–9.

[11] Supreme Court judgment dated 12 August 1999 in Civil Appeal 7719 of 1994 (*Parambikulam A.P.O. Association v. State of Tamil Nadu and Ors*).

[12] Supreme Court judgment dated 13 December 1996 in Writ Petition (Civil) 182 of 1996 (*M. C. Mehta v. Kamal Nath & Ors*).

full enjoyment of life'.[13] This judgment explains that access to clean water is an environmental good. This case highlighted the pollution of Bokaro River caused by the dumping of slurry and sludge in the river and on its banks. The Court concluded that river pollution caused by the non-implementation of laws had violated the Right to Life enshrined in Article 21 of the Constitution.

On 20 March 1996,[14] the Supreme Court took cognizance of groundwater depletion reported in an *Indian Express* news item titled 'Falling Groundwater Level Threatens City' which appeared two days earlier. Even though the news report was related to the situation in the National Capital Territory (NCT)-Delhi, the directions in this case resulted in the setting up of regulatory and institutional measures to assess, manage and monitor groundwater use across the country. The Court recommended the need for the central and state water resources management authorities back in 1996. These authorities were envisaged as overarching institutions to review, conserve, regulate and manage all surface water and groundwater sources, including wetlands. This was based on the recommendations of the National Environmental Engineering Research Institute (NEERI) with the purpose of 'holological approach of Water Resources Management'. This report emphasised the need to find technological solutions, draft regulations, prepare medium- and long-term land use plans and create awareness. The judgment laid out fifteen functions for such authorities that included assessment of irrigation practices, ensuring minimum flows in rivers, review of groundwater availability and conservation of traditional water retaining structures.[15] However, it took several more years for the institutions to emerge.

This chapter covers the following sections:

I.   *Legal Framework for Groundwater Extraction*
II. *Surface Water Utilisation and Challenges*
III. *Regulating Wetlands*
IV. *Water Policies and Planning*

---

[13] Supreme Court judgment dated 21 September 1991 in Writ Petition (Civil) 381 of 1998 (*Subhash Kumar v. State of Bihar*).

[14] Interlocutory Application (I.A.) No. 32 in Writ Petition (Civil) No. 4677 of 1985 (*M. C. Mehta v. Union of India and Others*) as decided on 10 December 1996.

[15] Supreme Court judgment dated 10 December 1996 in I.A. No. 32 in Writ Petition (Civil) No. 4677 of 1985 (*M. C. Mehta v. Union of India*).

I   LEGAL FRAMEWORK FOR GROUNDWATER EXTRACTION

India is the world's largest user of groundwater. It provides 80 per cent of the country's drinking water and two-thirds of our irrigation needs. The right of extraction and use of groundwater was historically governed by general provisions in the Indian Easement Act, 1882. As discussed earlier, this legislative framework allowed a person owning a piece of land, the rights to extract groundwater.[16] This right is now subject to several state and national legislations that set the terms and limits for such extraction.

Groundwater overexploitation and bad quality of aquifers affect 60 per cent of India's districts.[17] The depletion of groundwater has been the subject of detailed scrutiny by public planning institutions, courts and policy think tanks. In 1996, the Supreme Court[18] referred to several submissions and documents that showed that groundwater was declining at a rapid rate in India. One such affidavit was by Central Groundwater Board (CGWB) that stated the decline was occurring 'from 1962 onwards'. The judgment in this case quoted a report of NEERI stating 'that the main reason for gradual decline in the level of groundwater in certain areas of the country is over-exploitation'.

The legal and institutional framework for groundwater regulation covers three objectives:

- Assessing the availability of groundwater and its quality. This allows for a mapping of areas that are stressed or have abundance of groundwater.

- Distributing groundwater for competing demands such as irrigation, drinking water and industry through the process of issuing no objection certificates (NoCs).

- Taking enforcement actions in instances of violations such as drawing of groundwater without approval or extracting qualities in excess of permitted quantities.

These objectives were the basis of the 'Guidelines/ Criteria for evaluation of proposals/requests for ground water abstraction' issued by the Central Ground Water Authority (CGWA) in November 2015 and specific state laws for

---

[16] Vaidyanathan and Jairaj, 'Legal Aspects of Water Resource Management', 4.

[17] Ministry of Water Resources, River Development and Ganga Rejuvenation, *A 21st Century Institutional Architecture for India's Water Reforms*, 33.

[18] Supreme Court judgment dated 10 December 1996 in I.A. No. 32 in Writ Petition (Civil) No. 4677 of 1985 (*M. C. Mehta v. Union of India*).

regulating use of groundwater. The institutions engaged in these functions are outlined below.

Institutional Framework

- *The CGWB* is the oldest institution set up in 1970 and acts as a national apex agency to provide scientific inputs for assessment, management and augmentation of groundwater in the country. When first established, the CGWB was under the Ministry of Agriculture and renamed from its earlier formation, that is, Exploratory Tube Wells Organisation. It was merged with the groundwater wing of the Geological Survey of India in 1972. The CGWB has 18 regional offices, which carry out monitoring and present data on the status of groundwater availability and quality. These studies inform state governments and other regulatory agencies in planning and management of groundwater. The CGWB does not directly regulate groundwater use or pollution, except in specific instances when they are called upon to do so. According to a notice dated 12 December 2018, Regional Directors/Heads of regional offices of the CGWB are delegated the powers of granting NoCs for groundwater extraction by projects and industries up to 10 cubic metre per day.[19] The CGWB, whose role was so far limited only to assessing and monitoring the status of groundwater, would now play a regulatory role of granting NoCs to specific industries and projects located within their jurisdiction.

- *Central Ground Water Authority* was notified in 1997 under Section 3 (3) of the Environment Protection Act (EPA).[20] The 1996 Supreme Court judgment discussed above directed the setting of central- and state-level ground water authorities.[21] This judgment issued on 10 December 1996 directed that the existing CGWB be appointed as an authority with 'all the powers necessary to deal with the situation created by the depletion of groundwater levels, dwindling surface water resources, deterioration of surface and groundwater quality and haphazard land use'. Though set up under the EPA, the CGWA is under the administrative control of

---

[19] Central Ground Water Authority No. 26-1lCGWA/Notification-2302.

[20] Ministry of Environment and Forests Notification No. S.O. 38(E) dated 14.01.1997 under Subsection (3) of Section 3 of the Environment Protection Act, 1986 (29 of 1986), in pursuance of the order of the Hon'ble Supreme Court of India in I.A. No. 32 in Writ Petition (Civil) No. 4677 of 1985.

[21] Supreme Court judgment dated 10 December 1998 in I.A. No. 32 in Writ Petition (Civil) No. 4677 of 1985 (*M. C. Mehta v. Union of India*).

the water resources ministry and is expected to implement notifications, guidelines and orders for groundwater regulation made by the central water resources ministry. Besides undertaking some overarching functions, the CGWA regulates the use of groundwater in 23 states and UTs.[22] The CGWA issues NoCs for groundwater use. It also has the powers to take penal actions defined under Section 5 of the EPA.[23] The chairperson of the CGWB also chairs the CGWA.

In November 2018, the central water resources ministry issued a public notice[24] directing all units that have not taken NoCs to extract groundwater to do so by making an application before the CGWA. This public notice also referred to ongoing directions of the National Green Tribunal (NGT) where such violations had been reported. The notice mentioned that the violating units would be liable to pay fines, environmental compensations or additional payments to rectify damages as would be decided by the NGT. For projects which had already been convicted, ₹5,000 would need to be paid for each day that the violation continues. The time period granted for this was five months and was extended twice upto March 2020.

- *State groundwater authorities*: Thirteen state and UT governments have their own groundwater authorities through their laws enacted for the regulation of groundwater.[25] Since the 1970s, the central government has issued model groundwater bills encouraging state governments to adopt their own laws. Both the National Water Policy, 2002 and the 2005 model Bill pushed for this idea.[26] Today, Andhra Pradesh, Goa, Jammu and Kashmir, Karnataka, Himachal Pradesh, West Bengal, Telangana, Tamil

---

[22] These are Arunachal Pradesh, Assam, Bihar, Chhattisgarh, Gujarat, Haryana, Jharkhand, Madhya Pradesh, Maharashtra, Manipur, Meghalaya, Mizoram, Nagaland, Odisha, Punjab, Rajasthan, Sikkim, Tripura, Uttar Pradesh, Uttarakhand, Andaman and Nicobar Islands, Dadra and Nagar Haveli, Daman and Diu.

[23] Ministry of Water Resources, River Development and Ganga Rejuvenation Notification No. S.O. 6140(E) dated 12 December 2018.

[24] Central Ground Water Authority (CGWA) Notice No. 26-1/CGWA/PN/2018 dated 14 November 2018.

[25] National Water Policy, 2002.

[26] As discussed in Planning Commission of India, *Draft Model Bill for the Conservation, Protection and Regulation of Groundwater, 2011* (New Delhi: Planning Commission of India, 2011), available at http://www.planningcommission.nic.in/aboutus/committee/wrkgrp12/wr/wg_back.pdf, accessed on 9 August 2019. Since then the water resources ministry has updated the model bill in 2016.

Nadu, Kerala, NCT-Delhi, Chandigarh, Puducherry and Lakshwadeep have their own groundwater laws and authorities. Karnataka has had a law since 2003, which was replaced in 2011 and is now in operation. Laws and regulations such as the Tamil Nadu Groundwater (Development and Management) Act, 2003; the Karnataka Ground Water (Regulation and Control of Development and Management Act, 2011; Delhi NCT Groundwater Regulation Directions, 2010 and Goa Ground Water Regulation Act, 2002 allow for state governments to set up their groundwater authorities which have the power to restrict and regulate groundwater use in the state. The functions of State Groundwater Authority are similar to CGWA and adopted from the model bills drafted by the central government.

For instance, the Karnataka Groundwater Authority needs to keep several aspects in mind while appraising applications for extraction. The 2011 Act[27] states, 'No permit shall be given for water intensive crops like paddy, sugarcane in notified areas'. While appraising applications, the state authority needs to take into account the likelihood of whether the groundwater extraction would adversely affect the water availability of any drinking water sources in its vicinity. The Karnataka law also requires all groundwater users, without exception, to register themselves with the authority.

The 2003 law applicable in Tamil Nadu discusses the need for 'conjunctive use of surfacewater and groundwater' with the purpose of maximising the 'beneficial use' of the water resources in the state. For this, the Tamil Nadu Groundwater Authority may

> identify and notify suitable areas for conjunctive use of surfacewater and groundwater to stabilise the existing use or to improve or increase the use of water.

- *District collectors* have powers delegated by the CGWA to 'monitor compliance, check violations and seal illegal wells, launch prosecution against offenders etc. including grievance redressal related to ground water'. According to the 2015 CGWA guidelines, the district collectors can also grant NoCs for groundwater use.

---

[27] The Karnataka Ground Water (Regulation and Control of Development and Management) Act, 2011 (Act 25 of 2011).

The groundwater crisis has continued to escalate despite putting into place the above regulatory structure. In 2011, the Working Group set up by the Planning Commission as part of the process to prepare the 12th five year plan concluded that India is moving 'towards a serious crisis of groundwater overuse and groundwater quality deterioration'.[28] The above institutional framework shows that the regulation of groundwater cuts across central and state government jurisdictions as well as specific roles assigned to the ministries of agriculture, water and environment. The responsibility to monitor, regulate and take necessary steps on contamination and pollution of groundwater rests with the Central Pollution Control Board (CPCB) and State Pollution Control Boards (SPCBs). Chapters 2 and 4 discuss the role of the CPCB and SPCBs in detail.

ZONATION AND NOTIFIED AREAS

The CGWA adopted a zonation model to regulate groundwater use based on the assessments of groundwater availability carried out by the CGWB. This is to determine what kinds of uses can be permitted in specific stressed zones or notified areas. For instance, in highly stressed groundwater areas, high levels of extraction will not be permitted. This is to forewarn groundwater users that their application seeking NoCs will not be processed. The 2015 CGWA guidelines identified 162 'Notified Areas'. In such areas, no permission to abstract groundwater 'through any energized means' would be accorded unless for the purposes of drinking water. The statewide distribution of these areas shows Punjab had 46 such notified areas, followed by 45 in Rajasthan, 19 in Karnataka and 18 in Tamil Nadu.

All other areas were divided into four categories: safe areas, semi-critical areas, critical areas and overexploited areas. In each of these areas, the 2015 guidelines specified restrictions for groundwater withdrawal for industrial, mining and infrastructure projects. For instance, in overexploited areas, the NoC would be granted only on the condition that all industries 'fully recycle and reuse the waste water'. These measures were to be regulated through district administrative heads, that is, district collectors assisted by advisory committees that are to be specially constituted under the provisions of Section 4 of the EPA, 1986.

---

[28] Planning Commission of India, *Report of the Working Group on Sustainable Groundwater Management: As Input to 12th Plan* (New Delhi: Planning Commission of India, 2011), 8.

ENFORCEMENT OF GROUNDWATER REGULATIONS

Groundwater was treated as an exploitable resource between the 1950s and 1980s. It was only after this period that the need to regulate the use of groundwater was systematically considered by the government. However, the prevalent NoC model of regulation 'cannot police 30 million recorded groundwater structures'.[29] The regulatory set-up to manage groundwater resources involves several agencies. The multiplicity of ministries, institutions and laws at the state and central levels has challenged the enforcement of groundwater regulations. The lack of enforcement has further aggravated the problem of groundwater extraction and has provoked a number of legal cases including the Supreme Court case of 1996 discussed earlier. There is a range of litigation seeking restraining orders on excessive groundwater extraction, pushing directions against institutional failure and reporting illegal extraction without due approvals.

In April 2019, the Punjab and Haryana High Court ordered the district authorities of Gurugram in Haryana to file a fresh report on the status of groundwater. This direction came on 9 April 2019 in an 11-year-old public interest litigation (PIL) against inaction of the Environment Ministry and other authorities on illegal extraction of groundwater through borewells. The court had regularly given directions to the district authorities, and yet between 2014 and 2018, the groundwater level depleted by 2.5 metres in Gurugram district and by 3 metres in Gurugram city.[30]

In October 2018, in a case filed by 75 packaged drinking water companies seeking to quash a state government order on seeking NoCs prior to the drawal of water,[31] the Madras High Court[32] observed that water and other natural resources are national assets with the government as their custodians.

---

[29] Ministry of Water Resources, River Development and Ganga Rejuvenation, *A 21st Century Institutional Architecture for India's Water Reforms*, 8.

[30] *Hindustan Times*, 'High Court Seeks Report on Groundwater Extraction in Gurugram', 19 April 2019, available at https://www.hindustantimes.com/gurugram/high-court-seeks-report-on-groundwater-extraction-in-gurugram/story-EdR684eYjb1gWnDK38gPYI.html, accessed on 5 June 2019.

[31] Press Trust of India, 'Extracting Groundwater Illegally Amounts to Theft: HC', *Deccan Herald*, 3 October 2018, available at https://www.deccanherald.com/national/extraction-groundwater-695988.html, accessed on 5 June 2019.

[32] Madras High Court judgment dated 3 October 2018 in Writ Petition No. 28535 of 2014 (*M/S. Sarooja Agro Foods v. Government of Tamil Nadu & Ors*).

The High Court, while dismissing the case, concluded that extraction of groundwater illegally amounted to theft. The court's view was that those extracting groundwater illegally should be punished under Sections 378 and 379 of the Indian Penal Code related to theft.

The regulations for groundwater extraction have been shaped significantly by the orders from the NGT. The following cases have influenced the legal framework for groundwater regulation.

- On 15 April 2015, directions were issued to the CGWA to ensure that all groundwater extraction including for tube wells would follow only after an NoC is issued (in Original Application [O.A.] Nos. 204/205/206 of 2014).

- On 9 July 2015, all industrial units sending treated effluent to the Common Effluent Treatment Plants were directed to approach the CGWA to register their borewells and take a NoC (O.A. Nos. 34 and 37 of 2014).

- On 13 July 2017, the Tribunal directed every industry to pay a water extraction fee subject to which the NoC would be granted. The provision for Water Conservation Fee was introduced into the regulation after these directions (O.A. No. 200 of 2014).

- On 28 and 29 August, 2018, the Tribunal directed the Central Water Resources Ministry to review the existing mechanisms to regulate groundwater and take 'effective steps for conserving ground water resources' (O.A. Nos. 176 of 2015 and 59 of 2012).

GROUNDWATER GUIDELINES, 2018

A notification to replace the 2015 guidelines was issued on 12 December 2018 by the Ministry of Water Resources, River Development and Ganga Rejuvenation.[33] This notification included 'Guidelines to regulate and control Ground Water Extraction in India' that were to come into effect from 1 June 2019. The notification used 'the words "abstraction", "extraction", "drawal" and "withdrawal" … interchangeably'. The 2018 notification did not specify any notified areas. However, it retained the categories of safe, semi critical, critical and overexploited areas, to regulate groundwater extraction. The notification did not cite any reasons for the same.

---

[33] Notification No. S.O. 6140(E) of the Ministry of Water Resources, River Development and Ganga Rejuvenation dated 12 December 2018.

On 4 January 2019 the NGT restrained[34] the central government from enforcing these new guidelines as it had 'serious shortcomings' and that it 'worsened the situation by liberalizing the extraction of ground water even for commercial purposes in violation of spirit of order of the Hon'ble Supreme Court.'[35] In September 2019, the NGT also constituted a committee to prepare recommendations on steps that could be taken to prevent ground water depletion and develop a monitoring mechanism for unauthorised extraction and non-compliance with conditions. The committee was also to look into mechanisms for environmental compensations for ground water extraction. The committee submitted its report in March 2020.[36]

This 2018 notification stayed by the NGT had proposed four major changes:

- The notification did not specify the existing precautionary classification of 'notified areas' where only non-energised drinking water extraction was permitted. However, it retained the reference to safe, semi-critical, critical and overexploited areas.

- It brought into effect a set of activities, which would be exempt from seeking NoCs. These include drawal of groundwater through non-energised means (bucket and rope, hand pump, and so on) by individual households drawing/proposing to draw groundwater from a single dug well/bore well/tube well through a delivery pipe of up to 1 inch diameter, agricultural users and armed forces establishments during operational deployment or during mobilisation in forward locations.

- It introduced a system of levying a water conservation fee based on the quantum of groundwater to be extracted. The details of how the fee could be calculated has been specified in the 2018 guidelines. The idea of such a fee was introduced as part of the 2019 guidelines pursuant to the order of the NGT dated 13 July 2017 in O.A. No. 200/2014[37] related to various issues of River Ganga.

---

[34] Shinjini Ghosh, 'NGT Stays Notification on Groundwater', *Economic Times*, 4 January 2019.

[35] National Green Tribunal order dated 11 September, 2019 in Original Application (O.A.) No. 176 of 2015 (*Shailesh Singh v. Hotel Holiday Regency, Moradabad & Ors*).

[36] Notification No. S.O. 3289(E) of the Department of Water Resources, River Development and Ganga Rejuvenation, Ministry of Jal Shakti, dated 20 September 2020.

[37] National Green Tribunal judgment dated 13 July 2015 in O.A. No. 200 of 2014 (*M. C. Mehta v. Union of India—C. Writ Petition No. 3727/1985*).

- It specified the list of states for which the provisions of the notification will be applicable, and others where groundwater use will be regulated through the state government laws and regulations.

The proposed 2018 mechanism required submitting an online application form to the CGWA. A specified set of documents were to be submitted depending on the nature of use. The procedures specified the time period within which a decision was to be taken.

The 2018 guidelines were replaced by the Ministry of Jal Shakti notification dated 20 September 2020. The 2020 notification was challenged before the NGT in O.A. No. 218/2020.

II SURFACE WATER UTILISATION AND CHALLENGES

Water in rivers, streams, estuaries, creeks, reservoirs, ponds, lakes and wetlands are all part of the surface water system. They are seen as a resource for various needs such as industry, infrastructure and energy projects and municipal water supply for domestic uses including drinking water and for agriculture. There is no single or overarching legal framework to regulate access to surface water or no single institution to monitor the use of surface water in India. This makes the legal governance of surface water a complex federal issue.

State governments have exclusive powers to regulate aspects such as water supply, hydropower and fisheries. They are the main parties regulating irrigation, canals, embankments and drainage of surface water. State governments have enacted specific laws for irrigation, drinking water or fisheries management. These laws have prioritised water provisioning and are not set up for water conservation. Take, for instance, the laws for drinking water supply. These have been enacted with 'a single objective of providing and regulating water supply in the state or with a dual objective of water supply in the state and the setting up of corporations or boards for the same'.[38]

As stated in this chapter's introduction, the Supreme Court recognised the right to clean drinking water as a fundamental right. Subsequent legal challenges have broadened the scope of judgment to include continued supply

---

[38] Meena Panicker, *State Responsibility in the Drinking Water Sector: An Overview of the Indian Scenario* (Geneva: International Environmental Law Research Centre, 2007), available at http://www.ielrc.org/content/w0706.pdf, accessed on 10 August 2019.

of water, access to drinking water and access to safe drinking water.[39] These responsibilities are of the state governments. Many of these cases involved citizens challenging the lack of water supply from municipalities as a violation of Article 21. One such case is from the Allahabad High Court where a dispute between two villages Jethai and Pangchora was leading to the water supply not reaching one or the other village. The court ordered the subdivisional magistrate and other respondents to not just continue water supply but also 'take stringent steps if anyone tries/attempts to stop flow of water to either of the two villages'. The judgment also recorded that the survival of human beings and animals cannot be conceived without water.[40]

The drinking water crisis has been recognised at the highest levels of the government. In June 2018, the NITI Ayog, which replaced the Planning Commission, released a study which warned that India is facing its worst water crisis in history, and that the demand for potable water will outstrip its supply by 2030 unless special measures are taken.[41] There is an increasing pressure on state governments to ensure that authorities and corporations vested with this responsibility respond to this crisis.

The central government is in charge of inter-state rivers and has jurisdiction to legislate on the use of surface water for shipping and national waterways. The central government set up the CWC in 1945 to perform several overarching functions for surface water management, just like the CGWB does it for groundwater resources. The CWC is set up under the Central Water Resources Ministry.

The power to declare rivers as national waterways rests with the central government. In 2016, the National Waterways Act[42] was enacted to recognise 106 additional stretches of rivers as 'national waterways'. The Act aims to develop and regulate national waterways to enable their use for 'shipping

---

[39] Videh Upadhyay, 'Water Rights and the "New" Water Laws in India: Emerging Issues and Concerns in a Rights Based Perspective', in *India Infrastructure Report 2011: Water: Policy and Performance for Sustainable Development* (New Delhi: IDFC and Oxford University Press, 2011), 56–66.

[40] Allahabad High Court judgment dated 4 May 2000 in Civil Misc. Writ Petition No. 16424 of 1995 (*Diwan Singh and Another v. Sub Divisional Magistrate, Almora and Others*)

[41] Jacob Koshy, 'India Faces Worst Water Crisis: NITI Aayog', *The Hindu*, 14 June 2018.

[42] The National Waterways Act, 2016 (No. 17 of 2016).

and navigation'. The 2016 law amended the Inland Waterways Authority of India Act, 1985 and added to the list of five waterways notified under it. The Department-Related Parliamentary Standing Committee on Transport, Tourism and Culture recommended these amendments.[43]

This move complicates the already vexed surface water law framework that strides the centre and state jurisdictions. While the national government aims to use rivers for commercial and cargo transport, the right over the use of water, riverbed and the 'appurtenant' (or connected) land vests with the state governments. The implementation for any scheme would require collaboration with the state governments. A national waterway enclosed for tourism, recreation or shipping purposes by private or public sector companies could lead to conflicts with pre-existing rights or uses. The current legal framework does not provide much legal clarity on how this would be addressed.

One way to navigate the network of laws regulating surface water is to understand them through the three sectors that presently demand the highest amount of surface water. These are irrigation, industry and hydropower. The sectoral laws regulating these projects do not require them to disclose the environment and social impacts of their water use. The authorities in charge of surface water reviewing their applications are also not mandated to assess these aspects.

WATER FOR IRRIGATION

Farm irrigation has been one of the largest users of surface water in India. Post-independence, the construction of dams for irrigation was part of nation building. In 1947, India had 300 large dams. By the year 2000, that figure increased to 4,000. It is estimated that more than half of these dams were built between 1971 and 1989. Studies indicate that in 96 per cent of these, dams have been built to support irrigation.[44] Some of the largest dams in India like

---

[43] Press Information Bureau, 'Cabinet Approves Central Legislation to Declare 106 Additional Inland Waterways as National Waterways', Government of India, New Delhi, 9 December 2015, available at http://pib.nic.in/newsite/PrintRelease. aspx?relid=132900, accessed on 28 May 2019.

[44] Esther Duflo and Rohini Pande, 'Dams', *The Quarterly Journal of Economics* 122, no. 2 (May 2007): 601–646.

the Bhakra Nangal, Hirakud and, in more recent times, Sardar Sarovar and Polavaram are designed as irrigation dams. As of 2019, 'Madhya Pradesh has 899 big dams, followed by Gujarat (620), Chhattisgarh (248), Karnataka (230) and Rajasthan (209)'.[45] These dams have been hugely debated for their social impacts, ecological damage and rehabilitation-related issues[46] and the lack of water in rivers.[47]

State governments have their own laws for irrigation. Researchers have pointed out that until the 1990s, most states have followed the pattern of colonial irrigation laws with little innovation. These laws vest all powers with the state governments to determine whether surface water is to be used for irrigation and whether it would serve any public purpose.[48] Powers of the government also extend to determine the management and use of dam reservoirs and canals. For instance, The Karnataka Irrigation Act, 1965 (amended up to 2010) specifies its purpose as:

> An Act to make provisions relating to the construction, maintenance and regulation of irrigation works, the supply of water therefrom, obtaining labour in emergencies and certain other matters pertaining to irrigation in the State of Karnataka.

This Act lays out how water supply can be regulated from 'irrigation works' like all reservoirs, tanks, wells, ponds, canals, pipes, channels, aqueducts and sluices. It also includes embankments, structures, supply and escape channels connected with such irrigation works. The Act gives powers to irrigation officers to survey, investigate, execute and maintain irrigation and canal-related structures and operations.

The Gujarat Irrigation and Drainage Act, 2013 has been enacted with the purpose of managing

---

[45] Radheyshyam Jadhav, 'More and More Dams Planned, but Where Is the Water?' *Hindu Business Line*, 20 February 2019.

[46] Shekhar Singh, 'Social and Environmental Impacts of Large Dams in India', *The Ecologist Asia* 11, no. 1 (January–March 2003): 61–70.

[47] Radheyshyam, 'More and More Dams Planned, but Where Is the Water?'

[48] Philippe Cullet and Joyeeta Gupta, 'Evolution of Water Law and Policy in India', in *The Evolution of the Law and Politics of Water*, ed. Joseph W. Dellapenna and Joyeeta Gupta (Dordrecht: Springer Academic Publishers, 2009).

all the works constructed or maintained relating to irrigation by the State Government, State Government Institutions and Grant-in-Aid Institutions of the State and includes all services rendered thereof.

The law lays out detailed mechanisms through which the state government needs to ensure the maintenance of canal and drainage systems and regulate the use of canal water. Every person desirous of using canal water needs to make an application to a 'Canal Officer'. Such permission can be granted with conditions and restrictions related to how much water can be drawn and for what purpose. There is also a dedicated section related to compensation that can be claimed in case of 'substantial damage' caused by the exercise of powers under this Act. This includes stoppage of supply of water except when it is due for alterations, repair or 'any cause beyond the control of the state government'. It also does not include any damage that may occur due to 'deterioration of climate'.

Since the mid-1990s, at least 15 state governments have included Participatory Irrigation Management (PIM) and creation of water user areas and Water Users Associations as mechanisms in their legal frameworks. This was to include farmers dependent on irrigated agriculture into irrigation management and to ensure equity in distribution and sharing of water.[49] Critiques indicate that the legal frameworks in state laws give limited rights to farmer users. They continue to depend on state government authorities for both information and supply of water. Further, there is little in these laws in the form of remedies in case the agreements or legal requirements are not honoured.[50]

While the state governments implement irrigation projects, the central government has played a significant role in providing financial and technical assistance to the state governments. In 1996–97, the central government launched the 'Accelerated Irrigation Benefits Programme (AIBP) to provide Central Loan Assistance (CLA) to major/medium irrigation projects in the country'. The CWC plays an important role in designing, implementing and monitoring these schemes.[51]

---

[49] The central Water Resources Ministry credits this concept to an non-governmental organisation, SOPPECOM (note available at http://mowr.gov.in/sites/default/files/CADWM_Status_of_PIM_0.pdf, accessed on 8 May 2019).

[50] Upadhyay, 'Water Rights and the "New" Water Laws in India'.

[51] Discussion on the Accelerated Irrigation Benefits Programme on the website of the Central Water Commission, available at http://www.cwc.gov.in/aibp-projects, accessed on 12 August 2019.

WATER FOR INFRASTRUCTURE AND INDUSTRIAL OPERATIONS

Surface water is needed for various types of projects such as manufacturing, power plants, ports, highways and real estate construction. According to a 2011 study by the Federation of Indian Chambers of Commerce and Industry (FICCI), surface water contributes to 41 per cent of the demand by industries. This is followed by groundwater (35 per cent) and municipal water supply (24 per cent).[52] Yet there are no clear estimates of the quantum of fresh surface water consumed by industries. In a study by the Infrastructure Development Finance Company (IDFC), the figure ranges from 6 per cent according to the Ministry of Water Resources to 13 per cent by the World Bank.[53]

Since the supply of water is under state jurisdiction, industrial units usually have a Memorandum of Understanding with a state ministry of water resources that assures them unhindered supply of water. This agreement specifies the exact quantity and duration of water required from specific sources. However, the FICCI study also states that limited or non-availability of water is one of the greatest risk factors for industries.[54]

Typically, surface water for industrial purposes is sourced from rivers or streams, which are regulated through different laws and may be meant for different purposes. States have to amend these laws to reallocate water to industries from earlier uses like irrigation. Since the 1990s, most state governments, through their Department of Water Resources, have responded to applications for industrial water supply by amending state irrigation laws. The sources of water in these cases could be flowing rivers or reservoirs of dams constructed primarily for irrigating farm lands and power generation. For example, the 1993 amendment to the Odisha Irrigation Act, 1959 for 'Regulation and Use of Water from Government Water Sources' devised a mechanism to issue licences for industrial or commercial water use from any government water source. In this law, a government water source is defined as

---

[52] FICCI Water Mission, *Water Use in Indian Industry Survey* (New Delhi: FICCI Water Mission, 2011), available at http://ficci.in/Sedocument/20188/Water-Use-Indian-Industry-Survey_results.pdf, accessed on 10 August 2019.

[53] Aggarwal Suresh Chand and Surendra Kumar, 'Industrial Water Demand in India: Challenges and Implications for Water Pricing', in *India Infrastructure Report 2011: Water: Policy and Performance for Sustainable Development* (New Delhi: IDFC and Oxford University Press, 2011), 274–291.

[54] FICCI Water Mission, *Water Use in Indian Industry Survey*, 3–4.

any water source created naturally or otherwise by collection or deposit of water at a fixed place, any sub-soil water or water in a state of running such as rivers, nalas, springs, streams and the alike, which is other than an irrigation work and is the property of the Government.

Increasingly, state governments view reservoirs made for irrigation and hydropower as revenue earning assets as the demand for water from industrial, energy or municipal operations has increased. In 2016, the Odisha government amended the Orissa Irrigation Act, 1959 and Orissa Irrigation Rules, 1961 to facilitate ease of water supply to industrial, commercial and municipal entities in return for a licence fee. The notification lays down the procedures through which applications are to be made by concerned industries and the timeframe within which the state government would respond.[55] The 2016 amendment to the Odisha Irrigation Rules, 1961 adds a clause that all other users of water from dam reservoirs would sign supplementary agreements with the Odisha Hydro Power Corporation Limited. Any user, including buildings, brick kilns or for bulk supply to municipalities, would need to

compensate the loss of energy generation due to its drawal and the Odisha Hydro Power Corporation Limited, shall raise demands for compensation of loss of energy generation within first week of every month against the quantity of water drawn or allocated, whichever is higher.[56]

Such diversions of water meant for irrigation or power generation to industry has been critiqued. The Sardar Sarovar Narmada Nigam Limited's supply of water to industry has been criticised on the grounds that several tribals and farmers were displaced to create public goods out of public funds to supply water for irrigation and drinking water to the arid regions. The distribution of water from the Narmada Valley Development projects including the Sardar Sarovar project was determined by the Narmada Water Disputes Tribunal set up in 1969.[57] The purpose of this project was to generate power and supply water

---

[55] Notification dated 9 February 2016, No. 2017 WR lrr.-II-WRC-07/16 (procedure for permission/allocation of water to industrial, commercial and other establishments).

[56] Notification S.R.O. No. 450/2016 of the Odisha Revenue and Disaster Management Department dated 24 September 2016.

[57] Narmada Disputes Tribunal was set up under Section 4 of the Inter-State Water Disputes Act, 1956.

for irrigation to Gujarat. Out of its allocated share, the Gujarat government made a provision that '11% of the total allocated water, i.e. 3,582.17 mld (1.06 MAF or 1,307.5 MCM), would be used for drinking water and industrial use in 135 urban centres and 9,633 villages in the state, including villages in Kutch, North Gujarat and Saurashtra regions'.[58]

Research shows that as of 2016, more than 18 per cent of this irrigation project is used to provide water for industrial and other non-agricultural purposes[59]. All industries in Kutch show the Sardar Sarovar Dam as their main source of water, and most of them are drawing more than their allocated amount. While such diversion of water has been made legal, it has been questioned on the grounds of equity and social justice.[60]

As discussed earlier, every state government grants or rejects access to river or reservoir water on a case-to-case basis. Over-extraction of surface water beyond the approved limits and their impacts caused to communities dependent on river/reservoir water due to approvals granted by the government have been challenged in various high courts. For example, two PILs were filed in the Madurai bench of the Madras High Court[61] seeking restraining orders against the extraction of water from the Tamirabarani River. The state government approved the sale of 1.5 million litres of water per day at the rate of ₹37.50 for every 1,000 litres of water to two companies operating in the State Industries Promotion Corporation of Tamil Nadu (SIPCOT) area in Tirunelveli District. These units were packaging water for two large multinational companies. The state government also constructed a check dam near Seevalaperi village in order to facilitate this supply. These measures affected the availability of drinking water and irrigation water in the region.

The petitioners in the above PILs contended that the drawing of water from Tamirabarani River required the approval of Water Resources Control and Review Council. It was argued,

---

[58] Avantika Mehta, 'Drought-Hit Gujarat Has Water for Factories, but Not for Farmers', *IndiaSpend*, 16 April 2019, available at https://www.indiaspend.com/drought-hit-gujarat-has-water-for-factories-but-not-for-farmers, accessed on 28 May 2019.

[59] Ibid.

[60] Express News Service. 'Farmers' Body Claims Govt Plan to Divert Narmada Water for Industry', *Indian Express*, 14 January 2018.

[61] Madras High Court judgment dated 2 March 2017 in Writ Petition (MD) No. 20558 of 2016 (*M. Appavu vs The Chief Secretary to Government & Ors*) and Writ Petition. (MD) No. 22425 of 2015 (*D. A. Prabakar v. The State of Tamil Nadu & Ors*).

Any agreement among Government agencies involving the use of water for domestic and municipal water supply, irrigation, hydro power production, industrial or other commercial uses, water shed, coastal areas and environmental protection measures shall, in all cases, be subject to review and approval by the Council (WRCRC), as per G.O.Ms.No.1404.

The court dismissed these PILs on the grounds that these were private grievances and not maintainable in public interest.

WATER FOR HYDROELECTRICITY

Access to flowing water is essential for the purpose of generating hydroelectricity. According to Schedule VII of the Constitution, 'water storage' and 'water power' is under the jurisdiction of the state governments. Therefore, use of water for hydropower dams can be decided by the state. The central government also has a role to play when the hydropower generation involves inter-state rivers, which is the case with most major rivers in India.

The central government has a say in hydropower through the Central Electricity Authority (CEA). The CEA was set up under Section 3(1) of Electricity (Supply) Act, 1948. This section was replaced by Section 70 of the Electricity Act, 2003. The CEA identifies potential sites for potential hydropower plants. It also grants clearance to these schemes.

In 2006, the central government issued a notification for various categories of hydroelectric schemes that would not need to be submitted to the CEA for their concurrence. These schemes included projects with capital expenditure up to ₹25 billion, provided they were included in the 'National Electric Plan' notified by the CEA, and the site was allocated as per a transparent bidding process defined by the central government. [62] It also included sites which were allocated to a proponent following a transparent bidding process in accordance with central government's guidelines.

Specific details of the land requirement and technology for the project are subsequently developed through a Detailed Project Report (DPR).

---

[62] Notification No. S.O. 550(E) dated 18 April 2006 issued under Section 8(1) of the Electricity Act, 2003, as discussed in the Central Electricity Authority's 'Guidelines for Formulation of Detailed Project Reports for Hydro Electric Schemes, Their Acceptance and Examination for Concurrence, January 2015 (Revision 5.0)', available at http://www.cea.nic.in/reports/others/hydro/hpa1/guidlines_dpr_he_ver5.pdf, accessed on 15 September 2020.

The DPRs need to be prepared according to the CEA's 'Guidelines for Formulation of Detailed Project Reports for Hydro Electric Schemes, their Acceptance and Examination for Concurrence'. The checklist of information as per these guidelines requires the disclosure of environmental, forest use, land requirement and rehabilitation & resettlement aspects. These details are meant to be included in the applications sent to for environment clearance under the Environmental Impact Assessment (EIA) notification, 2006, which is discussed in Chapter 5 on environmental protection.

Hydropower projects planned in recent years have maximised the utilisation of river flows for generation of profitable hydroelectricity. Such projects have increased in scale and size all over the country and especially in the Himalayan region. The planning of hydropower dams has conventionally failed to account for the downstream uses of water. This has caused conflicts over water sharing between states in the case of inter-state rivers and between upstream and downstream regions.

Environmental flows (e-flows) of rivers have been discussed in several policy documents and regulatory directions by the central environment ministry. A 2015 judgment of the NGT elaborated on the concept of e-flows and its specific relevance to River Ganga[63].

The judgment dated 13 July 2015 defined *Environmental flows or E-flows* as

> the flow regime in a river that describe the temporal and spatial variations in the quantity and quality of water required by the river to perform its natural ecological functions and support the aquatic and terrestrial biodiversity, meet agricultural and consumptive needs and also support the spiritual, social and cultural activities that depend on the river ecosystem.

Elaborating on the need to introduce e-flows into water regulation, the judgment explained,

> Environmental Flows are the flow required for the maintenance of the ecological integrity of the rivers and their associated ecosystems. Environmental Flows are increasingly recognized as a vital contributor

---

[63] National Green Tribunal judgment dated 13 July 2015 in O.A. No. 200 of 2014 (*M. C. Mehta v. Union of India—C. Writ Petition No. 3727/1985*).

to the continuing provision of environmental goods and services upon which the livelihood of people depend.

This concept is an important corrective to the legal and institutional frameworks that incentivise river water extraction and diversion. The National Water Policy, 2002 (discussed later in this chapter) had observed, 'minimum flow should be ensured in the perennial streams for maintaining ecology and social considerations'. Several rivers across the country, especially peninsular rivers, practically do not exist during the post monsoon period due to excessive extraction and diversion through various means including dam structures. It took over a decade for the water resources and environment ministries to recognise the relevance of e-flows and incorporate it into regulatory practice. It was only after 2007 that the environment ministry included e-flows as a condition for the environment clearance (under EIA notification, 2006) of all river valley and hydroelectric projects. The actual percentage of flow to be considered sufficient to meet this has been undergoing change based on the recommendations of thematic expert committees at different points of time.[64]

The Central Water Resources Ministry has given special emphasis to ecological flows in the River Ganga through two notifications in 2016 and 2018. The first notification dated 7 October 2016[65] established The River Ganga (Rejuvenation, Protection and Management) Authority. This authority was required to 'determine the magnitude of ecological flow in the River Ganga and its tributaries' and devise a system for continuous monitoring of this flow. The second notification dated 9 October 2018[66] specifies the ' minimum environmental flows to be maintained at locations downstream of structures or projects meant for diversion of river flows for purposes like irrigation, hydropower, domestic and industrial and other requirements'.

---

[64] L. Anantha, S. Dharmadhikary and N. Bhadbhade, *E-flows in Indian Rivers: Methodologies, Issues, Indicators and Conditions—Learnings from Hasdeo Basin* (Pune: Forum for Policy Dialogue on Water Conflicts in India, 2017).

[65] Ministry of Water Resources, River Development and Ganga Rejuvenation Notification S.O. 3187(E) dated 7 October 2016.

[66] Ministry of Water Resources, River Development and Ganga Rejuvenation Notification S.O. 5195(E) dated 9 October 2018 in File Reference No. F. No. Estt.01/2016-17/111/NMCG (Vol. III).

INTER-STATE WATER SHARING

The central government has the responsibility to address issues arising out of inter-state water sharing. The legislation empowering the central government to do this is the Interstate River Water Disputes (ISRWD) Act, 1956.

Several cases related to inter-state sharing of water have been filed before tribunals set up under the ISRWD Act, 1956. The law allows for complaints to be filed by any state governments before the Ministry of Water Resources, River Development and Ganga Rejuvenation. The ministry can also directly intervene and set up special tribunals if state governments are unable to settle disputes between themselves. Presently, tribunals have been set up for several rivers including Godavari, Cauvery, Mahadayi, Vansandhara, Krishna, Ravi and Beas, Narmada, Godavari and Mahanadi.

One such dispute is related to the use of water of the Hirakud irrigation dam on the Mahanadi River that flows through the states of Odisha and Chhattisgarh. The Odisha government claims that the Chhattisgarh state government constructed dams and barrages on the Mahanadi upstream of Hirakud. It states that this has significantly reduced the water available in the Hirakud.[67] On 12 March 2018, the central government, based on the directions of the Supreme Court, constituted a Mahanadi Water Dispute Tribunal under Section 4 of the ISRWD, 1956.[68] The Tribunal has subsequently issued notices to the states of Odisha, Chhattisgarh, Jharkhand, Madhya Pradesh and Maharashtra[69] as they are in the catchment area of the Mahanadi River. A final resolution of this dispute was pending as of mid-2020.

It is important to note that the state of Chhattisgarh is diverting a substantial portion of this dammed water for industrial and coal power

---

[67] K. Panda Ranjan, 'Mahanadi: Looking Beyond Coal', Heinrich Boell Stiftung, New Delhi, 8 May 2019, available at https://in.boell.org/2019/05/08/mahanadi-looking-beyond-coal, accessed on 1 July 2020.

[68] Notification No. S.O. 1114(E) issued by the Ministry of Water Resources, River Development and Ganga Rejuvenation on 12 March 2018, available at http://164.100.24.220/loksabhaquestions/annex/14/AS384.pdf, accessed on 9 March 2021.

[69] Press Information Bureau, 'Notice Issued by Mahanadi Water Disputes Tribunal', Ministry of Water Resources, New Delhi, 30 May 2018, available at https://pib.gov.in/PressReleaseIframePage.aspx?PRID=1533851, accessed on 30 September 2020.

generation.[70] As discussed earlier, Odisha also has an elaborate system in place to divert surface water meant for irrigation to industrial use. This is also the case with the Hirakud dam, one of the first major multipurpose river valley projects initiated soon after India's independence in Sambalpur district of Odisha. The inter-state water dispute between Odisha and Chhattisgarh can be seen as a fallout of competitive industrialisation based on water extraction between the two states.

III REGULATING WETLANDS

Wetlands are land areas that are inundated by water either seasonally or throughout the year. They can be found in a variety of habitats that includes rivers, lakes, coastal lagoons, mangroves, peatlands and coral reefs. They also include human-made structures such as ponds, farm ponds, irrigated agricultural land, salt pans, dam reservoirs, gravel pits, sewage farms canals and sacred groves. They can be found in both urban and rural areas. Wetland conservation is increasingly being linked to water security.[71] An International Convention on Wetlands was held in Ramsar, Iran in 1971, to which 169 countries including India have signed up. In all, 37 wetlands from India are included in the convention's list of internationally important wetlands.[72]

The legal definition for wetlands applicable in India is specified in the Wetlands (Conservation and Management) Rules, 2017. The rules define a wetland as

> an area of marsh, fen, peatland or water; whether natural or artificial, permanent or temporary, with water that is static or flowing, fresh,

---

[70] Anuj Goyal, *Hasdeo Bango Project: Changing Allocation Pattern* (Pune: Forum for Policy Dialogue on Water Conflicts in India), available at http://waterconflictforum.org/lib_docs/INT_RPT_AG_Hasdeo-Bango.pdf, accessed on 6 June 2019.

[71] Richa Banka, 'Wetlands Authority to Preserve Water Bodies: Delhi Government Tells High Court', *Hindustan Times*, 5 July 2019, available at https://www.hindustantimes.com/delhi-news/wetlands-authority-to-preserve-water-bodies-delhi-govt-tells-hc/story-vu9RS8AwgfTsSEouehvG7K.html, accessed on 14 September 2020.

[72] Ramsar Sites Information Service, 'Annotateed List of Wetlands of International Importance: India, Ramsar Convention Secretariat, Switzerland', available at https://rsis.ramsar.org/sites/default/files/rsiswp_search/exports/Ramsar-Sites-annotated-summary-India.pdf?1600153527, accessed on 15 September 2020.

brackish or salt, including areas of marine water the depth of which at low tide does not exceed six meters, but does not include river channels, paddy fields, human-made water bodies/tanks specifically constructed for drinking water purposes and structures specifically constructed for aquaculture, salt production, recreation and irrigation purposes.

The government's initiatives on wetland conservation can be traced back to 1987 when the environment ministry set up the National Wetlands Conservation Programme. The objective of this programme was to develop policies for conservation and management of wetlands. As part of this initiative, the ministry prepared an inventory of wetlands and undertook conservation activities in 115 wetlands.[73]

In 2001, a writ petition[74] was filed in the Supreme Court seeking the protection of wetlands. It asked for wetlands to be inventorised and conserved. The Court directed all state and UT governments to file their responses and counter affidavits.[75] The case was repeatedly adjourned until January 2009, when the Court granted four weeks for responses to be filed. In March 2009, in a detailed order, the Court expanded the scope of the case linking conservation of wetlands to the acute water shortage in the country. The Court said,[76]

There is acute shortage of water in our country and one of the main reasons for that is that most of the water conservation bodies in our country such as ponds, tanks, small lakes etc. have been filled up in recent times by some greedy persons and such persons have constructed buildings, shops etc. on the same.

The Ministry of Science and Technology was directed to file a status report on water shortage in the country. The Court clarified that they will regularly

---

[73] MoEF (Ministry of Environment and Forests), 'Press Note: The Wetlands (Conservation and Management) Rules, 2010', Government of India, New Delhi, 2010.

[74] Writ Petition (Civil) No. 230 of 2001 (*M. K. Balakrishnan & Ors v. Union of India & Ors*).

[75] No orders between 9 September 2001 and 23 March 2006 are available in this case on the Supreme Court website as accessed on 27 July 2019.

[76] Supreme Court order dated 26 March 2009 in Writ Petition (Civil) No. 230 of 2001 (*M. K. Balakrishnan & Ors v. Union of India & Ors*).

monitor the progress of this case and asked the government to set up a committee to address the issues raised by the Court in the March 2009 order. The case was not heard between January 2010 and April 2013,[77] following which there were no substantive hearings.

During the pendency of this case, wetlands were brought under the legal ambit of the environment ministry by the Wetlands (Conservation and Management) Rules in December 2010.[78] The preamble of these rules specified the importance of the wetland ecosystem and clearly laid out that there is a need to legally protect wetlands to respond to threats. The 2010 Rules say,

> many wetlands are seriously threatened by reclamation through drainage and landfill, pollution (discharge of domestic and industrial effluents, disposal of solid wastes), hydrological alterations (water withdrawal and inflow changes) and over exploitation of their natural resources resulting in loss of biodiversity and disruption in goods and services provided by wetlands.

In September 2014, the issue of the effectiveness of the 2010 Rules was argued before the Supreme Court in the abovementioned writ petition. In an order dated 10 September 2014, the Court asked the environment ministry to respond to whether the Wetlands Authority that was to be set up under the 2010 Rules was functional. The ministry was also asked to report on the status of the inventory of wetlands, which was also legally mandated under the 2010 rules. The Court recorded that a National Wetland Inventory and Assessment Project, sponsored by the Environment Ministry, was underway. In 2010, the Space Applications Centre, Indian Space Research Organisation, based in Ahmedabad, undertook the task of making an inventory of all the wetlands in the country. There was no progress in this case until the end of 2016. On 1 December 2016, the Court referred to the September 2014 order and gave another six weeks for responses to be filed. On 12 July 2017, the Supreme Court fined the ministry ₹50,000 for not complying with its directions. The ministry was given until 30 June 2018 to identify and inventorise 201,503 wetlands at a national level. During this time, the Environment Ministry had

---

[77] As per orders available on the Supreme Court website as accessed on 27 July 2019.

[78] Vide G.S.R. 951(E) dated the 4 December 2010.

already issued draft amendments to the wetland conservation rules in March 2016 that were finally notified in 2017 (discussed below).[79]

In November 2015, two cases related to the non-implementation of the 2010 Rules were filed in the NGT.[80] These matters continued to be heard even as the draft and new rules were notified. According to the website of the NGT, the two cases and a Miscellaneous Application[81] filed in 2016 are 'pending' final decision in the Supreme Court case.

Today, the Wetlands (Conservation and Management) Rules, 2017 are applicable to all wetlands, other than those falling in areas under the ambit of Indian Forest Act, 1927; the Wild Life (Protection) Act, 1972; the Forest (Conservation) Act, 1980; the State Forest Acts and the Coastal Regulation Zone Notification, 2019. All these laws have been discussed in earlier chapters of this book.

The 2017 legal framework is primarily designed to achieve three outcomes:

- *Identification and inventorisation of wetlands* need to be carried out by Wetland Authorities to be set up in every state.

- There are several activities that are *prohibited in or around wetlands*. This includes the conversion of an identified wetland into a non-wetland use. It prohibits setting up of industries, waste dumping (construction or hazardous), discharge of effluents, any permanent construction (except boat jetties 50 metres away from wetlands) and poaching. State governments are allowed to add additional activities to the central government's list of prohibitions.

- State governments are required to demarcate and declare a *zone of influence* around every notified wetland along with list of activities that are permissible or restricted in the influence areas. The State Wetland Authority would be the body to enforce this. There is no specification of how large or small this influence zone needs to be.

The lacunae in the implementation of both wetland rules and the limitation of the 2017 framework have been critiqued by several ecologists and

---

[79] Draft Wetlands (Conservation and Management) Rules, 2016, vide number G.S.R. 385 (E) dated 31 March 2016.

[80] National Green Tribunal O.A. No. 501 of 2015 (*Anand Arya v. Union of India*) and O.A. No. 560 of 2015 (*Pushp Jain v. Union of India & Ors*).

[81] National Green Tribunal Miscellaneous Application 1212 of 2016 in O.A. 501 of 2015 (*Anand Arya v. Union of India*).

environmentalists. Critics have argued that the new rules are vague and do not provide a clear institutional accountability for the conservation of wetlands. They fear that the definition of wetlands in the 2017 Rules is too broad and runs the risk of leaving out important wetland types from the wetland inventory. These include salt pans, wetlands in river floodplains or lagoons like the Chilika in Odisha that are covered by the Ramsar Convention. The 2017 Rules also omit criteria such as natural beauty, ecological sensitivity, genetic diversity or historical value that are priority areas for wetland conservation. One specific critique was regarding the phrase 'wise use' in the objectives of the new rules. As per the 2017 Rules, wetlands are to be managed by applying the Principle of Wise Use. Resolution IX.1 Annex A (2005) of the Ramsar Convention defines Wise Use of wetlands as

> the maintenance of their ecological character, achieved through the implementation of ecosystem approaches, within the context of sustainable development.

Critics warn that these words provide discretion to state governments and wetland authorities in prioritising the use of wetlands, which may have negative implications.[82]

IV   WATER POLICIES AND PLANNING

Policies on how to manage and use water have been drawn up both at the central and state levels. At the national level, the Water Resources Ministry prepares water policies. However, each state government has a specific designated institution. For instance, in Odisha, the Orissa Water Planning Organization is the nodal agency for planning water resource management in the state, under the Department of Water Resources. This agency prepared the State Water Plan, 2004. It took 2001 as the base year to determine water usage and project the total water demand in the state until year 2051.

---

[82] Neha Sinha, 'Reconsider the Rules: On 2017 Wetland Rules', *The Hindu*, 21 December 2017, available at https://www.thehindu.com/opinion/op-ed/reconsider-the-rules/article22085813.ece, accessed on 14 September 2020; Mayank Aggarwal, 'Government's Draft Wetland Rules Draw Flak from Environmentalists',*Mint*,11 April 2016, available at https://www.livemint.com/Politics/dL1y5lPG6Hw3rZqJpgIQMI/Governments-draft-wetland-rules-draw-flak-from-environmenta.html, accessed on 14 September 2020.

In the last two decades, the Water Resources Ministry has issued two national water policies in 2002 and 2012. A comparison of the two is given below.

Table 7.1 Comparison of National Water Policies of 2002 and 2011

	National Water Policy, 2002	National Water Policy, 2012
Uses of Water	Six priority areas for water allocation including drinking water, irrigation, hydropower, ecology, agro-industries and non-agricultural industries and navigation. However, the priorities could be modified or added based on region-specific requirements.	Utilisation in all diverse uses of water should be optimised and an awareness of water as a scarce resource should be fostered. Specifically suggests rivers and other water bodies should be considered for development of navigation as far as possible.
National Water Framework Law	No discussion on a national law.	A new water law framework should present guiding principles for state level legislations. It should be based on an understanding that water is not just scarce but is life sustaining. Groundwater must be managed as a community resource.
Climate Change	No discussion on climate change or related topics.	Specific emphasis on planning and management strategies for combating and adapting to climate change.
Approaches for Water Conservation and Use Efficiency	Recommends Water Users' Associations (WUA) and the local bodies such as Municipalities and gram panchayats to be involved in the operation, maintenance and management of water infrastructures and facilities.	Scientific conservation of rivers, river corridors, water bodies and infrastructure implemented with community participation. Removal of encroachments emphasised.
Water Pricing	No discussion on water pricing. Polluter Pays principle emphasised in case of illegality.	Pricing of water to be introduced and managed through a Water Regulatory Authority. Suggests the concept of WUAs to manage pricing locally.

(Contd)

(*Contd*)

	National Water Policy, 2002	National Water Policy, 2012
Flood and Drought Management	Master plan for flood control and management for each flood prone basin. Strict regulation of settlements and economic activity in the flood plain zones.	Efforts to avert flood and drought occurrence by several measures including forecasting and management of natural drainage systems.
Water Disputes	Suggests a review of the Inter-State Water Disputes Act, 1956. Also suggests the need for appropriate river basin organisations for conflict resolution.	Suggests a permanent Water Disputes Tribunal at the national level.

Source: Authors.

8

Land Acquisition

INTRODUCTION

India's laws for land acquisition are based on the principle that governments have the power to acquire private property for public use. These powers arise from the Doctrine of Eminent Domain, which allows governments to take such action in exchange of due compensation to the property owner. Legal researchers have argued that the payment of a compensation 'changes the nature of this take-over from appropriation to acquisition'.[1]

The legal history of land acquisition can be traced back to the eighteenth century when the British colonial administration enacted laws[2] to 'facilitate the acquisition of land and other immovable properties for roads, canals, and other public purposes by paying the amount to be determined by the arbitrators'.[3] This includes the Land Acquisition (LA) Act, 1894. These laws outline a procedure for acquisition, limits of consent and mechanisms to compute compensation. The legal requirement to ensure rehabilitation of affected families has been part of the legal framework for land acquisition only from 2013.

---

[1] Usha Ramanathan, 'On Eminent Domain and Sovereignty', *Seminar*, no. 613, September 2010, available at http://www.india-seminar.com/2010/613/613_usha_ramanathan.htm, accessed on 28 July 2019.

[2] These are the Bengal Regulation I of 1824, Bombay Act No. XXVIII of 1839, Bombay Act No. XVII of 1850, Madras Act No. XX of 1852 and Madras Act No. 1 of 1854. Bengal, Bombay and Madras were three presidencies of the British colonial administration.

[3] Supreme Court judgment dated 15 April 2011 in Civil Appeal No. 3261 of 2011 (*Radhey Shyam (D)Thr. Lrs & Ors v. State of U.P. & Ors*), arising out of Special Leave Petition (Civil) No. 601 of 2009.

Today, the acquisition of land is a complex issue that is governed by a multiplicity of laws enacted by both state and central governments. Land is an item in the State List of the Seventh Schedule of the Constitution.[4] However, 'acquisition and requisitioning of property' lies in the Concurrent List,[5] giving powers to both central and state governments to enact laws. For matters on the Concurrent List, no state government can override an existing central government law, except with Presidential assent.[6]

The legal framework in operation today is the outcome of this distribution of powers. The most discussed law for land acquisition is the Right to Fair Compensation and Transparency in Land Acquisition, Rehabilitation and Resettlement (RFCTLARR) Act, 2013. However, key provisions of this law have been replaced or exempted in state level laws enacted by some state governments. These provisions have also been read down in some state rules notified under the central law. These changes were preceded by efforts of a newly elected government to amend the 2013 law through a series of ordinances.[7] The implications of this have been discussed further in this chapter. The 2013 law also has limited applicability for land acquired under thirteen other national laws related to coal, railways, highways, atomic energy and other sector specific acquisition.

Laws for acquisition are important to include in the study of environmental laws. Land acquisition legislations come into play when the government requires private land for a variety of purposes, which can include defence, infrastructure, industrial, disaster management or rehabilitation projects. The acquisition could impact not only private property but also livelihoods, water sources, existing government infrastructure and can involve the physical displacement of communities. According to India's Draft National Land Utilisation Policy of 2013, changes in land use are likely to pose critical challenges for sustainable development. It said that infrastructure expansion, industrial acceleration and urbanisation are likely to increase the pressure on

---

[4] Entry 18, List II, Seventh Schedule, Constitution of India, 1950.

[5] Entry 42, List III, Seventh Schedule, Constitution of India, 1950.

[6] Article 254 (2) of the Constitution of India, 1950.

[7] Kanchi Kohli and Debayan Gupta, 'Mapping Dilutions in a Central Law', Occasional Paper, Centre for Policy Research, New Delhi, 2017, 5.

existing land uses.[8] A 2015 report of the World Bank predicted 'enormous strains' on land governance in India. The report stated that this could lead to several conflicts including social dislocation, food insecurity and environmental degradation.[9]

The land acquisition process also intersects with environment regulation. The amount of land required for a project, its administrative break up and status of approvals have to be revealed while seeking environment and forest clearances (discussed in Chapters 3 and 5). Land acquisition laws have also been discussed in environmental litigation when land use change was carried out before compensation was paid[10] or when more land was acquired than what was immediately required by a project or activity.[11] Finally, rehabilitation and resettlement (R&R) is not just a requirement under land acquisition laws, but also a condition of environment approvals granted under the Environment Impact Assessment (EIA) notification discussed in Chapter 5. For these reasons, it is critical to study the legal processes for land acquisition along with the laws that seek to protect the environment and regulate land uses.

This chapter has four sections:

I.   *The 1894 Law*
II.   *Legal Framework of the 2013 Law*
III. *Rehabilitation and Resettlement*
IV.   *State-Level Amendments to the Central Law*

I   THE 1894 LAW

THE COLONIAL LAW

The colonial government enacted a law for land acquisition in 1870[12] based on pre-existing legislations on acquisition of land for private companies and

---

[8] Department of Land Resources, *National Land Utilisation Policy* (New Delhi: Government of India, 2013), 1.

[9] World Bank, *India: Land Governance Assessment (National Synthesis Report)* (Washington, DC: World Bank, 2015), 1.

[10] National Green Tribunal judgment dated February 2016 in Appeal No. 48 (THC) of 2012 (*Puran Singh & Anr v. State of Himachal & Ors*).

[11] National Green Tribunal judgment dated 30 March 2012 in Appeal No. 8 of 2011 (*Prafulla Samantaray & Anr v. Union of India & Ors*).

[12] Land Acquisition Act X of 1870 passed by the Governor General of India in Council on January 4, 1870.

public purpose. It gave responsibility to the collector to conduct a proper valuation of the land acquired and assess the compensation to be paid. In case a land holder refused to accept the compensation amount, the case would be referred to a civil court. The law envisaged that special assessors would be appointed to assist the court. In instances where the court and assessors disagreed, the decision would rest before the high court.

This law gave way to the LA Act, 1894.[13] The first version of the 1894 law did not provide any person aggrieved by the acquisition the opportunity to raise an objection against the intent for acquisition or the amount of compensation. This became possible when the law was amended in 1923.[14] The objects and reasons of this amendment specified that the primary purpose for enacting the new law was to introduce mechanisms through which objections could be raised before local governments.

The LA Act, 1894 continued to be operational after India's independence. It was replaced by the RFCTLARR Act, 2013. However, it is important to understand the procedures under the old law for two reasons:

- There are ongoing legal cases related to the 1894 law that are either pending final decision in courts or awaiting final payment of compensation dues.

- Section 24 in the new 2013 law provides for proceedings of the 1894 law to either lapse or fresh compensation amounts to be calculated in certain circumstances.

The above issues are discussed in this chapter.

Five aspects of the 1894 Act allow us to understand the framework of this law:

- *Preliminary notification and objections*: This was the first step of the law where the district collector issued a preliminary notification specifying the government's intention to acquire land. This was popularly known as the 'Section 4 order' or notification. It allowed officers or 'workmen' of the government to survey, inspect and mark boundaries. Any person who was 'interested in any land' could make objections in writing and in person to the collector.

---

[13] This Act was amended several times with the last amendment being the Land Acquisition (Amendment) Act, 1984 (68 of 1984).

[14] Act No. 38 of 1923.

- *Declaring intent to acquire for public purpose*: Once all responses were examined and the state government was satisfied that the land required was for public purpose, the government would publish a 'declaration' in the official gazette and in two local newspapers. A public notice would also be issued by the district collector(s). This declaration was seen as 'conclusive evidence that the land is needed for a public purpose or for a company'. Following this declaration, the district collector could initiate the proceedings for land acquisition which include demarcating the land and issuing public notices. Any person who had prior 'interest' in the land to be acquired could stake a claim for compensation. This claim needed to be made in writing.

- *Public purpose*: The 1894 law's definition of public purpose included planned development in or of village sites, towns, cities and specific government schemes. It also included land required for residential purposes in areas affected by natural calamities, especially for poor and landless families. Land could be acquired for rehabilitation of displaced persons, carrying out any educational, housing, health or slum clearance schemes of the government. Public purpose was also applicable for creating premises or buildings for locating a public office. This did not include land for offices or buildings for private companies.

- *Determining compensation*: The 1894 law did not provide for any formula based on which compensation could be calculated. It was the responsibility of the district collector to compute and issue 'awards' for compensation. This was based on the true area of the land. The compensation was to be determined on the collector's 'opinion' of what amount should be paid and to whom based on available information. Once the awards were 'received' by claimants, the collector could take possession of the land to be acquired.

- *Acquiring land for companies*:[15] The law allowed the government to acquire land for companies if they were convinced that it is for work that is 'likely to prove useful to the public' or for a company which is engaged in 'work which is for a public purpose'.

Other than the 1894 law, there are several other central and state statutes which allowed governments to acquire land. This includes laws that existed during the precolonial times and newer ones enacted post-independence. Several of these laws are in operation even today and lay out their own procedures for land acquisition. The R&R provisions for some of them are according to the Right to Fair Compensation and Transparency in Land

---

[15] Part VII: Acquisition of Land for Companies of the Land Acquisition Act, 1894.

Acquisition, Rehabilitation and Resettlement (RFCTLARR), 2013, discussed later in this chapter.

Land acquisition has been contested both in and outside the court. A 2017 study[16] reveals that challenges to acquisitions were made under fifteen central- and eighteen state-level statutes. The database of cases fought in the Supreme Court show that 87 per cent of 1,269 cases studied were related to acquisitions under the 1894 law. The research states that there is a high likelihood of litigation related to other central laws and the specific state laws in the district and high courts. The same study also points out that 63.4 per cent of the total land acquisition related cases in the Supreme Court are about petitioners seeking enhanced compensation.

LAND ACQUISITION AND ENVIRONMENTAL CLEARANCES

The acquisition of land is often dealt with separately or in parallel to the project approvals related to environmental impacts, pollution and forest diversions. However, there are major overlaps in these processes when applied to the construction and operation of any industrial, infrastructure or mining project. These overlaps between the land acquisition process and project approvals have been carried over to the post-2013 legal framework for land acquisition.

The legal trajectory of the 400 megawatt (MW) Maheshwar Dam in Madhya Pradesh is an important case to understand how the environment clearance of a project is directly linked with processes of land acquisition. The case also illustrates the conflicts arising out of the pendency of R&R after land acquisition.

The Maheshwar dam is located in Nimad, the south-western region of the state of Madhya Pradesh. The project is part of the Narmada Valley Development Plan[17] under which 30 large and 135 medium-sized dams have been planned in

---

[16] N. Wahi, A. Bhatia, D. Gandhi, S. Jain, P. Shukla and U. Chauhan, *Land Acquisition in India: A Review of Supreme Court Cases from 1950 to 2016* (New Delhi: Centre for Policy Research, 2017), 12.

[17] The Narmada Valley Development Plan (NVDP) is the largest interstate, multipurpose river development scheme of India. The project was launched in 1975 by the Narmada Valley Development Authority (NVDA), a state government-run organisation constituted in 1985 for planning and implementation of the NVDP. The NVDA, under the NVDP, has prepared a detailed plan for exploiting water of Narmada and its tributaries for irrigation, power generation, navigation and other purposes (see http://www.mp.gov.in/en/web/guest/narmada-valley-development1).

the Narmada valley. The project impacts 61 villages. The dam would submerge 5,697 hectares of which 1,060 hectares is private land. Of these villages, 13 would be fully submerged by the dam, 9 would be partially submerged and the remaining 39 villages would lose their agricultural land.

The land acquisition for this project was initiated in the late 1980s. However, it was only in 1997 that some villagers received notices under the 1894 law.[18] In the same year, the project authorities initiated construction activities. The situation escalated into an opposition against the land acquisition where affected people led a series of rallies, protests and demonstrations under the banner of the Narmada Bachao Andolan. The first mass protest against the dam was reportedly on 11 January 1998 and more than 25,000 people were present.[19]

These protests led to the government constituting a Task Force in January 1998[20] and eventually a formal review of the entire project. The Task Force concluded that the dam should not be constructed without addressing the issues related to displacement and livelihood loss. It also recorded that the number of families affected is 4,000 and not 2,264 as the project authorities disclosed earlier. Since a proper and humane approach to R&R was required, the constitution of another committee was recommended to ensure this. It also suggested that the project should resume construction only after a cost benefit analysis with fresh computation and a comprehensive plan for rehabilitation according to the state R&R policy was done.[21] However, the construction activity continued without complying with the Task Force's recommendations.[22]

---

[18] Kanchi Kohli, Meenakshi Kapoor, Manju Menon and Vidya Viswanathan, *Midcourse Manoeuvres: Community Strategies and Remedies for Natural Resource Conflicts in India* (New Delhi: Centre for Policy Research, 2018), 101.

[19] C. Palit and A. Vanaik, 'Monsoon Risings', *Counter Currents*, 11 June 2003, available at https://www.countercurrents.org/en-narmada110603.htm, accessed on 1 July 2020.

[20] Shripad Dharmadikary, 'Implementing the Report of the World Commission on Dams: A Case Study of the Narmada Valley in India', *American University International Law Review* 16, no. 6 (2001): 1591–1630, accessed from https://digitalcommons.wcl. american.edu/cgi/viewcontent.cgi?referer=https://search.yahoo.com/&httpsredir=1& article=1259&context=auilr, accessed on 15 September 2020.

[21] Ibid.

[22] Sanjay Sangvi, 'Role of State in the Maheshwar Project', *Economic and Political Weekly* 33, no.37 (1998): 1686–1688.

The completion of R&R was also one of the conditions of the environment clearance letter issued to the project. This approval was issued under the EIA notification, 1994 (See Chapter 5) in favour of Narmada Valley Development Authority (NVDA). In 2001, this approval was transferred to a private developer which was a joint venture between the Madhya Pradesh State Government and a private company. One of the conditions of the environment clearance was that

> M/s SMHPCCL would be fully responsible for ensuring satisfactory resettlement and rehabilitation. Full arrangements should be made for resettlement and rehabilitation of project affected families from 61 villages (13 fully submerged, 9 partially submerged and 39 where only land is getting submerged) getting affected due to maximum water level at 165.80 mt.

The R&R was to be completed by December 2003 or six months prior to the commencement of submergence, whichever was earlier.

Financial irregularities reportedly led to construction work being stalled for five years between 2001 and 2006. During this time, there was no movement on R&R. When the work resumed in 2006, after financial restructuring of the company and the property attached to the loan was released by the state government, there was no R&R plan in place.[23] The work on the dam was started without submitting the R&R plan with the environment ministry as stipulated in its environment clearance (EC). In 2006, the Environment Ministry issued a stop work order to the project authorities. The construction had to be suspended till the time the R&R plan was finalised and approved by the monitoring committee.[24]

In February 2009, the monitoring committee, constituted as part of the environment clearance process, carried out a site inspection. The committee reported the non-compliance of the R&R conditions. Between October 2009 and April 2011, the environment minister and Chief Minister of Madhya

---

[23] L. Bavadam, 'Private Project, Public Fears', *Frontline* 23, no. 13 (14 July 2006), available at http://www.frontline.in/static/html/fl2313/stories/20060714003502700. htm, accessed on 31 July 2017.

[24] *The Hindu*, 'Centre Suspends Work on Maheshwar Project', 13 June 2006.

Pradesh corresponded on the non-compliance of the R&R condition. The monitoring committee's report to the site in December 2009 and February 2010 also observed that there was no progress on R&R. However, the construction activity had reached an advanced stage. On 23 April 2010, the ministry issued a stop work order recording negligible progress on the R&R.[25]

The political tensions grew between February and May 2011 when the Chief Minister of Madhya Pradesh wrote to the Prime Minister's Office against the stop work order. The letter also raised questions on the environment ministry's locus standi to raise issues of R&R. In this regard, the environment minister, Jairam Ramesh responded,

> R&R proceeding pari passu with dam construction was a key condition of the May 2001 environmental clearance and the MoE&F is well within its rights to be concerned with this issue, as can be seen from the environmental clearance letter of 1 May 2001.

The Environment Ministry continued to be dissatisfied with the state government's response that 70 per cent of the R&R was complete. However, stop work order was lifted in May 2011.[26] The note by the environment minister concluded,

> I have no option but to agree to the lifting of the stop-work order on the construction of the last five spillway gates. However, these gates shall not be lowered until satisfactory completion of R&R and its review. The filling up of the reservoir up to 154 mtrs will be considered after the R&R work has been completed.

The affected people approached different benches of the High Court of Madhya Pradesh as the work continued on the project without R&R measures being in place. One case in the Jabalpur Bench went on until 2018 when it was finally disposed of with an order that the petitioners can approach the

---

[25] Detailed note by Jairam Ramesh, Minister, Ministry of Environment and Forests, on the Maheshwar Hydroelectric Project, Madhya Pradesh, dated 6 May 2011.

[26] Ibid.; V. Venkatesan, 'A Tale of Two Dams', *Frontline* 25, no. 12 (17 July 2011), available at http://www.frontline.in/static/html/fl2812/stories/20110617281203200.htm, accessed on 15 September 2020.

Grievances Redressal Authority constituted on 3 October 2017 specifically for 'relief of rehabilitation of the oustees' of the project.[27]

In 2012, a series of cases were filed in the Indore bench of the High Court of Madhya Pradesh. These were combined in a single writ.[28] The petitions challenged the notification issued under the LA Act, 1894. The farmers argued that no compensation was paid to them and they were still in possession of their land. They demanded that the land acquisition proceedings be discontinued. The case was pending even as the new law for land acquisition was enacted in 2013, which legally provided for return of land in certain circumstances (discussed further). On 17 March 2016, the court ruled in favour of the petitioner to say that the proceedings for land acquisition should start afresh under the 2013 law. The court relied on the precedence set in a previous judgment of the same court.[29] In the present case, the court concluded,

> the proceedings initiated by respondents deserves to be quashed and are accordingly quashed. The land owners are already in possession of the land in question and they shall continue to remain in possession of the land in question, however, respondents shall be free to take appropriate action in accordance with law under the Act of Right to Fair Compensation and Transparency in Land Acquisition Rehabilitation and Resettlement Act, 2013, if they so desire.

Another case was filed before the National Green Tribunal (NGT) in 2011.[30] The main grievance here was the Environment Ministry's decision to allow the government to raise the height of the dam to 154 metres. Since this would increase the submergence and impact area, the compensation and rehabilitation issues had to be addressed. In 2012 the petition was disposed of with directions. However the tribunal continued to monitor the implementation of the R&R plan. In April 2014, the NGT restrained the storage of water in the reservoir till R&R was completed. The petition

---

[27] High Court of Madhya Pradesh at Jabalpur, Writ Petition No. 1359 of 2009 (*Narmada Bachao Andolan v. the State of Madhya Pradesh*).

[28] High Court of Madhya Pradesh at Indore, Writ Petition (Civil) 217 of 2012 (*Kamalchand v. State of M.P. and Others*).

[29] High Court of Madhya Pradesh judgment dated 2 February 2016 in Writ Petition No. 1263 of 2012 (*Indore Hygiene Product & Ors v. Indore Development Authority & Ors*).

[30] National Green Tribunal judgment dated 9 August 2012 in Appeal No. 26 of 2012 (*Antarsingh Patel & Anr v. Union of India & Ors*).

demanded that compensation should be paid as per the 2013 law. In July 2014, the NGT put the onus on the state government to address this issue. In October 2015,[31] the state government was directed not to open the gates of the dam till R&R was complete. The NGT stated,

> We may not go into the issue of the release of funds but we can certainly reiterate our orders that there shall be no lowering/closure of the gates causing submergence of the area without the R & R being completed and the order is reiterated accordingly.

The R&R for the Maheshwar project remained incomplete.[32] In April 2020, the Madhya Pradesh government cancelled the project by terminating the power purchase, escrow and rehabilitation agreements.[33]

THE PUSH FOR A NEW LAW

High-level committees of the Parliament and other specialised bodies have made recommendations on the need to revise the colonial law. Networks of social movements and civil society also made recommendations.[34] One specific recommendation was the need to combine R&R provisions with a law for land acquisition. The 1894 law had dealt only with the award of compensation. There was no comprehensive national or sector specific legislation for R&R. The National Advisory Council (NAC) that was set up in 2004 advised the government to combine acquisition, compensation and R&R into one single law.[35] The NAC stated that acquisition and R&R 'are two sides of the same coin'.

---

[31] National Green Tribunal Order dated 28 October 2015 in Appeal No. 26 of 2012. 2012 (*Antarsingh Patel & Anr v. Union of India & Ors*).

[32] *Counterview*, 'Narmada Dam Oustees of 3 States to Begin Protest March on May 27, Reach Bhopal for Public Hearing on June 4', 26 May 2018, available at https://www.counterview.net/2018/05/narmada-dam-oustees-of-3-states-to.html, accessed on 1 July 2020.

[33] Lyla Bavadam, 'Victory and Vindication', *Frontline*, 5 June 2020, available at https://frontline.thehindu.com/the-nation/article31657615.ece, accessed on 30 September 2020

[34] Ashirbani Dutta, *Development-Induced Displacement and Human Rights* (New Delhi: Deep & Deep Publications, 2008), 184.

[35] Chapter II of the Right to Fair Compensation and Transparency in Land Acquisition, Rehabilitation and Resettlement Act (RFCTLARR Act), 2013.

The 2007 report of the Standing Committee on Rural Development also highlighted some contentious issues.[36] This committee observed that from some individuals and communities, land is acquired more than once. The committee recommended the new land acquisition Bill proposed in 2007 should ensure that benefits received are doubled every time land is acquired for the second or subsequent time. The committee also recommended that the government should make a list of all the land that is acquired and has remained unutilised.

In February 2009, the lower house of the Indian Parliament passed two separate bills. These were the Land Acquisition (Amendment) Bill 2009 (LAA 2009) and Resettlement and Rehabilitation Bill, 2009 (R&R 2009).[37] However, these bills lapsed[38] after they could not be passed in the Rajya Sabha, the upper house of the Parliament.

In 2011, the NAC's working group on Land Acquisition, Resettlement and Rehabilitation recommended the enactment of a single comprehensive legislation titled the 'National Development, Land Acquisition, Resettlement and Rehabilitation Act'. The NAC's May 2011 press release[39] observes,

> A single comprehensive law which would discourage forced displacement, and minimise adverse impacts on people, habitats, environment, food security and biodiversity.

Securing agricultural lands from acquisition was an important focus of the NAC's recommendations, which said,

> The law should ensure that all possible options of more barren, less fertile and waste lands have been explored before acquiring agriculture land.

---

[36] Department-Related Standing Committee on Rural Development, *39th Report on the Land Acquisition (Amendment) Bill, 2007*, (New Delhi: Lok Sabha Secretariat, 2008), 1–2.

[37] Bill No. 97-C of 2007 and Bill No. 98-C of 2007.

[38] Puja Mehra, 'No Man's Land', *Business Today*, 7 August 2011, available at https://www.businesstoday.in/magazine/features/land-acquisition-protests-halt-india-projects-greater-noida/story/17153.html, accessed on 2 August 2019.

[39] Press release dated 25 May 2011 related to the thirteenth meeting of the National Advisory Council, available at https://www.prsindia.org/sites/default/files/bill_files/NAC%20press%20release_Land%20Acq%20and%20others.pdf, accessed on 15 September 2020.

This set in motion the process of the introduction of The Draft National Land Acquisition and Rehabilitation & Resettlement Bill, 2011[40] and subsequently the enactment of the RFCTLARR Act in 2013.

II   LEGAL FRAMEWORK OF THE 2013 LAW

The RFCTLARR Act, 2013 came into force on 1 January 2014. This is one of the several national and state-level laws under which land acquisition is undertaken. This section discusses the legal framework prescribed under the 2013 law and how it interfaces with some of the other laws under which acquisition is carried out.

The Preamble of the RFCTLARR, 2013, requires

a humane, participative, informed and transparent process for land acquisition for industrialisation, development of essential infrastructural facilities and urbanisation with the least disturbance to the owners of the land and other affected families and provide just and fair compensation to the affected families whose land has been acquired or proposed to be acquired or are affected by such acquisition.

It also seeks to

make adequate provisions for such affected persons for their rehabilitation and resettlement and for ensuring that the cumulative outcome of compulsory acquisition should be that affected persons become partners in development leading to an improvement in their post acquisition social and economic status.

The law is different from the 1894 version in the following ways:

- It brings together the process of land acquisition and rehabilitation in a single comprehensive legislation.
- It extends the provisions of compensation and rehabilitation to not just land owners but all those who are likely to lose livelihoods and use rights over the land being acquired.

---

[40] Ministry of Rural Development, *The Draft National Land Acquisition and Rehabilitation & Resettlement Bill, 2011* (New Delhi: Government of India, 2011).

- It introduces the provision of seeking consent of the landowners, which was missing in the earlier law.

- It introduces the concept of Social Impact Assessment (SIA) to understand who are the people affected by the acquisition and inform the rehabilitation plan.

- It introduces conditions for the lapse of land acquisition proceedings and repatriation of land unutilised from prior acquisitions.

- It does not distinguish the acquisition of land for government and for companies. Land can be acquired for private companies as long as it satisfies public purpose.

- It has special safeguards for tribal dominated Fifth and Sixth Schedule Areas and to ensure food security concerns prior to land acquisition.

- It presents a mechanism to calculate compensation rates for the loss of land.

This law defines those affected by acquisition as family units. An affected family is one whose land or immovable property is to be acquired. It also includes all families who don't own land but use or are dependent on such a land. This includes agricultural labourers, tenants, sharecroppers and those who hold usufruct rights like seasonal graziers. The law narrows this down to families whose 'primary source of livelihood' is dependent on the land at least for three years prior to the acquisition. Artisans like potters, basket weavers and makers of fishing nets who would be dependent on the materials available on the land being acquired can also be included under the definition of those affected. All rights recognised under the Scheduled Tribes and Other Traditional Forest Dwellers (Recognition of Forest Rights) Act, 2006 (See Chapter 3) are also covered by the definition. Any livelihood that satisfies the definition would have access to compensation and rehabilitation under this Act.

Purpose and Scope of Acquisition

The RFCTLARR allows a wide ambit of purposes for which the government can acquire land. It covers acquisition for government departments, public sector undertakings and the private sector, provided it is for public purpose as defined in the law.

Public purpose includes:

- *Strategic and defence purposes* related to national-level naval, military, armed forces, paramilitary forces, state police and any work related to 'national security'.

- *Housing infrastructure* related to housing for specific income groups or 'weaker sections' as specified by government at any given point of time. The projects for residential purpose can also include those for poor or landless persons residing in areas affected by natural calamities or displaced by the implementation of any government scheme.

- *Area development projects* for 'planned development' or 'improvement of village sites' or urban areas.

- *Infrastructure projects* including a wide range of activities in the 2009 Department of Economic Affairs notification which covers mining, power generation, industry, roads, pipelines, hotels, industrial parks, special economic zones, educational facilities, etc.[41] It does not include private hospitals, private educational institutions and private hotels. It also includes projects for sports, healthcare, tourism or transportation.

Areas predominantly inhabited by constitutionally recognised scheduled tribes have been granted special governance and protection status in India. This is through the Schedules V and VI as prescribed under Article 244 of the Constitution of India. While the Schedule V covers 10 states in India,[42] special provisions apply to the VI Schedule Areas demarcated in the states of Assam, Meghalaya, Tripura and Mizoram.[43] The 2013 law affords special protection for scheduled tribes. It emphasises that 'as far as possible no acquisition shall

---

[41] Department of Economic Affairs (Infrastructure Section) number 13/6/2009-INF dated 27 March 2012.

[42] Ten states, namely, Andhra Pradesh, Chhattisgarh, Gujarat, Himachal Pradesh, Jharkhand, Madhya Pradesh, Maharashtra, Odisha, Rajasthan and Telangana have Fifth Scheduled Areas (available at http://pesadarpan.gov.in/en_US/fifth-schedule-areas/-/asset_publisher/LmZ9LplaCh7b/content/fifth-schedule-are-2?inheritRe direct=false&redirect=http%3A%2F%2Fpesadarpan.gov.in%2Fen_US%2Ffifth-schedule-areas%3Fp_p_id%3D101_INSTANCE_LmZ9LplaCh7b%26p_p_lifecycle%3D0%26p_p_state%3Dnormal%26p_p_mode%3Dview%26p_p_col_id%3Dcolumn-1%26p_p_col_count%3D1, accessed on 26 March 2018).

[43] Details of the Sixth Schedule are available on the website of Ministry of External Affairs at https://www.mea.gov.in/Images/pdf1/S6.pdf, accessed on September 30 2020.

be made in Scheduled Areas'. The law states that if land is to be acquired in such areas, it is to be done only as a 'demonstrable last resort'.

The Act also has a special provision related to maintaining food security for the country. Therefore, irrigated multi-cropped land cannot be acquired, except under exceptional circumstances and also as a demonstrable last resort.

LEGAL PROCESS FOR ACQUISITION OF LAND

The process of acquiring land under the 2013 law can be understood through the following four major steps. These need to be read along with the modifications and amendments carried out by various state governments discussed in Section 4. These steps will also differ if the land is acquired under one of the 13 central laws listed in Schedule IV of the RFCTLARR, 2013.[44]

The process includes:

- *Prior consent*: Prior consent is a mandatory step in any land acquisition process initiated under the 2013 law. If land is being acquired for a private company, consent of at least 80 per cent of the affected families as defined above is mandatory. If acquisition is for a private company under the public–private partnership (PPP) model or collaboration with a public agency, consent of at least 70 per cent of affected families must be taken. This process has to be initiated by the district collector along with the SIA.

    If land is being acquired in Fifth and Sixth Scheduled Areas, consent is mandatory irrespective of which national law is being used to acquire the land. This is because Section 41,[45] which grants this special protection to Scheduled Areas, (including requirement of consent) is under Chapter V on Rehabilitation and Resettlement Award. The 2013 law mandates that even though land may be acquired under the 13 national laws listed in the

---

[44] The 13 central laws are as follows: the Ancient Monuments and Archaeological Sites and Remains Act, 1958; the Atomic Energy Act, 1962; the Damodar Valley Corporation Act, 1948; the Indian Tramways Act, 1886; the Land Acquisition (Mines) Act, 1885; the Metro Railways (Construction of Works) Act, 1978; the National Highways Act, 1956; the Petroleum and Minerals Pipelines (Acquisition of Right of User in Land) Act, 1962; the Requisitioning and Acquisition of Immovable Property Act, 1952; the Resettlement of Displaced Persons (Land Acquisition) Act, 1948; the Coal Bearing Areas Acquisition and Development Act, 1957; the Electricity Act, 2003 and the Railways Act, 1989.

[45] Sections 41 and 41 (3) to be read with Section 105 of the RFCTLARR Act, 2013.

Fourth Schedule of the law, the R&R benefits will need to be processed as per the 2013 law. This implies that even though land is being acquired under the National Highways Act, 1956 or the Coal Bearing Areas Act, 1957, all benefits related to Chapter V will be applicable.

• *Notification for determining public purpose and SIA*: The first step in any acquisition proceedings under the 2013 law is to issue a notification stating the intent and purpose for land acquisition and for the commencement of consultations for SIA. The objective of the SIA is to understand the social impacts of the project and also to determine whether acquisition is within the ambit of the definition of 'public purpose'.

The notification has to be made available in the local language to the panchayat, municipality or municipal corporation and in the area where land is being acquired. A copy of this notification is also to be made available in the offices of the district collector, subdivisional magistrate and block level revenue office (*tehsil/taluka*). It also needs to be uploaded on the state government's website. The SIA has to be carried out in consultation with the abovementioned local authorities. The funds for SIA have to be provided by the 'requiring body' that is seeking acquisition. It also needs to include the Social Impact Management Plan 'listing the ameliorative measures required to be undertaken for addressing the impact of the project'.[46]

In case an EIA is required for the activity for which an SIA is being prepared, a copy of the SIA has to be made available at the time of the EIA appraisal (see Chapter 5). The 2013 law exempts SIA for all irrigation projects that are under the purview of the EIA notification, 2006.[47]

• *Preliminary notification for acquisition* can be issued only after the determination of public purpose and prior consent are complete. This notification needs to specify the area to be acquired, a statement for public purpose and the details of the 'Administrator' responsible for R&R. The notification needs to be published in the official gazette of the government and in two daily newspapers, which are circulated in the locality where acquisition is taking place. One of these needs to be a regional language newspaper. It also needs to be made available in the local language to the concerned panchayat, municipality or municipal corporation and be

---

[46] Details of the social impact assessment (SIA) process are prescribed under the RFCTLARR (Social Impact Assessment and Consent) Rules, 2014.

[47] Section 1 (C) of the schedule ('List of Projects or Activities Requiring Prior Environmental Clearance') under the EIA notification.

available in the offices of the district collector, the subdivisional magistrate and the block (tehsil/taluka) office. The notification has to be uploaded on the state government's website and circulated in the affected area through specific mechanisms prescribed under the SIA or respective state rules.

- *Restriction on altering land use*: Once this notification is issued, there can be no transaction such as sale, lease or renting made with the land that is going to be acquired. The law also states that no person should cause any 'encumbrances on such land'. The notification allows for carrying out surveys on the land, marking out boundaries and digging of bore wells.

- *Determining R&R*: The process related to R&R comes into play after issuance of this preliminary notification, which is to be facilitated by the district collector. The district collector, following the completion of the R&R process, issues the final public notice for acquisition. It also provides the opportunity to the person whose land is being acquired to raise objections. This process is discussed in Section 3.

RETURN OF UNUSED LAND

The RFCTLARR deals with the issue of repatriation of unused land acquired under the 2013 Act and the 1894 Act. It does not address similar instances of land acquired under any other central or state laws.

If the land acquired under the 2013 law is not utilised for five years, then the central law allows for two possibilities. This procedure is laid out under Section 101 of the Act. The land can either be returned to the original owner or reverted to a land bank created by the state government. The land bank allows state governments to use the land once acquired at any given point of time without initiating fresh land acquisition proceedings.

If acquisition proceedings were initiated under the 1894 law at least five years before the commencement of this law and no compensation has been paid or if physical possession is still with the land owner, the proceedings are deemed to lapse. Fresh land acquisition proceedings have to start under the 2013 law including the process of consents and SIA. This clause is in Section 24 (2). Section 24 deals with all cases related to land acquisition proceedings initiated under the LA Act, 1894 and is discussed in detail in the next section.

Section 24 (2) can be understood through a 2017 judgment of the Gujarat High Court.[48] A group of farmers from Jamnagar approached the court

---

[48] High Court of Gujarat at Ahmedabad judgment dated 11 November 2017 in Special Civil Application No. 20362 of 2015 with Special Civil Application No. 12012 of 2014 to Special Civil Application No. 12023 of 2014.

seeking return of land as their land was still in their possession and they had not received compensation. The private company responded saying that the responsibility to pass on the compensation was that of the state government which had acquired the land on their behalf. The company's contention was that while they had transferred the compensation amount to the government, the farmers refused to accept it. The area in question covers 5 per cent of the company's special economic zone.

In response to this case, the company challenged the constitutional validity of Section 24(2) as it was 'unreasonable'. All these matters were heard together, and a common judgment was issued on 11 November 2017. The company argued that the 1894 law did not specify any timelines within which the land had to be taken into possession. The company also stated that the inclusion of Section 24 in the 2013 law was too sudden and did not give any transition time. Therefore, a private company should not be penalised for the lethargy of the government, which was responsible for securing possession of the land.

The central government countered this challenge and argued that Section 24(2) is 'constitutionally valid and not ultra vires'. However, they agreed that possession had not been taken from the appellant farmers.[49]

The judgment of the High Court interpreted that Section 24(2) had to be applied differentially to acquisition of land by government for its own use and for private companies. The judgment records,

> Even after the deposit of compensation as determined in the awards, it is
> an obligation on the part of the State to take possession from the owners
> of the land and to hand over the same to the beneficiary company in
> terms of the agreement.

It is for this and other reasons cited in the judgment that Section 24(2) was read down by the Gujarat High Court. The judgment observed,

> [T]he provision under section 24(2) of the Right to Fair Compensation
> and Transparency in Land Acquisition, Rehabilitation and Resettlement

---

[49] Express New Service, 'Centre Opposes Reliance Plea against New Land Act Clause', *Indian Express*, 19 February 2016, available at http://indianexpress.com/article/india/india-news-india/centre-opposes-reliance-plea-against-new-land-act-clause/, accessed on 3 August 2019.

Act, 2013 is to be read down by holding that the said provision cannot be applied to the acquisition of lands covered by Part-VII of the Land Acquisition Act, 1894, where award is passed and compensation is deposited by the beneficiary Company as per the terms of the award.

This High Court judgment was challenged in the Supreme Court.[50] The proceedings in this case need to be read along with the developments discussed in the next section.

Legal Challenges to the Retrospective Application of RFCTLARR

Section 24 of the 2013 Act deals with all cases related to the acquisition of land under the 1894 Act. Section 24 (1) addresses two situations. One in which an award under Section 11 of the LA Act, 1894 has been made and the other in which a final award has not been made. In the first case, the land acquisition proceedings would continue under the LA Act, 1894, and in the second instance, compensation would be calculated as per the provisions of the RFCTLARR, 2013.

Section 24(2) (as discussed earlier) deals with cases where a final award has been made, but either the compensation has not been paid or the physical possession has not been taken for five years. In such cases, acquisition proceedings will deem to have lapsed and fresh acquisition proceedings have to be initiated under the RFCTLARR, 2013. Section 24, more particularly subsection 2, has been a subject of litigation since the Act came into force in 2014.

A five-judge constitutional bench of the Supreme Court passed a detailed judgment interpretating the applicability of Section 24.[51] The bench was constituted following two conflicting interpretations of the Supreme Court related to Section 24, the details of which are discussed further.

The judgment of 2014 (known as the Pune Municipal Corporation case) dealt with the question of when section 24(2) can be invoked to claim that

---

[50] Supreme Court Special Leave Petition (C) Nos. 000474–000476 of 2018 (*Narbheshanker Prabhashanker Khetia & Ors v. Union of India & Ors*).

[51] Supreme Court of India judgment dated 6 March 2020 in Special Leave Petition (Civil) No(s). 9036–9038 of 2016 (*Indore Development Authority v. Manohar Lal & Ors*).

'compensation has not been paid'.[52] The Court held that this clause cannot be invoked if the compensation amount has been submitted in court (as per the requirement under Section 31 of the LA Act, 1894) or has been given to the affected landowners.

This conclusion relied upon several cases of the Supreme Court, one of which is a 2016 judgment.[53] In this judgment, the Court held that the five-year period, after which the existing acquisition proceedings lapse, is inclusive of the time delays in compensation or possession due to court interference. In its conclusion, the judgment said that Section 24(2)

> has been prescribed with a view to benefit the land-losers and the period spent in litigation due to challenge to the award or the land acquisition proceedings cannot be excluded.

This interpretation was questioned in another case being heard by a two-judge bench of the Supreme Court.[54] So, the matter was referred to a larger, three-judge bench of the Court.

In February 2018, this three-judge bench in what is called the *Indore Development Authority* case held that the *Pune Municipal Corporation* judgement had failed to consider certain aspects pertaining to the understanding of Section 24. The three-judge bench was of the opinion that the word 'paid' in Section 24(2) would also include the unconditional tendering of the compensation amount even if such an amount is not deposited in court or given to the affected landowners. They also were of the opinion that the periods spent in litigation not be included within the five-year time period to determine the lapsing of the acquisition proceedings.[55]

The question before the three-judge bench was regarding what orders could be given. Two judges were of the opinion that the interpretation of the

---

[52] Supreme Court of India judgment dated 24 January 2014 in Civil Appeal No. 877 of 2014 (arising out of Special Leave Petition [Civil] No. 30283 of 2008) (*Pune Municipal Corporation and Another v. Harakchand Misirmal Solanki and Ors*).

[53] Supreme Court of India judgment dated 10 September 2014 in Civil Appeal No. 8700 of 2013 (*Sree Balaji Nagar Residential Association v. State of Tamil Nadu*).

[54] Supreme Court judgment/order dated 12 January 2016 in Special Leave to Appeal (Civil) No. 10742 of 2008 (*Yogesh Neema v. State of Madhya Pradesh*).

[55] Supreme Court judgment dated 8 February 2018 in Civil Appeal No. 20982 of 2017 (*Indore Development Authority v. Shailendra (Dead) thr. Lrs. & Ors*).

Pune Municipal Corporation case should be considered as *per incuriam* (that is, through lack of care). The third judge suggested that this entire matter should be referred to a larger bench.

In a few weeks after the preceding judgment, another three-judge bench of the Supreme Court was apprised of situations where several high courts were relying upon the *Indore Development Authority* case to decide on matters. This three-judge bench ordered that no benches of the Supreme Court or the high courts would take decisions in cases where similar matters were pending until the issue was addressed by a larger bench of the Supreme Court.[56] The order states,

> Insofar as cases pending in this Court are concerned, we request the concerned Benches dealing with similar matters to defer the hearing until a decision is rendered one way or the other on the issue whether the matter should be referred to larger Bench or not. Apart from anything else, deferring the consideration would avoid inconvenience to the litigating parties, whether it is the State or individuals.

Thereafter, two separate division benches of the Supreme Court[57] referred the matter to the Chief Justice of India requesting him to constitute a 5-judge constitution Bench to deal with the matter. The constitutional bench issued its judgment in March 2020 [58] with a detailed interpretation of the retrospective application of the 2013 law. One of the most significant directions was:

> The word 'or' used in Section 24(2) between possession and compensation has to be read as 'nor' or as 'and'.

This implies that land acquisition proceedings initiated under the 1894 law would apply only if compensation has not been paid and the land is in the possession of the land owner. The judgment also interpreted which instances would fall within the ambit of compensations as not paid.

---

[56] Supreme Court judgment/order dated 21 February 2018 in Special Leave to Appeal (Civil) No. 8453 of 2017 (*State of Haryana v. Goenka Tourism Corporation Limited and Anr*).

[57] Supreme Court judgment/order dated 6 March 2018 in Special Leave to Appeal (Civil) No(s). 9798–9799 of 2016 (*Indore Development Authority v. Shyam Verma*); Supreme Court judgment/order dated 22 February 2018 in Civil Appeal No. 4835 of 2015 (*State of Haryana v. Maharana Pratap Memorial Trust*).

[58] Supreme Court judgment dated 6 March 2020 in Special Leave Petition (Civil) No(s). 9036–9038 of 2016 (*Indore Development Authority v. Manohar Lal & Ors*).

III   REHABILIATION AND RESETTLEMENT

The 1894 law did not provide for a specific section on R&R. There were different national- and state-level policies through which this process was undertaken. If land was acquired through any national or state laws like the Coal Bearing Areas Act, 1957, the guidelines for R&R provided under that law or relevant state R&R policy would be applicable.

MOVING TOWARDS A NATIONAL REHABILITATION AND RESETTLEMENT POLICY

In the early 2000s, some attempts were made to enact a national rehabilitation policy, which could serve as the guidance for rehabilitation across the country. A National Policy on Resettlement and Rehabilitation for Projected Affected Families, 2003, was gazetted on 17 February 2004. The first draft of this policy was brought out in 1993 and was revised a number of times. One of biggest critiques of the 2003 policy was its definition of 'affected people' that was limited to only landowners whose land was being acquired. This meant many others dependent on the land or those living downstream of a project would not be compensated for change of land use. It was argued that communities living downstream of hydropower or mining projects are often affected by obstruction of river flow, sedimentation or mine overruns.[59]

Another critique was that the policy was relevant only for projects displacing more than 500 families in the plains and 250 families in hilly areas. Both the methodology and the misuse of setting such a lower limit were questioned. It was argued that this clause was likely to be misused, and projects or government departments requiring land could under-report the number of affected families.[60]

With the enactment of the 2013 law, land acquisition and R&R were part of one legal regime. The R&R provisions of this law apply to three contexts:

- *Incomplete proceedings under the 1894 law*: The new law specified that in instances where the final award for land acquired under the 1894 law was

---

[59] Manju Menon, *Resettlement and Rehabilitation: Moving from an Inadequate Policy to a Bad One*' (Pune: Centre for Communication and Development, 2003), available at http://www.infochangeindia.org/environment/151-environment/analysis/5743-resettlement-and-rehabilitation-moving-from-an-inadequate-policy-to-a-bad-one, accessed on 4 August 2019.

[60] Ibid.

pending, R&R provisions of the 2013 law would apply (Section 24(1)
(a)). In such cases, irrespective of when the acquisition proceedings were
initiated, compensation benefits would be provided under the 2013 law. If
final awards had been issued, the compensation proceedings would need
to be carried out according to the provisions of the 1894 law.

- *R&R for land acquisition by laws listed in the Fourth Schedule*: The
  RFCTLARR Act, 2013 clarifies that if land is acquired under any of
  the thirteen laws listed in its Fourth Schedule after 2013, then the R&R
  benefits of the new law would apply. This was specified in Section 105
  of the Act and for which a notification was issued. The notification was
  issued as RFCTLARR (Removal of Difficulties) Order, 2015.[61]

- *R&R under the 2013 law*: The law provided a detailed procedure by which
  compensation is to be calculated and R&R provisions are to be executed.
  This is discussed further.

INSTITUTIONAL ARRANGEMENT FOR REHABILITATION AND RESETTLEMENT

The law puts into place an elaborate institutional mechanism through which
an R&R scheme would need to be prepared, executed and monitored. The
first step entails the appointment of an administrator and a commissioner for
R&R by the state government.

The administrator should not be below the rank of a collector, additional
collector or deputy collector, and if not, an equivalent official of the Revenue
Department. The administrator is responsible for 'the formulation, execution
and monitoring of the Rehabilitation and Resettlement Scheme....'

The commissioner for R&R should not be below the rank of a commissioner
or secretary and is responsible for 'supervising the formulation of rehabilitation
and resettlement schemes or plans and proper implementation of such schemes
or plans' and 'for the post-implementation social audit in consultation with the
Gram Sabha in rural areas and municipality in urban areas'.

In case the land to be acquired is more than hundred acres, an R&R
Committee has to be set up under the chairpersonship of the district
collector. The role of this committee is 'to monitor and review the progress of
implementation of the Rehabilitation and Resettlement scheme'.

---

[61] Ministry of Rural Development Notification S.O. 2368 (E) on 28 August 2015.

The district collector can pass individual land acquisition awards once the commissioner approves the R&R scheme with specific entitlements. No land use change of the acquired area is permitted if the R&R is not complied with.

The law also provides for setting up of a Land Acquisition, Rehabilitation and Resettlement Authority, which is headed by the District Judge or a qualified legal practitioner with at least seven years' experience. All appeals related to R&R procedure and awards can be filed before this adjudicating authority.

COMPENSATION AND FINAL AWARD

It is the district collector's responsibility to compute compensation. As a first step, the collector needs to determine the market value of land using any of these three criteria:

- The market value of land is determined through registration of sale deeds or agreements under the Indian Stamp Act, 1899. In many urban areas, this is known as circle rates.

- The average sale price of similar land situated in the nearest village or vicinity to where the land is being acquired.

- In case of private companies or public–private partnership projects, the amount mutually agreed upon at the time consent is being sought for acquisition.

The compensation amount is arrived at based on the following criteria:

- Market value of the land.

- Adding a multiplier amount depending on where the land is located. In urban areas, the multiplier is 1 and, in rural areas, it is fixed based on a sliding scale from 1 to 2 depending on the radial distance from urban centres (See Section 4 for amendments).

- Compensation for the loss of assets like buildings, trees, plants, crops and any other damages.

In addition to this, the final award needs to include a 'solatium' of 100 per cent on the entire amount. The 'solatium' amount is a compensation to ameliorate the pain of forcible acquisition.

REHABILITATION AND RESETTLEMENT PACKAGE

The 'Rehabilitation and Resettlement Package' is discussed in the Second and Third Schedules of the RFCTLARR Act, 2013.

The Schedule II provides for the essential elements for any R&R scheme, which needs to be prepared by the district collector. This includes both landowners and families whose livelihoods were dependent on the land acquired.

The schedule presents eleven elements of the R&R scheme like provision of housing in case of displacement and offer for land against loss of land. In case the land is acquired for 'urbanization purposes', 20 per cent of the land that would be developed needs to be reserved for land owning project affected families. The affected family also needs to be offered employment or annuity.

If families are displaced, they are to be provided ' a monthly subsistence allowance equivalent to three thousand rupees per month for a period of one year from the date of award'. The law also requires the payment of a one-time transportation cost of ₹50,000 for shifting of families along with building materials, belongings and cattle.

The Fourth Schedule explains how R&R is to be dealt with for livelihood losses of artisans, small traders or self-employed persons not owning land. The law provides:

> Each affected family of an artisan, small trader or self-employed person or an affected family which owned non-agricultural land or commercial, industrial or institutional structure in the affected area, and which has been involuntarily displaced from the affected area due to land acquisition, shall get one-time financial assistance of such amount as the appropriate Government may, by notification, specify subject to a minimum of twenty-five thousand rupees.

The costs for the preparation of the R&R scheme and its execution need to be borne by the institution or company who requires the land. In cases where a 'requiring body' needs additional land for the purposes of the R&R, the district collector needs to be informed. The district collector can also take a view that the R&R can be quantified in monetary terms. In this case, the payment secured from the 'requiring body' can be transferred to the families whose land is being acquired.

The Third Schedule is related to infrastructural entitlements in the area where the affected families are being resettled. The schedule lists minimum requirements for such resettlement. It includes construction of all-weather roads with easement rights for all resettled families. Proper drainage and sanitation plans have to be executed before physical resettlement.

The resettlement site needs to have one or more 'assured sources of safe drinking water' but does not specify whether it needs to be piped supply or access to natural water source. Drinking water supply and grazing land are also required for cattle. In addition, infrastructural facilities, such as *Panchayat Ghars*, fair price shop, burial and cremation grounds, electrical facilities, health centre and playgrounds for children are needed. The Third Schedule requires that separate land must be earmarked for traditional tribal institutions.

IV   STATE-LEVEL AMENDMENTS TO THE CENTRAL LAW

The central government introduced substantial amendments to the RFCTLARR Act, 2013 through an ordinance promulgated by the President of India on 31 December 2014.[62] This ordinance introduced changes in the applicability of the consent and SIA clauses to almost all purposes for which acquisition can take place as per the law. It also extended the scope of 'public purpose' to include private hospitals and private educational institutions among other projects, which was otherwise exempted. Subsequently, the RFCTLARR (Amendment) Bill 2015 was introduced in the Lok Sabha on 24 February 2015 and passed on 10 March 2015.[63]

The passage of this Bill was held up in the Rajya Sabha, and the issue of the proposed amendments was placed before a Joint Parliamentary Committee (JPC) in May 2015 chaired by Shri. S. S. Ahluwalia, Member of Parliament. This committee had over 24 sittings and also called for public views and suggestions between May and June 2015.[64] The report of this JPC is yet to be finalised.[65]

During the pendency of the Bill, two more ordinances, namely the RFCTLARR (Amendment) Ordinance, 2015 and the RFCTLARR (Second Amendment) Ordinance, 2015 were issued on 2 April 2015 and 30 May 2015,[66] respectively. At the same time, there were huge protests from farmers,[67] social

---

[62] The RFCTLARR (Amendment) Ordinance, 2014 (No. 9 of 2014).

[63] The RFCTLARR (Second Amendment) Bill, 2015 (Bill 152 of 2015).

[64] ANI, 'Government Invites Public Suggestion on Land Acquisition Bill by June 8', *Financial Express*, 23 May 2015, available at https://www.financialexpress.com/economy/government-invites-public-suggestions-on-land-acquisition-bill-by-june-8/75293/, accessed on 15 September 2020.

[65] Details are available at http://164.100.47.194/Loksabha/Committee/Committee Information.aspx?comm_code=67&tab=1, accessed on 15 September 2020.

[66] The RFCTLARR (Amendment) Ordinance, 2015 (No. 4 of 2015) and the RFCTLARR (Amendment) Second Ordinance, 2015 (No. 5 of 2015).

[67] Jayashree Nandi, 'Farmers to Hold Mass Rally in Delhi against Land Ordinance', *Times of India*, 24 January 2015, available at http://timesofindia.indiatimes.com/city/delhi/Farmers-to-hold-mass-rally-in-Delhi-against-land-ordinance/articleshow/46003736.cms, accessed on 23 August 2017.

activists[68] and the political parties, especially the Indian National Congress, that considered this law as one of its flagship legislations.[69]

In August 2015, the third ordinance was allowed to lapse and a 'statutory order' was issued[70] to extend the benefits related to compensation, infrastructure amenities and R&R to acquisitions under thirteen central acts listed in the Fourth Schedule of the RFCLARR, 2013.

The lapse of the ordinances, pendency of the Bill and JPC report did not restrain state governments from carrying out their own changes to the land acquisition laws. They did this through state rules under the central Act or by enacting new land acquisition laws.

The RFCTLARR, 2013 allows states to create rules under the Act. Using the power given under Section 109 of the Act, the central and several state governments passed certain rules. Andhra Pradesh, Assam, Bihar, Himachal Pradesh, Jharkhand, Sikkim, Tamil Nadu, Telangana, Tripura and Uttar Pradesh passed rules after the central government.[71]

As discussed in the introductory section, acquisition of property is part of the concurrent list of the constitution and allows both central and state governments to make laws. State governments cannot bypass a central law like the RFCTLARR, 2013 except with presidential assent as provided for under Article 254(2) of the Constitution of India. This has allowed several state governments

---

[68] TNN, 'Land Acquisition Ordinance: Anna Hazare Vows to Fight Modi Govt's "Anti-farmer" Bill', *Times of India*, 23 February 2015, available at http://timesofindia. indiatimes.com/india/Land-acquisition-ordinance-Anna-Hazare-vows-to-fight-Modi-govts-anti-farmer-bill/articleshow/46340475.cms, accessed on 23 August 2017.

[69] PTI, 'Congress Launches Protest against Land Ordinance in Bhatta Parsaul', *The Hindu*, 14 January 2015, available at http://www.thehindu.com/news/national/other-states/congress-launches-protest-against-land-ordinance-in-bhatta-parsaul/article6789097.ece, accessed on 23 August 2017.

[70] Ministry of Rural Development Order No. S.O. 2368 (E) dated 28 August 2015 (F. No. 13011-01/2014-LRD).

[71] The Andhra Pradesh RFCTLARR Rules, 2014 (20 November 2014); Assam RFCTLARR Rules, 2015 (31 July 2015); Bihar RFCTLARR Rules, 2014 (27 October 2014); Himachal Pradesh RFCTLARR (SIA and Consent) Rules, 2015 (9 April 2014); Jharkhand RFCTLARR Rules, 2015 (30 March 2015); Sikkim RFCTLARR Rules, 2015 (2 March 2015); Telangana RFCTLARR Rules, 2015 (19 December 2015); Tripura RFCTLARR Rules, 2015 (22 April 2015; and Uttar Pradesh RFCTLARR Rules, 2015 (12 December 2015).

to pass their own laws incorporating specific amendments proposed in the 2014 and 2015 ordinances. Some of the states that have done so include Tamil Nadu, Gujarat, Maharashtra, Telangana, Rajasthan and Jharkhand.[72]

Tables 8.1 and 8.2 highlight the changes made in the land acquisition proceedings through 15 State Amendment Acts or State Rules under the 2013 law.[73]

Table 8.1 Exclusions made by state land acquisition amendment Acts

RFCTLARR, 2013	Amendment			
	Gujarat	*Maharashtra*	*Tamil Nadu*	*Telangana*
Requirement of SIA and Consent.	Exemptions from SIA and consent for a range of projects. For example, those important for national security or defence, rural infrastructure, affordable housing for poor, industrial corridors and other infrastructural projects, including projects under PPPs.	Only private projects will be subject to the provisions of SIA and consent. Public–private partnerships excluded from requirement.	Acquisitions carried out under the four state laws are exempt from the provisions of the RFCTLARR. These laws don't require SIA and consent. Laws include those for Harijan Welfare Schemes, Acquisition for Industrial Purposes and Highways.	Exemptions from SIA and consent for a range of projects. For example, those important for national security or defence, rural infrastructure, affordable housing for poor, industrial corridors and other infrastructural projects, including projects under PPPs.

Source: Authors.

---

[72] Some examples of state-level Acts are the RFCTLARR (Tamil Nadu Amendment) Act, 2014; the RFCTLARR (Gujarat Amendment) Act, 2016; and the RFCTLARR (Telangana Amendment) Act, 2017.

[73] These tables are adapted from Kohli and Gupta, 'Mapping Dilutions in a Central Law'. After this publication of Kohli and Gupta in 2017, the states of Andhra Pradesh and Karnataka have enacted state amendment legislations. However, copies are not available in the public domain as of February 2020.

Table 8.2 Changes to the central land acquisition law through state rules

Clause in central rules, 2014	Dilutions				
I. Social impact assessment					
	Andhra Pradesh	Jharkhand	Sikkim	Telangana	Uttar Pradesh
Notice period for ensuring participation in public hearing: 3 weeks	1 week	2 weeks	2 weeks	1 week	1 week
Time period for conducting the SIA and submitting the report: 6 months	No amendment	No amendment	No amendment	No amendment	2 months
Requirement to determine nature of the land, the size of holdings, ownership patterns, land prices and changes in ownership	Present	Present	Present	Present	Absent

(Contd)

II. Consent		
Clause in the central rules	*Amendments*	
	Himachal Pradesh	*Sikkim*
Quorum of gram sabha for obtaining consent: 50%	The quorum of the *gram sabha* is to be fixed according to provisions of either the Himachal Pradesh Panchayati Raj Act, 1994 (one-third of the total members) or the Himachal Pradesh Municipal Corporation Act, 1994 (half of the total members) or the Himachal Pradesh Municipal Act, 1994 (half of the total members)	No amendment
Requirement of having one-third woman representation.	Absent	Present
Time period for displaying the terms and conditions of the acquisition: 3 weeks	No Amendment	15 days

(Contd)

(Contd)

III. Return of unused land as acquired under 2013 law

Clause in central rules	Amendments					
	Jharkhand	*Karnataka*	*Odisha*	*Tamil Nadu*	*Telangana*	*Tripura*
If any land is lying unused for a period of 5 years or more, the land may be returned either to the original owners and their heirs or be reverted to a government land bank	No provision for returning land back to the original owners or their heirs. The land will revert back to a government land bank	Before the land is returned to the original owners or their heirs, they must deposit the current market value of the land minus the compensation paid at the time of acquisition	No provision for returning land back to the original owners or their heirs. The land will revert to a government land bank.	If the DC is satisfied that the land in question may be used for some other purpose, then he can pass a notice reverting the land to a government land bank	The 5-year period has been increased from either 5 years or the time period required for setting up the project, whichever of the two is higher	Before the land is returned to the original owners or their heirs, they must deposit the current market value of the land minus the compensation paid at the time of acquisition

IV. Calculating compensation

Clause in central rules	Amendments			
	Assam	*Chhattisgarh*	*Haryana*	*Tripura*
Multiplying factor for land in urban areas: 1.0	Multiplying factor for land in urban areas: 1.5	Multiplying factor for land in urban area: 1.0	Multiplying factor for land in urban areas: 1.0	Multiplying factor for land in urban area:1.0
Multiplying factor for land in rural areas: 2.0	Multiplying factor for land in rural areas: 2.0	Multiplying factor for land in rural areas: 1.0	Multiplying factor for land in rural areas: 1.0	Multiplying factor for land in rural areas: 1.0

(Contd)

(Contd)

Clause in central rules	Amendments		
	Chhattisgarh	Jharkhand	Madhya Pradesh
The limit on acquisition of irrigated multi cropped land: Not more than 1% of the total irrigated land in the state	The limit on acquisition of irrigated multi cropped land: Not more than 2% of the total irrigated land in the state	The limit on acquisition of irrigated multi cropped land: Not more than 2% of the total irrigated land in the state	No change
The limit on acquisition of other agricultural land: Not more than 5% of the total net sown area in the state	The limit on acquisition of other agricultural land: Not more than 1% of the total net sown area in the state	The limit on acquisition of other agricultural land: Not more than one-fourths of the total net sown area in the state	The limit on acquisition of other agricultural land: Not more than 50% of the total net sown area in the state

V. *Provisions related to food security*

Source: Authors.

9

Climate Litigation and Policy Frameworks

INTRODUCTION

Climate change has emerged as the most pressing environmental concern of global, national and local significance. The main cause of climate change is greenhouse gases (GHGs) that include carbon dioxide, methane, nitrous oxide and fluorinated gases emitted from a variety of industrial processes.[1] Carbon emissions from various industrial sectors form the largest percentage of GHGs. High levels of GHGs in the atmosphere create a gaseous blanket around the earth, packing in the earth's radiation and preventing the escape of heat. It also disrupts the earth's water and weather cycles that connect the oceans and land. Most parts of the world have witnessed unpredictable and intense weather events affecting food productivity, human health and wild biodiversity. This wide spectrum of effects is increasingly being linked to climate change. According to a 2019 assessment by Morgan Stanley, the global economic cost of climate disasters over three years was USD 650 billion.[2] The high carbon-emitting economic activities also involve changes in local land use that cause deforestation, land degradation and environmental pollution.

The international discussions on global warming and climate change date back to the mid-1970s. However, it was only in 1988 that the Intergovernmental

---

[1] Website of the United States Environment Protection Agency (USEPA), available at https://www.epa.gov/ghgemissions/overview-greenhouse-gases, accessed on 22 August 2019.

[2] Tom DiChritopher, 'Climate Disasters Cost the World $650 billion over Three Years—Americans Are Bearing the Brunt: Morgan Stanley', CNBC, 14 February 2019, available at https://www.cnbc.com/2019/02/14/climate-disasters-cost-650-billion-over-3-years-morgan-stanley.html accessed on 30 September 2020.

Panel on Climate Change (IPCC) was established under the aegis of the United Nations (UN) with the objective of collating scientific evidence related to climate change.[3] The IPCC's first assessments in 1990 and 1992 presented conclusive evidence on rising temperature and attributed it to rising GHG emissions.[4] Since then, the 1992 United Nations Framework Convention on Climate Change (UNFCCC) and the Kyoto Protocol (1997) have recognised 'that developed countries are principally responsible for the current high levels of GHG emissions in the atmosphere as a result of more than 150 years of industrial activity'.

The UNFCCC is one of the three framework conventions of the Rio Earth Summit, the other two being the Convention on Biological Diversity and the UN Convention to Combat Desertification. These are elaborated in Chapter 1. The UNFCCC is the overarching international framework binding all its signatories, including India, to commit to a reduction in GHG emissions based on the principle of 'common but differentiated responsibilities (CBDR)'.[5] This principle was central to the discussions at the Earth Summit held in Rio de Janeiro in 1992. It recognised that while some countries were more responsible for climate change, the entire international community had to contribute to reduce emissions and address the challenges posed by climate change.[6] The convention requires countries to act 'on the basis of equity and in accordance with their common but differentiated responsibilities and respective capabilities'.[7] In fact, the CBDR has been a pivot around which the international climate negotiations have coalesced.

---

[3] The Intergovernmental Panel on Climate Change, (undated), About the IPCC, at https://www.ipcc.ch/about/, accessed on 16 September 2020.

[4] Intergovernmental Panel on Climate Change (IPCC), *Climate Change: The 1990 and 1992 IPCC Assessments* (Canada: World Meteorological Organisation and UNEP).

[5] Principle 7 of the Rio Earth Summit (1992) declared,
   In view of the different contributions to the global environmental degradation, State have common but differentiated responsibilities. The developed countries acknowledge the responsibility that they bear in the international pursuit of sustainable development in the view of pressures their societies place on the global environment and the technologies and financial resources they command.

[6] Website of the United Nations Framework Convention on Climate Change (UNFCCC), available at https://unfccc.int/kyoto_protocol, accessed on 22 August 2019.

[7] United Nations, *United Nations Framework Convention on Climate Change* (New York: United Nations, 1992).

The first phase of international climate negotiations did not result in much because of the classic problem of collective action. In most countries, the domestic economic growth and developmental imperatives such as poverty alleviation and growing energy demands took precedence over mitigating GHG emissions. The 1997 Kyoto Protocol organised the international community into various groupings to increase their bargaining positions and leverage support for their proposed actions. The commitments made as part of the UNFCCC were unenforceable and non-binding on the parties of the convention. Bidwai states that the Kyoto Protocol bound a small group of countries to emissions reduction targets.[8] Countries such as the United States of America (USA), India and China could not reach an agreement on specific targets. However, scholars claim that the Protocol provided crucial lessons for the next steps in the global efforts to deal with climate change.[9]

The present phase of global efforts is framed by the Paris Agreement which came into force in 2016. Following the challenges of the Kyoto Protocol, this agreement sought to establish climate action targets at the national level. The approach shifted to what domestic governments could do to develop while

> holding the increase in the global average temperature to well below 2°C above pre-industrial levels and pursuing efforts to limit the temperature increase to 1.5°C above pre-industrial levels, recognizing that this would significantly reduce the risks and impacts of climate change.

To achieve progressive actions on the issue, the commitments made in the Paris Agreement are voluntary until 2020 after which the committed targets would become binding. As part of the Paris Agreement, all signatory countries, including India, have developed what are known as Intended Nationally Determined Contributions (INDCs), which are to be enforced nationally

---

[8] Praful Bidwai, *The Politics of Climate Change and the Global Crisis* (New Delhi: Orient BlackSwan, 2012), xiii.

[9] Daniel Bodansky, 'The Legal Character of the Paris Agreement', *Review of European, Comparitive and International Environmental Law* 25, no. 2 (2016): 142–150, available at http://climatechange.moe.gov.lb/Library/Files/Uploaded%20Files/The%20Legal%20Character%20of%20the%20Paris%20Agreement.pdf, accessed on 30 August 2019.

and reported to the UNFCCC secretariat. National governments also need to ensure monitoring, reporting and verification mechanisms for the INDCs.[10]

The Paris Agreement pushed national governments to introduce new laws and policies or revisit, revise and strengthen their existing ones to enable national actions on climate change. The three decades of international climate talks have led several countries to develop climate policies, enact new laws and amend their domestic environmental regimes in response to the international policy discourses on climate change. National governments have established ministries and departments dedicated to implement programmes and schemes arising out of climate policies and plans. Today, there are more than 1,500 laws and policies worldwide that address climate change or transition to a low carbon economy, 106 of which have been introduced since the Paris Agreement.[11]

There has also been an increasing global trend to litigate on climate change issues. The USA and Europe have witnessed several cases filed to push government and private sectors on climate actions. As per a global database on climate change litigation, there are 276 cases across 25 countries excluding the USA as of 2018.[12]

- Seventy-seven per cent of cases are primarily concerned with emissions reduction or mitigation-related grievances. Australia has cases on adaptation, mostly dealing with coastal planning and risks from climate change hazards.[13]

---

[10] United Nations Environment Programme, *The Status of Climate Change Litigation: A Global Review* (Nairobi: UNEP, May 2017), available at http://columbiaclimatelaw. com/files/2017/05/Burger-Gundlach-2017-05-UN-Envt-CC-Litigation.pdf, accessed on 22 August 2019.

[11] Michal Nachmany and Joana Setzer, *Policy Brief: Global Trends in Climate Change Legislation and Litigation: 2018 Snapshot* (London: Grantham Research Institute on Climate Change and the Environment and the Centre for Climate Change Economics and Policy, 2019), available at http://www.lse. ac.uk/GranthamInstitute/wp-content/uploads/2018/04/Global-trends-in-climate-change-legislation-and-litigation-2018-snapshot-3.pdf, accessed on 22 August 2019.

[12] Website of the Climate Change Litigation Databases, available at http://climatecase chart.com, accessed on 22 August 2019.

[13] Mitigation actions are those taken to reduce emissions or enhance sinks to stabilise levels of greenhouse gases (GHGs) in the atmosphere. Adaption actions are needed to deal with the consequences of emissions already in the atmosphere.

- Corporations take most matters to court; they bring 40 per cent of cases to court, 90 per cent of which are against governments. These cases seek to overturn administrative decisions denying licences on grounds of climate change impacts of activities.

- Cases have also been filed to challenge allocation of allowances under emissions trading schemes or governmental schemes to specific sectors or companies.

International climate policy frameworks look to go beyond the international climate regime set by the terms of the Paris Agreement. These frameworks involve three approaches: multi-level governance, sub-national actions[14] and co-benefits.[15] The first approach recognises that while the push for formulating policies arise from international agreements, they are to be implemented by institutions based at regional, national and sub-national levels. The second approach recognises that local and regional governments have an important role to implement the Paris Agreement and to take actions to address climate change.[16]

The third approach of co-benefits is seen as a 'win–win' strategy to get ambitious climate actions. This strategy is 'aimed at capturing both development and climate benefits in a single policy or measure',[17] and it allows climate mitigation and adaptation actions to be tied to domestic developmental outcomes. This framing aims to make climate actions less intractable by viewing

---

[14] Kirsten Jörgensen, Anu Jogesh and Arabinda Mishra, 'Multi-level Climate Governance and the Role of the Subnational Level', *Journal of Integrative Environmental Sciences* 12, no. 4 (2015): 235–245.

[15] Navroz Dubash and Lavanya Rajamani, *Rethinking India's Approach to International and Domestic Climate Policy (Policy Challenges 2019–2014)* (New Delhi: Centre for Policy Research, 2019), 31.

[16] Pasquale Capizzi, Emily Castro, Ingrid Gonzalez, Jakob Lindemann, Patricia Lizarazo, Tangmar Marmon and Maryke van Staden, *Enabling Subnational Climate Action through Multilevel Governance* (Bonn: GIZ, 2017), available at https://www.international-climate-initiative.com/fileadmin/Dokumente/2017/GIZ_ICLEI_UNHabitat_2017_EN_Enabling_subnational_climate_action.pdf, accessed on 30 August 2019.

[17] Akiko Miyatuska and Eric Zusman, *What Are Co-benefits: Fact Sheet No. 1, Asian Co-benefits Partnership* (Kanagawa: Institute for Global environmental Strategies, 2019), available at https://pub.iges.or.jp/system/files/publication_documents/pub/nonpeer/2393/acp_factsheet_1_what_co-benefits.pdf, accessed on 30 August 2019.

them as developmental actions and see them for their potential to 'green' local development rather than an imposition of limits by the global community. India's National Action Plan on Climate Change (NAPCC) discussed in the following section drew substantially on this approach. The objective of these action plans is not only quicker and effective reduction of carbon emissions but also channelising international funds available for climate mitigation and adaption to implement mainstream developmental programmes and projects in India.

This chapter discusses India's climate policy framework and efforts to litigate on climate change issues in India and other countries. It is arranged in the following sections:

I.   Policy Framework on Climate Change in India
II.   Climate Litigation

Scholars have argued that a robust legal and policy framework for climate change is important for India which is 'a country with 7,500 km of coastline, extensive tracts of low-lying areas, high population density, poor infrastructure and continued reliance on agriculture for livelihoods'.[18]

While India has made forays into climate litigation, as discussed in Section II of this chapter, it does not have a specific overarching law that frames its responses to the problem of climate change. The litigation efforts largely draw upon existing environment law and regulation.

I   POLICY FRAMEWORK ON CLIMATE CHANGE IN INDIA

India's policy framework on climate change mitigation and adaptation can be understood through the three approaches in international climate policy discussed earlier: multilevel governance, sub-national actions and co-benefits. These are brought together in the elements that form the rubric of India's policy framework: India's INDCs, the NAPCC and the State Action Plans.

INTENDED NATIONALLY DETERMINED CONTRIBUTIONS

As discussed in the introduction, the UNFCCC's Paris Agreement required national governments to finalise their specific national commitments that will

---

[18] Navroz Dubash and Lavanya Rajamani, *Rethinking India's Approach to International and Domestic Climate Policy (Policy Challenges 2019–2014)* (New Delhi: Centre for Policy Research, 2019), 28–32.

help achieve the terms of the agreement. Foremost in these terms were efforts to keep the global temperature rise below 2 degree Celsius. Towards this, India finalised its INDC document in October 2015.[19] The document specified the national context which requires the national government to balance development priorities with environmental protection[20]. It clarifies that all actions proposed are required to be viewed through the lens of equity and CBDR and says,

> Given the development agenda in a democratic polity, the infrastructure deficit represented by different indicators, the pressures of urbanization and industrialization and the imperative of sustainable growth, India faces a formidable and complex challenge in working for economic progress towards a secure future for its citizens.

There are eight commitments in the INDC document. Some of these are values that can be adopted by citizens and others are concrete targets.

- The first two commitments encourage healthy and sustainable living based on traditions and values as well as adopting a climate-friendly path in economic development.

- The third, fourth and fifth commitments are specific contributions to be achieved by 2030. These include a commitment to reduce the emission intensity of the country's gross domestic product by 33–35 per cent with 2005 levels as the baseline, derive 20 per cent of India's cumulative installed electricity capacity from non-fossil fuel based energy sources and to carry out a massive afforestation drive to create a carbon sink that can store 2.5 to 3 billion tonnes of carbon.

- The last three contributions are about investments and funds. This includes enhancing investments in sectors and regions such as agriculture, water resources, the Himalayan region, coastal areas, health and disaster management that are considered vulnerable to climate change. For this, there is a call for new and additional funds from both domestic sources and the international community to help bridge the resource gap for joint and collaborative research and development on new climate technologies.

---

[19] Vidya Venkat, 'India to Cut Emissions Intensity', *The Hindu*, 2 October 2015, available at https://www.thehindu.com/news/national/india-sets-ambitious-goals-to-tackle-climate-change/article7715679.ece, accessed on 16 September 2020

[20] MoEFCC, *India's Intended Nationally Determined Contribution: Working towards Climate Justice* (New Delhi: Government of India, 2015).

NATIONAL ACTION PLAN ON CLIMATE CHANGE

The Prime Minister's Council for Climate Change (PMCCC) was established in 2007 with the objective of coordinating an action plan to assess impacts of climate change and implement measures for climate adaptation and mitigation. In June 2008, the council launched the NAPCC[21] which acknowledged that the plan was designed on the approach of co-benefits. The document clarified,

> The National Action Plan on Climate Change identifies measures that promote our development objectives while also yielding co-benefits for addressing climate change effectively. It outlines a number of steps to simultaneously advance India's development and climate change-related objectives of adaptation and mitigation.

The 2008 plan presented eight national missions which would direct the government's activities on climate adaptation and mitigation. According to the NAPCC, this would be a mix of ongoing government initiatives, programme revisions and a new set of actions which would allow for a 'multi-pronged, long-term and integrated strategies for achieving key goals in the context of climate change'.

The NAPCC's implementation is dependent on India's domestic environment laws and policy frameworks. It is useful to read these missions and their objectives along with the legal and policy discussions presented in different chapters of this book. For example, the afforestation drive of the Green India Mission is linked to the compensatory afforestation mechanisms mentioned in Chapter 3 on forest reservation and conservation. Similarly, the water mission talks about water pricing as a conservation tool, and this ties in with the concept of the water conservation fee introduced in ground water laws, which has been discussed in Chapter 7 on ground and surface water extraction.

The eight missions are as follows:

- *National Solar Mission* which looks to increase the share of solar power in the overall energy mix of the country. The ultimate objective is to make the solar power competitive with coal and other fossil fuel-based energy. It emphasised on international cooperation for technology and increasing domestic manufacturing capacity including through government funding.

---

[21] Prime Minister's Council on Climate Change, *National Action Plan on Climate Change (NAPCC)* (New Delhi: Government of India, 2008).

- *National Mission for Enhanced Energy Efficiency* introduces four new mechanisms. This includes a market-based certification mechanism to enhance cost effectiveness and improvements in efficiency of energy-intensive industries and facilities, pushing for a shift in energy efficient appliances, financing of demand side management in all sectors requiring energy and development of fiscal instruments to promote energy efficiency.

- *National Mission on Sustainable Habitat* aims to promote energy efficiency as a central component of urban planning. This is to be achieved in three ways: developing the Energy Conservation Building Code for energy efficiency in large and new commercial buildings, improving public transport and introducing urban waste recycling into urban planning.

- *National Water Mission* is directed towards encouraging integrated water resource management, which would allow for conserving water, minimising waste and ensuring equitable distribution within and across states. This mission set a goal of 20 per cent improvement in water use efficiency through pricing and other measures, so that the issue of water scarcity can be addressed.

- *National Mission for Sustaining the Himalayan Ecosystem* had a special focus on conserving the Himalayan glaciers. The NAPCC sees this as a joint effort of climatologists, glaciologists and other experts. It also presented the need to set up a monitoring network for assessing freshwater resources and overall health of the ecosystem.

- *National Mission for Green India* has an explicit mandate of creating carbon sinks. It proposes to carry out afforestation of 6 million hectares of degraded forest lands along with expanding the forest cover from 23 per cent to 33 per cent in the country. This mission also committed funds available with the Compensatory Afforestation Management and Planning Authority (CAMPA) to commence this work. The CAMPA has been discussed in detail in Chapter 3.

- *National Mission for Sustainable Agriculture* aims to make Indian agriculture more resilient to climate change by developing new varieties of crops. It also focused on improving the productivity of rainfed agriculture that was outside the irrigation network. This is to ensure alternative cropping patters that are capable of withstanding weather extremes.

- *National Mission on Strategic Knowledge for Climate Change* aims to promote high-quality research on climate change. It proposes to set up Climate Science Research Fund and also to invite the private sector to develop technologies for adaptation and mitigation of climate change through venture capital funds.

The NAPCC received mixed responses. On the one hand, the NAPCC's emphasis on the co-benefits approach was appreciated, but with a caveat. The plans were critiqued for their lack of methodological clarity. One suggestion was for the government to adopt a multi-criteria analysis to prioritise actions and trade-offs.[22] Bidwai critiqued the action plans for being fragmented, hasty and relying only on technological fixes. He argued that India would require much more than the NAPCC if a low-carbon pathway is to be achieved without compromising on the objectives of eliminating poverty, security of the underprivileged and improving standards of living.[23] Another review pointed to that the NAPCC did not address new threats to climate change such as monopolies on new technologies through intellectual property rights.[24]

In 2014, the PMCCC was reconstituted, and a year later, there were 4 more missions reportedly added to the NAPCC taking the total count to 12. These new missions were the National Wind Mission, Waste-to-Energy Mission, Coastal Mission and Health Mission.[25] However, these were still pending final cabinet approvals as of 2018.[26] A National Adaptation Fund for Climate Change as a central sector scheme was the next step taken to financially support climate mitigation and adaptation activities. The sectors eligible for this fund are agriculture, animal husbandry, water, forestry and tourism. The government appointed the National Bank for Agriculture and Rural Development as the National Implementing Entity.[27]

The NAPCC performance has been evaluated by parliamentary committees[28] and non-governmental organisations.[29] They point to lack of

---

[22] Navroz Dubash, Raghunandan D., Girish Sant and Ashok Sreenivas, 'Indian Climate Change Policy: Exploring a Co-benefits Based Approach', *Economic and Political Weekly* 47, no. 22 (June 2013): 47–61.

[23] Bidwai, *The Politics of Climate Change and the Global Crisis*, 9.

[24] Committee on Estimates, *Performance of the National Action Plan on Climate Change (NAPCC): Thirtieth Report* (New Delhi: Lok Sabha Secretariat, 2018).

[25] Vijeta Rattani, *Coping with Climate Change: An Analysis of India's National Action Plan on Climate Change* (New Delhi: Centre for Science and Environment, 2018).

[26] Committee on Estimates, *Performance of the National Action Plan on Climate Change (NAPCC)*.

[27] NABARD, 'National Adaptation Fund for Climate Change', ix, available at https://www.nabard.org/content.aspx?id=585, accessed on 21 September 2020.

[28] Ibid.

[29] Rattani, *Coping with Climate Change*.

coordination between different government departments and an ineffective and lackadaisical pace of implementation of the projects and activities. The Parliamentary Committee on Estimates in its 2018 report observes that 'against the allocation of ₹350 crore for two years, only ₹212.3 crore i.e. around 60 percent has been released which indicates slow and tardy pace of implementation of the projects'. The slow movement on missions related to sustainable habitat, water, agriculture and forestry, which are multi-sectoral and require interdepartmental coordination, has attracted the attention of critics. These have been compared with the National Solar Mission which has progressed well to achieve its targets. These evaluations also pointed out that the overall monitoring of the missions has not been effective as the PMCCC had reportedly met only once since its constitution in 2014.

STATE ACTION PLANS

Keeping with the objective of sub-national planning, Prime Minister Manmohan Singh called upon state and union territory (UT) governments in 2009 to devise their own climate action plans. Even though this call was to prepare these plans in line with the principles and missions identified at the national level, it was seen as an attempt to decentralise planning. As the Chhattisgarh State Action Plans on Climate Change (CSAPCC)[30] specifies in its introductory section,

> Given that the impacts of climate change will vary across states, sectors, locations, and populations, there can be no 'one-size-fits-all' climate change strategy. India recognises this, and that approaches will need to be tailored to fit specific sub-national contexts and conditions. As such, all Indian States have been asked to prepare State Action Plans for Climate Change (SAPCCs) in line the NAPCC.

This was particularly important as some subjects such as water and agriculture are those in which state governments have a larger role in governance. As discussed in Chapter 7 on water, most of the planning and management roles lie with the state governments except when it comes to using national waterways and interstate disputes. Similarly, forests are a concurrent subject of the constitution, giving both the centre and states a role in forest governance. A State Action Plan on Climate Change (SAPCC)

---

[30] Government of Chhattisgarh, *State Action Plan on Climate Change: Inclusive Growth for Improved Resilience* (Raipur: Government of Chhattisgarh, 2014).

laying down similar objectives as the National Mission for Green India would only facilitate centre–state coordination on climate policy and action.

As of 2018, 32 states and UTs had prepared their SAPCC or Union Territory Action Plan on Climate Change.[31] These action plans sought to reconcile ongoing development and natural resource-related programmes while formulating new schemes for mitigating and adapting to climate change. The Uttarakhand state action plan[32] states,

> The challenge for the state is to holistically converge these existing initiatives and make additional efforts to integrate climate concerns and response measures into all aspects of the development process, from policy and planning to implementation. The state has adopted this as the underlying principle in the formulation of the Uttarakhand Action Plan for Climate Change (UAPCC) and aims to become a green and carbon neutral state by 2020.

Reviews[33] of the SAPCCs point to the following concerns:

- Climate policy continues to be top-down with the push for preparing sub-national plans and their implementation coming from the central government.

- Differences in priorities and variation in institutional frameworks across the action plans throw up challenges in monitoring by the environment ministry.

- The focus of several state and UT action plans was on adaptation than mitigation of climate change. The CSAPCC states, '"Adaptation" will be the predominant philosophy and component of the climate response strategy of Chhattisgarh, while at the same time leveraging opportunities for mitigation co-benefits.'

- Constraints in implementation due to allocation of resources from state and the central governments as a well as capacity of the state-level institutions to implement these plans.

---

[31] Committee on Estimates, *Performance of the National Action Plan on Climate Change (NAPCC)*.

[32] Government of Uttarakhand, *Uttarakhand Action Plan on Climate Change: Transforming Crisis into Opportunity* (Dehradun: Government of Uttarakhand, 2014), 25.

[33] Parul Kumar and Abhayraj Naik, 'India's Domestic Climate Policy Is Fragmented and Lacks Clarity', *Economic and Political Weekly* 54, no. 7 (February 2019), available at https://www.epw.in/node/153770/pdf, accessed on 15 September 2020; Rattani, *Coping with Climate Change*.

- There was reluctance in adopting central government proposals, one example of which is the Conservation Building Code, which was adopted only by about half the states and UTs even after its revision in 2017.[34]

India's climate policy framework made up of the INDCs, national and state action plans and missions operates through institutions located at different geographic scales. The nodal ministry for India's commitments to the UNFCCC is the MoEFCC. However, the various climate missions are anchored within different central government ministries of which the environment ministry as the coordinating body. The environment ministry also works with state governments for implementation of specific state action plans and national policies that require state-level enforcement.

India's climate policy framework has also been critiqued by many scholars. Kumar and Naik state that India's actions need to go beyond the broad promises and numbers on afforestation, emissions reduction and other targets and attend to the economic systems in a more substantive manner.[35] There are trenchant critiques of the implications of market-based mechanisms such as emissions trading schemes ('cap and trade')[36] which do not take into account the political and economic inequities.[37] A recent paper by Dubash and Rajamani emphasises the need for climate policy and India's INDCs to be developed through a

---

[34] Twesh Mishra and Garima Singh, 'Most States Still Lukewarm to Green Building Norms', *Hindu Business Line*, 5 June 2018, available at https://www.thehindubusinessline.com/news/most-states-still-lukewarm-to-green-building-norms/article24089963.ece, accessed on 16 September 2020.

[35] Kumar and Naik, 'India's Domestic Climate Policy Is Fragmented and Lacks Clarity'.

[36] According to the USEPA,

Emissions trading, sometimes referred to as 'cap and trade' or 'allowance trading', is an approach to reducing pollution that has been used successfully to protect human health and the environment. Emissions trading programs have two key components: a limit (or cap) on pollution, and tradable allowances equal to the limit that authorize allowance holders to emit a specific quantity (e.g., one ton) of the pollutant. This limit ensures that the environmental goal is met and the tradable allowances provide flexibility for individual emissions sources to set their own compliance path. Because allowances can be bought and sold in an allowance market, these programs are often referred to as 'market-based'.

Available at https://www.epa.gov/emissions-trading-resources/what-emissions-trading, accessed on 30 August 2019.

[37] Bidwai, *The Politics of Climate Change and the Global Crisis*, 44–48.

consultative process and be particularly attentive to the human rights impacts arising out of any action or inaction related to climate change.[38] A target-based approach of these plans and their implementation have created and could increase conflicts with existing livelihoods and land uses.[39] Finally, the INDCs and the government missions rely on the existing architecture of environmental laws, policies and institutions which are themselves mired in difficulties and challenges and are unable to deal with environmental non-compliance.[40]

II CLIMATE LITIGATION

Environmental disputes have traditionally been centred around personal injury or public harms caused by certain actions. These actions may include an act amounting to a breach of law or an omission to perform an action mandated by law. Most environmental cases in India, including those concerning torts, nuisance or other criminal acts (as discussed in Chapter 1), have sought effective judicial and administrative response. Projects that were refused permissions or were issued executive directions for compliance have also approached courts for redress.

Climate litigation cases also follow a similar logic. They can be understood as new cases involving environmental disputes in which climate change arguments are at the centre of a grievance or a remedy being sought. A 2012 paper defines climate change litigation as one where 'the party filings or tribunal decisions directly and expressly raise an issue of fact or law regarding the substance or policy of climate change causes and impacts'.[41]

---

[38] Dubash and Rajamani, *Rethinking India's Approach to International and Domestic Climate Policy*, 30.

[39] Kanchi Kohli and Manju Menon, *Banking on Forests: Assets for a Climate Cure?* (Pune: Kalpavriksh and Heinrich Boll Foundation, 2011), 32–39; Karthikeyan Hemalatha, 'Why India's Solar Push Could Kill the Livelihood of Pastoral Communities', *IndiaSpend*, 5 August 2019, available at https://www.indiaspend.com/why-indias-solar-push-could-kill-the-livelihood-of-pastoral-communities/, accessed on 15 September 2020.

[40] Manju Menon and Kanchi Kohli, 'Regulatory Reforms to Address Environmental Non-compliance', in *Policy Challenges (2019–2024): Charting a New Course for India and Navigating Policy Challenges in the 21st Century* (New Delhi: Centre for Policy Research, 2019), 32–38.

[41] David Markell and J. B. Ruhl, 'An Empirical Assessment of Climate Change in the Courts: A New Jurisprudence or Business as Usual?' *Florida Law Review* 64, no. 15 (2012): 15–27.

Such cases have been witnessed in increasing numbers in Europe and the USA. Climate litigations are being pursued by a variety of stakeholders, including private corporations in the fossil fuel sector, that aim to challenge administrative decisions and cancellation of licences which were made on the basis of climate change arguments.

Cases have also been filed by citizens against their governments to demand accountability and greater climate action. One such case was filed in November 2015 by a law student in New Zealand, who took the government to court for their insufficient climate ambitions. The case was heard in court in June 2017. On 2 November 2017, the Wellington High Court held that climate change presents significant global risks and that the government is legally accountable for its actions to address climate change. The court determined that the New Zealand minister for climate change had acted unlawfully by failing to review the country's climate change targets for 2050 after the publication of the most recent IPCC Assessment Report. The court refrained from issuing an order against the government as the newly elected government took up office in October 2017 and committed itself to a target of CO_2-neutrality in 2050.[42]

In another case from Colombia, 25 young people in the 7–25 year age group brought a lawsuit against the Colombian national government, several local governments and a number of corporations. The petitioners claimed that climate change along with the government's failure to reduce deforestation and meet its 2020 zero-net Amazon deforestation target threatened their fundamental rights to a healthy environment, life, health, food and water. In April 2018, the Supreme Court ruled in favour of the young people, recognising the Colombian Amazon as having its own rights and ordered the government to make and carry out action plans to address deforestation in the Amazon.[43]

Cases have also been filed by large collectives. In 2015, the Urgenda Foundation along with 900 Dutch citizens[44] sued the Netherlands government for its inaction on climate change and for not doing enough to prevent harm

---

[42] High Court of New Zealand, Wellington Registry Civ 2015-485-919 [2017] NZHC 733 (*Sarah Thomson and the Minister for Climate Change Issues*).

[43] *Dejusticia*, 'In Historic Ruling, Colombian Court Protects Youth Suing the National Government for Failing to Curb Deforestation', 5 April 2018, available at https://www.dejusticia.org/en/en-fallo-historico-corte-suprema-concede-tutela-de-cambio-climatico-y-generaciones-futuras/, accessed on 22 August 2019.

[44] Available at https://elaw.org/nl.urgenda, accessed on 27 February 2020.

to ordinary people. The case relied on the government's duty of care, as well as on Articles 2 and 8 of the European Convention on Human Rights. On 14 April 2015, the court ordered the government to make more stringent GHG reductions by 2020. The appeal was heard on 28 May 2018 by the court of appeal. On 9 October, the court upheld the 2015 groundbreaking decision.[45] In December 2019, the Netherlands's Supreme Court also upheld this decision and observed that governments have a duty to protect citizens from the impacts of climate change. This judgment requires the Dutch government to reduce its GHG emissions by 25 per cent by the end of 2020 (as compared with the 1990 levels).[46]

In several countries, national governments or citizens have taken private corporations to court for their past negative effects on the environment by arguing that those have contributed to climate change. Cases are also filed by environmental organisations and community groups against the government's approvals to new fossil fuel projects which can exacerbate climate impacts in the future. In such cases, climate change may be presented as a peripheral argument and its function could be to improve the judiciary's understanding of the ways in which climate-sensitive decisions can influence laws and address environmental disputes.

STRATEGIC LITIGATION

In several parts of the world, including in India, strategic cases have been filed to develop case law that would examine linkages between climate change and protection of constitutional rights. Even though the 2015 Paris Agreement is not enforceable, these strategic cases are seeking to bind GHG emitters and government regulators accountable to the international climate commitments. For instance, there are cases where petitioners or appellants typically seek compensation from major carbon emitters for climate change-related damages. These could be coal mining, coal-based power producers or

---

[45] *Urgenda Foundation v. the State of the Netherlands*, (2015) HAZA C/09/00456689 (24 June 2015); aff'd on 9 October 2018 the District Court of The Hague and The Hague Court of Appeal (on appeal), available at https://elaw.org/nl.urgenda, accessed on 12 February 2020.

[46] Isabella Kaminsky, 'Historic Urgenda Climate Ruling Upheld by Dutch Supreme Court', *Climate Liability News*, 19 December 2019, available at https://www. climateliabilitynews.org/2019/12/20/urgenda-climate-ruling-netherlands-supreme-court/, accessed on 12 February 2020.

other industrial activity responsible for high levels of emissions. Some cases have also been filed to influence government policy to devise mechanisms and programmes for mitigation, adaptation and compensation.

A landmark strategic case is from Pakistan. A Pakistani farmer, brought a climate change case against the Pakistani government for failing to implement its national climate change law and policy. In 2015, the Green Bench of the Lahore High Court upheld the claim, invoking the right to life and the right to dignity. A significant decision of 4 September 2015 which precedes the main judgment observes,

> Environment and its protection has taken a center stage in the scheme of our constitutional rights. It appears that we have to move on. The existing environmental jurisprudence has to be fashioned to meet the needs of something more urgent and overpowering i.e., Climate Change. From Environmental Justice, which was largely localized and limited to our own ecosystems and biodiversity, we need to move to Climate Change Justice.

Finding that the government had done little to carry out its national climate law, the court in its judgment dated 25 January 2018 directed each government ministry to nominate a focal point to ensure implementation and present a list of action points. The court also created a Climate Change Commission, mandated to monitor the government's progress.[47]

According to a global database on climate change litigation, strategic climate litigation in the United States has involved a broad set of themes to achieve a number of environmental actions. These include resisting deregulation, ensuring federal government transparency, tackling opposition to climate action, filing municipality-led suits against fossil fuel companies, strengthening state-led efforts to decarbonise the electricity sector and ascertaining liability for failure to adapt. Litigants have also made arguments for climate action based on the public trust doctrine, which assigns responsibility to the state to manage a nation's public resources for future generations. They have raised issues regarding an individuals' fundamental rights and intergenerational equity, as well as concerns about the balance of powers among the judicial, legislative and executive branches or functions of governments.

---

[47] Lahore High Court, Lahore Judicial Department Case No. Writ Petition No. 25501 of 2015 (*Asghar Leghai v. Federation of Pakistan*).

CLIMATE LITIGATION IN INDIA

There are a few instances of strategic or climate change-focused cases in India. Most cases have invoked domestic environmental regulation to seek redress related to non-implementation of emission norms or failure of compensatory afforestation measures. Such cases do not mention climate change even peripherally in their arguments.

Some cases have creatively applied climate change arguments in social justice matters. These are discussed in a 2013 working paper by Lavanya Rajamani.[48] One such case is from the Delhi High Court where the petitioners stated that the 'limit, fixed by the respondent—Delhi Municipal Corporation (hereinafter referred to as "MCD") on the issuance of cycle rickshaw licenses is illegal and void'. The petitioners placed a collection of arguments, seeking the court's intervention to revoke policies that place limits on cycle rickshaw licences. The judgment[49] which upheld the petitioner's plea to remove limits on cycle rickshaw licenses records

> that the 4th Assessment Report of the year 2007 of the International Governmental Panel on Climatic Change (IPCC) emphasized the need for policies that encourage use of more fuel-efficient vehicles, hybrid vehicles, non-motorized transport, (such as cycling and walking), and better land-use and transport planning, to minimize rising pollution levels.

There are two cases of the National Green Tribunal (NGT) in India where climate change has been an important focus.

In the first case, the NGT took *suo moto* cognizance of degradation of the Rohtang Pass and made important observations on its vulnerability due to global warming and climate change. In February 2014, the NGT ordered authorities in the state of Himachal Pradesh to undertake several measures to remediate various environmental harms being caused due to excessive tourism in Rohtang Pass. The Rohtang Pass in the Himalayas is considered to be the

---

[48] Lavanya Rajamani, 'Rights Based Climate Litigation in the Indian Courts: Potential, Prospects and Potential Problems', Working Paper 2013/1 (May), Centre for Policy Research, New Delhi, 2013.

[49] High Court of Delhi judgment dated 10 February 2010 in Writ Petition (Civil) No. 4572 of 2007 (*Manushi Sangathan v. Govt. of Delhi & Ors*).

'crown jewel of Himachal Pradesh and had seen excessive degradation due to the pressures and mismanagment of tourism'.[50]

While issuing a range of directions related to afforestation, reducing pressure of tourism and pollution mitigation, the NGT also observed that the ecologically sensitive Rohtang Pass is vulnerable due to rise in global temperatures:

> Global warming has its impacts in other parts of the world, as in the Indian sub-continent. It is likely to affect the glaciers. There will be early and untimely melting of ice resulting in various environmental issues. Rohtang Pass being one of the eco-sensitive and fragile areas of the glacier, is likely to get affected more than other areas. Thus, there is a need for evolving schemes and mechanism to take greater care of the glacier in the interest of environmental and ecological balance.

The NGT concluded that the government of Himachal Pradesh had violated its obligations under Articles 21, 48A and 51A[51] by failing to restrict development and road and pedestrian traffic in and around the increasingly touristed area accessible via the Rohtang Pass. Melting of regional glaciers and deforestation led the list of environmental impacts noted by the NGT, which identified the emission of black carbon from vehicle traffic as a chief cause of the glacial melt.

The NGT described how GHG emissions cause global warming and affirmed the applicability of the 'polluter pays' principle to the respondents in this case. It only ordered the government of Himachal Pradesh to impose a host of restrictions on traffic and to undertake a programme of reforestation, both of which would be overseen by a monitoring committee that would submit quarterly reports to the NGT.

The second case is a strategic climate litigation filed by an Indian student in March 2017. The petition argues that the public trust doctrine, India's commitments under the Paris Agreement and India's existing environmental laws and climate-related policies oblige greater action to mitigate climate change. The petition also argued that the term 'environment', as used in the

---

[50] National Green Tribunal judgment dated 6 February 2014 in Application No. 237 (THC)/2013 (CWPIL No. 15 of 2010) (*Court on Its Own Motion v. State of Himachal Pradesh & Ors*).

[51] Discussed in Chapter 1.

Environment Protection Act, 1986, necessarily encompasses the climate. The case raised these claims invoking the NGT's powers to adjudicate on 'substantial question relating to the environment'. In addition, the petition cited the principles of sustainable development, precaution and intergenerational equity. It also referred to several international judicial decisions and judgments that have been passed on similar grievances.

The petition noted that India is the third-largest national emitter of GHGs (after China and the USA) and among those countries that are most susceptible to adverse climate change impacts. It sought directions for 'effective, science-based action to reduce and minimize the adverse impacts of climate change in the country'. The petition asked the court to order the national government to undertake a variety of measures, including but not limited to, the inclusion of climate change in the domestic environment laws related to air pollution, environmental impact assessments and forest diversion. It also sought that climate change concerns drive the preparation of a national GHG inventory and a national carbon budget against which project emissions and impacts can be assessed. The court observed,[52]

> There is no reason to presume that Paris Agreement and other international protocols are not reflected in the policies of the Government of India or are not taken into consideration in granting environment clearances.

The case was dismissed by the NGT one and a half years after it was filed.[53] In February 2021, the Supreme Court of India admitted an appeal against this dismissal.

---

[52] National Green Tribunal Original judgment dated 15 January 2019 in Application No. 187 of 2017 (*Ridhima Pandey v. Union of India & Ors*).

[53] Sourav Roy, 'The Kids Are Alright', *Deccan Herald*, 16 February 2020, Available at https://www.deccanherald.com/sunday-herald/sh-top-stories/the-kids-are-alright-804575.html, accessed on 27 February 2020.

10

Contemporary Environmental Law Reforms

INTRODUCTION

Indian environmental laws have seen tremendous changes since the 2000s. The predominant goal of successive governments in India has been to enhance economic growth, to provision for the needs of India's large population and to manage the global and local expectations of sustainable development. In order to achieve these goals, the central government has undertaken consistent and systematic processes to review and overhaul several environmental laws in the country and to streamline them with these governmental objectives. Amendments to environmental laws have also been brought on by court directions and parliamentary committee reviews. The erstwhile Planning Commission and the present-day NITI (National Institution for Transforming India) Aayog have also played a decisive role in recommending policy and legal changes. Citizen engagement with environment laws has also increased during this period due to the creation of specialised legal forums and information availability. They have litigated on violations of law and demanded accountability from regulatory institutions. The nature, scope and role of environmental regulation have undergone major shifts as courts, civil society and social movements are focused on the effects of environmental laws on the ecology, economy and society.

By the early 2000s, environmental scrutiny of most infrastructures, energy and real estate projects was a legal prerequisite. One set of pressures for amendments and revisions to environmental laws has come through an 'investment reforms' agenda. The primary mandate of these reforms was to reconcile environmental law frameworks with economic priorities such as expanding opportunities for international investments and growth of domestic businesses. The environmental law reform proposals in this phase were driven by the narratives that environment regulation caused delays, bottlenecks and hurdles for much needed investments. These reforms were executed through setting up of high-level committees by ministries such as coal, commerce, power and finance. These special committees also carried out consultations across ministries at the state and central levels and with economic actors such as corporations and industrial and investor associations.

Another set of changes came through courts and governments interested in finding solutions for the implementation challenges faced by environmental laws and regulations. Courts, the central government and various state governments suggested legal and policy changes to institute long-term monitoring committees, financial offset mechanisms, and grievance redressal and appellate mechanisms. The decades of institutional experience in pollution control and environmental management laws were reviewed, and measures were recommended to purportedly make environmental governance through institutions and enforcement actions more effective and efficient.

The Supreme Court in its 2011 Lafarge judgment[1] directed the need to set up a national environmental regulator (discussed in Chapter 5) to deal with questions of environmental regulation and enforcement.

The third imperative for reforms came from international environmental conventions and treaties to which India was a signatory. These legal discourses placed importance on transparency, public participation and consent, and these aspects needed to be built into the domestic environment governance regimes.

This chapter lays out the main sets of reform proposals that have had a significant and lasting influence on environmental laws, rules, notifications and guidelines since 2000. The reform proposals that stemmed from within the environment ministry's own processes and those that were driven from outside the environment ministry converged on certain themes. These include the multiplicity of approvals required by projects, devolution of power to states and local units to approve projects, need to change penal mechanisms, and better technological mechanisms to monitor projects' environmental performances. These processes picked up momentum from 2014 when

---

[1] Supreme Court of India judgment dated 6 July 2011 in Interlocutory Application (I.A.) Nos. 1868, 2091, 2225–2227, 2380, 2568 and 2937 (*Lafarge Umiam Mining Pvt. Ltd*) in Writ Petition (Civil) No. 202 of 1995 (*T.N. Godavarman Thirumulkpad v. Union of India & Ors*).

the environment ministry set up high-level committees to review major environmental laws.

This chapter has two sections:

I.   *Reforms Taken Up by the Environment Ministry*
II. *Reforms Suggested by Other Higher Level Committees*

The first section discusses reforms taken up by the central environment ministry, the highest nodal institution in charge of environmental regulation in India. The second section discusses the steps taken up by bodies above the environment ministry. However, it is important to bear in mind that both types of reforms were responding to international shifts in environmental thinking as well as domestic economic and environmental challenges.

I REFORMS TAKEN UP BY THE ENVIRONMENT MINISTRY

In 2014, two committees were set up to reform environmental laws. These were the T. S. R. Subramanian Committee and the Shailesh Nayak Committee. The time frame given to both these committees was short, and the task at hand was to cumulatively review eight environmental laws. The recommendations of these two committees formed the basis of sweeping changes in the framework of environmental laws. Several legal amendments since 2014 that have been discussed in this book draw their justification from these committee reports.

This section also discusses an important change in the government's view on the management of forests and the inclusion of the private sector in forestry.

'High Level' Review of Environmental Laws

In August 2014, the environment ministry set up a high-level committee under the chairmanship of former Cabinet Secretary T. S. R. Subramanian[2] to review six major environment laws: the Environment Protection Act (EPA), 1986; the Forest Conservation Act, 1980; the Air Prevention and Control of Pollution Act, 1981; the Water Prevention and Control of Pollution Act, 1974; the Wild Life Protection Act, 1972 and the Indian Forest Act (IFA), 1927. The committee submitted its report in November 2014, 10 days prior to the expiry of its term. The composition of this committee, the short time

---

[2] Ministry of Environment, Forest and Climate Change (MoEFCC) Office Memorandum (OM) No. 22-15/2014-IA. III dated 29 August 2014.

frame to review six major environment laws and what was finally proposed came under widespread scrutiny and critique.[3] The committee in its final report acknowledged the time limitations to say that it 'could not address all the laws, regulations, rules and executive instructions comprehensively within the time span available to it'.

The laws under review formed the backbone of environmental regulation in India, and any reform should have ideally dealt with the complex design and implementation challenges of these laws and the numerous conflicts that they created or ignored. The committee narrowed the problems with the environmental legal framework to the following three issues:

- There was a multiplicity of laws and, therefore, a need to clarify and consolidate the legal framework including parent Acts and subordinate legislations.

- The environmental and forest approval process for projects suffered from bad data, time 'delays', insufficient expertise and arbitrary decisions.

- The compliance and enforcement of laws and environmental management measures were poor.

The recommendations of the Subramanian Committee aimed to tackle these problems by adopting a framework involving minimum governmental interference to business and a bigger role for qualified technical experts. The committee's report foregrounded the principle of 'utmost good faith' and suggested that approvals should be granted and compliance should be assessed according to pre-stated terms and commitments made by project proponents. The specific suggestions included the following:

- Creating a single-window approval system and the merging of the Air Act, the Water Act and the EPA.[4]

---

[3]   Megha Bahree, 'Indian Govt Sets Up Committee to Review Environmental Laws: Not Everyone Is Happy with It', *Forbes Asia*, 16 October 2014, available at https://www.forbes.com/sites/meghabahree/2014/10/16/indian-govt-sets-up-committee-to-review-environmental-laws-not-everyone-is-happy-with-it/#7cc74c3f4deb, accessed on 15 September 2020.

[4]   PRS Legislative Research, *Report Summary: High Level Committee to Review Acts Administered by the Ministry of Environment, Forest and Climate Change* (New Delhi: Institute for Policy Research Studies, 2014), 1.

- Tackling the reasons for the supposed delay of projects by limiting the scope of public participation, exemptions to certain projects from public hearings and removing the requirement of completion of forest rights recognition process prior to the first-stage permission for forest diversion.

- Good data, technology for environmental management, technical experts, higher penalties and restoration costs were emphasised. To enforce these, the committee recommended the setting up of new, full-time technical institutions such as National Environment Management Authority at the centre and State Environment Management Authority at the states as the primary institution for environmental matters.[5]

- Setting up of appellate boards at the centre and other locations as the first forum of appeal against regulatory decisions, prior to those being challenged before any judicial or quasi-judicial body such as the National Green Tribunal (NGT). (The next section discusses the Environment Law Amendment Bill (ELAB) as a follow up to this recommendation.)

The Department-Related Parliamentary Standing Committee on Science and Technology and Environment and Forests reviewed the Subramanian Committee report and questioned the process adopted as part of this review.[6] It suggested that the report not be accepted and the 'views/ opinion and objections raised by stakeholders including environmental experts' should be considered. It also recommended setting up of another committee 'comprising of acclaimed experts in the field who should be given enough time to enter into comprehensive consultations with all stakeholders so that the recommendations are credit worthy and well considered which is not the case with the recommendations of High Level Committee under review'.

However, even as the review of the parliamentary committee was pending, the environment ministry put out a 93-page request for proposals for the

---

[5] The design of these institutions was similar to the National Environment Regulatory Authority as discussed in Chapter 5.

[6] Department-Related Parliamentary Standing Committee on Science and Technology and Environment and Forests, *Two Hundred and Sixty Third Report: High Level Committee Report to Review Various Acts Administered by Ministry of Environment, Forest and Climate Change* (New Delhi: Rajya Sabha Secretariat, 2015), 7–11. The committee was chaired by Shri Ashwani Kumar, Member of Parliament.

preparation of a project report and drafts of proposed legislations suggested in the Subramanian Committee report, in consultation with the ministry.[7]

ENVIRONMENT LAW AMENDMENT BILL

The Subramanian Committee proposed the need for a new institution for monitoring the compliance of projects and adjudicating civil penalties against damages. As a follow up to this recommendation, the Ministry of Environment, Forests and Climate Change put out a draft Environment Law Amendment Bill (ELAB) for public comment. One of the primary purposes of this bill was to put in place a mechanism where damages that accrued from violation of laws could be adjudicated and civil penalties could be calculated. It presented the design of an administrative adjudication system as the first avenue of appeal before matters are challenged before the NGT, set up by the ministry in 2010.

The draft released on 7 October 2015 gave a period of 15 days to send in comments. The sudden release of this document and the short time period drew criticism from several groups. The objections were threefold: the justification for this new law, what it proposed to do and the process of how it was drafted, including the short public comment period.[8] The following are the substantive issues with the ELAB:

- *Should there be a hierarchy of environmental damages?* The bill proposed a new taxonomy of environmental damages or impacts of environmental violations as 'minor', 'substantial' and 'non-substantial'. This classification seemed vague and discretionary. In the proposed form, it would not serve the purpose of providing environmental remedies.

- *Are limits on penalties useful?* The bill suggested 'caps' or upper limit of damages to be paid but did not establish a minimum or 'floor' limit. Without a tight definition, there would be a tendency to list all violations as 'minor', especially given the incremental nature of most project-related violations. Encroachments, destruction of mangroves,

---

[7] MoEFCC, *Request for Proposal (RFP) (Amended) for Selection of Technical Consultant for Supporting the Implementation of Agreed Recommendations of High Level Committee* (New Delhi: Government of India, 2015).

[8] Dhwani Mehta, 'The Government's Environment Laws Amendment Bill May Transfer More Power to the Executive and Weaken the NGT', *Caravan Magazine*, 18 January 2016, available at https://caravanmagazine.in/vantage/environment-laws-amendment-bill-transfer-power-to-executive-weaken-ngt, accessed on 15 September 2020.

river pollution and dust accumulation are all cases of violation that can progress on an incremental basis, and small fines such as the proposed one time, 'on the spot penalty' of ₹1,000–₹10,000 may not serve as adequate deterrents.

- *Risks of too much discretion*: The ELAB was also critiqued on the grounds that the definition of damages and the methodology to institute penalties was discretionary in nature. Both environmental groups and industry representatives argued that the current definition gave excessive powers to an undefined authority. In case the damages were labelled as minor, they would not be adjudicated upon.

- *Should environmental impacts be monetised?* The primary mechanism for securing remedies in the proposed law was to monetise environmental impacts and create a 'fund'. The experience of penal funds such as the Compensatory Afforestation Fund, which existed since 2002, were flagged as being reasons why such a fund may not be an effective mechanism to reduce instances of violation or damages.

- *Should environmental penalties be pre-fixed?* Another question raised was whether the amendment bill took a very limited view of the 'polluter pays' principle by pre-fixing amounts in case of violations and by insitutionalising it as a routine process of handling environmental violations. It was argued that if penal amounts were fixed in advance, projects could even pass them on to consumers.

The environment ministry has not made any draft or final proposals for creating the ELAB as of January 2021. However, the environment ministry has set up special expert committees to deal with violations under the EIA (discussed in Chapter 5) and CRZ notifications.

REVIEW OF COASTAL REGULATION ZONE NOTIFICATION

In June 2014, the environment ministry set up a committee 'to review the issues relating to the Coastal Regulation Zone Notification, 2011'.[9] It was chaired by Shailesh Nayak, former Secretary, Ministry of Earth Sciences. The process was set up to primarily carry out consultations with governments of all coastal states, with no scope for public consultations or inputs into the review exercise. The committee submitted its report in early 2015, but was

---

[9] MoEFCC Office Order No. F. No.19/112/2013/IA/3 dated 17 June 2014.

never made public by the ministry,[10] even though a judgment[11] of the Central Information Commission directed it.[12]

The committee's key recommendations in its 110-page report include the following:

* Special focus on ecologically fragile areas on the coasts, especially with dense mangroves, coral reefs and nesting grounds that are currently recognised as Coastal Regulation Zone (CRZ) IA (see Chapter 5).

* Devolution of powers to state and union territory (UT) governments for regulating activities that do not attract the provisions of the Environment Impact Assessment (EIA) notification.

* Easing regulation for building and tourism sector and promoting housing infrastructure and slum redevelopment activities.

* Reduction of the No Development Zone where certain activities were previously prohibited.

Between 2014 and 2018, the environment ministry issued at least eight amendments to the Coastal Regulation Zone (CRZ) notification, which drew extensively[13] from the Shailesh Nayak Committee report.[14] The first such

---

[10] See DHNS, 'Govt Receives CRZ Review Report', *Deccan Herald*, 3 April 2015, available at http://www.deccanherald.com/content/469386/govt-receives-crz-review-report.html, accessed on 15 September 2020; Lakshmi Iyer, 'It's State vs Centre on Coastal Road', *Mumbai Mirror*, 15 April 2015, available at http://www.mumbaimirror.com/mumbai/civic/Its-state-vs-Centre-over-coastal-road/articleshow/46938075.cms, accessed on 15 September 2020; TNN, 'Coastal regulation Zone Norms: Central Team to Submit Report in October', *Times of India*, 24 August 2014, available at http://timesofindia.indiatimes.com/city/kochi/Coastal-regulation-zone-norms-Central-team-to-submit-report-in-October/articleshow/40824178.cms, accessed on 15 September 2020.

[11] Central Information Commission judgment dated 13 May 2019 in CIC/SA/A/2016/000209 (*Kanchi Kohli v. PIO, M/o Environment & Forest*).

[12] The Central Information Commission was set up under the Right to Information Act as an appellate body for appeals against denial of information or receipt of partial information by central government public information officers.

[13] Centre for Policy Research, 'Behind the Scenes: Story of CRZ Revamp from Within the Ministry', CPR Blog, 22 November, 2018 available at https://www.cprindia.org/news/behind-scenes---story-crz-revamp-within-ministry accessed on 30 September 2020

[14] MoEFCC, *Report of the Committee to Review the Issues Relating to the Coastal Regulation Zone, 2011* (New Delhi: Government of India, 2015).

draft amendment came soon after the first presentation of this committee to the ministry, with preliminary findings. The new CRZ Notification, 2019, cites the Shailesh Nayak Committee report as one of its basis.

PRIVATE SECTOR PARTICIPATION IN FORESTRY

Another aspect of the environment law reforms observed during the last two decades relates to the involvement of private sector in managing forests. Even through discussions on this were initiated in the late 1990s, specific high-level recommendations and policy interventions were upscaled from 2014 onwards when the environment ministry formulated 'Guidelines for Participation of Private Sector in Afforestation of Degraded Forests' with the objective of improving productivity and quality of forests.[15] The guidelines justified this move by saying,

> There is a need for finding ways to improve productivity and quality of forests for realization of full ecological potential which seems unlikely from the current level of investment in the forestry sector. Around 40% of forest cover is under the category of Open Forests i.e. canopy density is <40%. A part of this open forest along with scrubs may be considered as degraded forests.

Private sector participation in forestry has been seen as a measure to increase forest cover in the country for over two decades. In the 1990s, the idea of leasing degraded forest land to the private sector was rejected by high-level planning bodies.[16] However, afforestation programmes on 'degraded' forest lands have gradually opened the space for inviting bids from the private sector to directly engage in plantations and harvesting of trees. This was also recorded in a 2004 report of the Planning Commission, which lays down the selection criteria for inviting private sector participation in forestry.[17]

In 2010, public–private partnerships for raising plantations on degraded farmland was encouraged as a mechanism to combat climate change. This was

---

[15] MoEFCC F.No.7-8/2014-FP (undated).

[16] Planning Commission of India, *Report of the Working group of Planning Commission of India on 'Leasing of Degraded Forest Lands'*, chaired by N. C. Saxena (New Delhi: Government of India, 1998).

[17] Planning Commission of India, *Report of the PPP Sub-group on Social Sector: Public Private Partnership* (New Delhi: Government of India, 2004), 48.

recorded in the Green India Mission document finalised by the environment ministry as part of the Prime Minister Manmohan Singh led Council on Climate Change.[18] In 2015, the environment ministry finalised the guidelines for the 'Participation of Private Sector in Afforestation of Degraded Forests'. These guidelines justified private sector participation in forestry due to the availability of 'limited public funds and low investment, resulting in poor productivity and thereby serious ecological consequences'.[19] The Draft National Forest Policy, 2018[20] (see Chapter 3), states,

> Public private participation models will be developed for undertaking Afforestation and reforestation activities in degraded forest areas and forest areas available with Forest Development Corporations and outside forests.

The proposed changes to the IFA in 2019 include specific amendments to attract private persons and the private sector in forest management including carrying out plantations. Comments were sought from state governments and the public on the proposed amendments.[21] The amendments are yet to be finalised.

II REFORMS SUGGESTED BY OTHER HIGHER LEVEL COMMITTEES

This section contains four critical sets of reforms that came through influential committees of the government. These are only illustrative of how the highest levels of the government, such as the Union Cabinet and the Planning Commission, viewed environmental regulations and also the influence that

---

[18] Ministry of Environment and Forests (MoEF), *National Mission for a Green India: Under the National Action Plan on Climate Change Draft Mission Document* (New Delhi: Government of India, 2010), 18.

[19] Suresh Ghattamaneni, 'Plan to Privatise 40% of Forests Will Undermine Law Giving Adivasis Control of Their Habitats', *Scroll.in*, 17 September 2015, available at https://scroll.in/article/756055/plan-to-privatise-40-of-forests-will-undermine-law-giving-adivasis-control-of-their-habitats, accessed on 12 July 2019.

[20] MoEFCC, *Draft National Forest Policy, 2018*, vide Notification F. No. 1-1/2012-FP (Vol. 4) (New Delhi: Government of India, 2018), available at https://smartnet.niua.org/sites/default/files/resources/draft_national_forest_policy_2018.pdf, accessed on 1 July 2020.

[21] MoEFCC, *Proposed Indian Forest (Amendment) Act, 2018* (New Delhi: Forest Policy Division, Government of India, 2019). This document was sent to all state governments on 7 March 2019 (F. No. 2-1/1997-FP (Vol. 6)) and circulated for public comments thereafter.

they had in shaping environmental regulation frameworks. The reforms suggested by committees, set up by these bodies, were basically concerned with addressing 'delays' in project approvals and ensuring timely and uninterrupted access to natural resources.

GOVINDARAJAN COMMITTEE ON INVESTMENT REFORMS

Environmental approvals for industrial and infrastructure projects were instituted only in the 1990s. Within a period of a few years, these approvals became unpopular with the industry sector. One of the first committees which influenced the trajectory of environmental regulations was set up by the Cabinet Secretariat. The Govindarajan Committee was set up for 'Reforming Investment Approval and Implementation Procedure'. Part II of the committee's report submitted in 2002 informed the process for amending the procedures for environmental approvals.[22]

The report listed the problems as:

1. Cumbersome procedure for environmental clearance and public hearing.

2. Submission of incomplete information and poor quality of EIA/EMP necessitating additional information.

3. Reopening of technical issues at every stage and raising of fresh issues during the course of grant of environmental clearance.

4. Delays in the meetings of the Expert Committee and site visit, if considered necessary, before clearance.

This informed a series of measures taken by the environment ministry. The 2004–05 annual report of the ministry states that statutory amendments and simplification of EIA-related processes have been undertaken to ensure 'expeditious decision on project clearances'.[23] India's National Environment Policy, 2006,[24] states,

---

[22] Cabinet Secretariat, *Report on Reforming Investment Approval and Implementation Procedure: Part II* (New Delhi: Government of India, 2002).

[23] MoEF, *Annual Report (2004–2005)* (New Delhi: Government of India, 2005), 41.

[24] MoEF, *National Environment Policy* (New Delhi: Government of India, 2006), 17.

The recommendations of the (the Govindarajan Committee) which identified delays in environment and forest clearances as the largest source of delays in development projects, will be followed for reviewing the existing procedures for granting clearances and other approvals under various statutes and rules.

The recommendations of this committee formed the basis of the 're-engineering' of the EIA Notification, 2006, and the multiple amendments to the CRZ notification, both of which are discussed in Chapter 5 on environmental protection.

HODA COMMITTEE ON NATIONAL MINERAL POLICY

A high-level committee was set up to review the National Mineral Policy. The Planning Commission constituted this committee under the chairmanship of Anwarul Hoda, Member, Planning Commission and is known as the Hoda Committee.[25] The overall objective of this committee was to examine the challenges in boosting mining in India and to suggest measures on how extraction could be increased. One of the tasks assigned to this committee was to

review the procedures for according clearance to mineral exploration and mining projects under the Forest (Conservation) Act, 1980 and the Environment (Protection) Act, 1986 and suggest ways for speeding them up.

The December 2006 report of the committee had six pages of suggestions on how to reduce delays in forest diversion and environment approvals for mining projects. This included recommendations such as single-window approval rather than separate approvals under different environment laws. There were suggestions for delegating powers to state mining departments to approve projects after in-principle approvals are granted by the ministry, staggering the payments due from the mining projects for compensatory afforestation over the lease period for a mine, reducing the number of days within which an environment clearance is processed and limiting public participation in

---

[25] Planning Commission, Government of India (GOI), Order No. I&M-25(3)/2005 dated 14 September 2005.

specific instances where the mine lease area is 'less than 50 hectares and also for renewal of leases'.

CHAWLA COMMITTEE REPORT ON STREAMLINING NATURAL RESOURCE ALLOCATION

In 2011, the Cabinet Secretariat constituted a committee 'to deliberate on measures required for enhancing transparency, effectiveness and sustainability in utilization of natural resources'.[26] This committee was set up under the chairmanship of former Finance Secretary, Ashok Chawla.[27]

The 'natural resources' studied by this committee included coal and other minerals, natural gas, water, forests, petroleum, spectrum and land. The premise of the exercise was that these resources are natural assets and, therefore, had to be allocated in a way that involves minimum corruption and maximum utilisation. The committee recognised that while some of these natural resources could be entirely governed by market processes, others would need different mechanisms. According to the committee, 'union government has some role to play in the allocation and pricing decisions for all of these natural resources'.

Regarding transparency in forest diversions, the committees report said,

Not all decisions can however be made through market processes, e.g., the decision as to whether or not forest clearance should be granted for a mining project. This decision is necessarily an administrative determination based on the facts of the case. However, in the interests of transparency, it is incumbent on those that are responsible for making decisions to make clear the process by which decisions are reached. At various points in this report, this Committee has recommended that minutes of meetings be detailed and reasoned and available in the public domain to convey the process by which a particular decision was arrived at.

The report presented several arguments and recommended the adoption of open, transparent and competitive mechanism for allocation of all natural

---

[26] Cabinet Secretariat OM dated 31 January 2011 (vide Order No. 483/1/1/2011-Cab), xviii.

[27] Cabinet Secretariat, *Report of the Committee on Allocation of Natural Resources* (New Delhi: Government of India, 2011).

resources. It also suggested greater disclosure in existing approval processes, including those related to environment and forest diversion.

The report was finalised in May 2011. It was accepted by the Group of Ministers headed by the then Minister of Finance, Pranab Mukherjee, at its meeting on 15 October 2011.[28] News reports indicate that almost all the recommendations of this committee were accepted, including ways in which delays in forest clearances for mining projects can be minimised.[29]

This 2011 report was not attributed in specific legal changes. However, some legal and policy changes that were made after the completion of this report mirror the recommendations of the committee. For instance, the committee had introduced the idea of creating inviolate forest areas with a 'view to improving the predictability of clearances for diversion of forest land'. This recommendation was similar to what the environment ministry had set out with its policy on 'go and no-go areas' or 'inviolate areas'[30] to determine forest diversions for coal mining.

The committee also discussed the importance of Free Prior Informed Consent (FPIC) to ensure 'social sustainability' in natural resource allocation. It observed that the forest diversion process needs to respond to the complexity and importance of incorporating FPIC in legal procedures. This is because 'practices like mining frequently have adverse consequences for local communities'. The committee's report also emphasised,

> Recognition that project affected persons (PAP) are not just people who lose land, but also those whose current livelihoods are severely impaired.

---

[28] Sujai Mehudia, 'Ashok Chawla Committee Report on Allocation of Natural Resources Gather Dust', *The Hindu*, 16 April 2012, available at https://www.thehindu.com/business/ashok-chawla-committee-report-on-allocation-of-natural-resources-gather-dust/article3320826.ece, accessed on 15 September 2020.

[29] Vikas Dhoot, 'Ashok Chawla Committee Punctures Manmohan Singh's Claims on Coal Allocation', *Economic Times*, 19 September 2012, available at https://economictimes.indiatimes.com/news/news-by-industry/indl-goods/svs/metals-mining/ashok-chawla-committee-punctures-manmohan-singhs-claims-on-coal-allocation/articleshow/16227832.cms, accessed on 15 September 2020.

[30] MoEF, *Report of the Committee to Formulate Objective Parameters for Identification of Inviolate Forest Areas* (New Delhi: Government of India, 2012).

The new land acquisition law of 2013 discussed in Chapter 8 sets out to do this.

The report also recommended the urgent need for a comprehensive national legislation on water, which could be done either by bringing water on the Concurrent List of the constitution or by obtaining state-wide consensus for a central 'framework law' on water. For groundwater management, the committee suggested the need for aquifer-level mapping, hydrological studies and pilot projects. These have been included in several model groundwater laws at central and state levels, as well as in the National Water Policy, 2012.

LAW REFORMS FOR A GREEN TRIBUNAL

One of the most significant legal reforms for environmental regulation came out of the 186th report of the Law Commission of India finalised in September 2003.[31] From 2001 to 2003, Justice (Retd) M. Jagannadha Rao headed the commission. The law commission recommended the setting up of specialised environmental courts. Its report states,

> In view of the involvement of complex scientific and specialized issues relating to environment, there is a need to have separate 'Environment Courts' manned only by the persons having judicial or legal experience and assisted by persons having scientific qualification and experience in the field of environment.

This was the first proposal for setting up of the National Environment Tribunal and regional environment tribunals, which were later discussed in the draft note for the cabinet in 2006.[32] This note proposed a structure for the tribunals. The note was shared by the environment ministry with all other ministries, seeking their comments and responses. In 2008, the Department-Related Parliamentary Standing Committee on Science and Technology and Environment and Forests in its 192nd report spoke of the need to establish

---

[31] Law Commission of India, *One Hundred Eighty Sixth Report on Proposal to Constitute Environment Courts* (New Delhi: Government of India, 2003), 164.

[32] Draft note to the Cabinet. F No. J (18)/2003-PL dated 31 October 2006.

environmental courts in each state and UT.[33] The committee's report on the NGT Bill, 2009, relied on the report of the Law Commission of India and its own recommendations to justify the need for the tribunal.[34]

The NGT was enacted in 2010.[35] The NGT accepted 31,730 cases until 30 November 2019 and disposed 28,761 while 2,969 of them were pending.[36] Until early 2018, the NGT was fully functional with two principal benches in New Delhi and four others in Bhopal, Kolkata, Pune and Chennai.[37] For most of 2019 it was limited to a single bench functional from New Delhi. Since 2020, the regional benches have become functional.

---

[33] Department-Related Parliamentary Standing Committee on Science and Technology and Environment and Forests, *One Hundred and Ninety Second Report on Functioning of Central Pollution Control Board* (New Delhi: Rajya Sabha Secretariat, 2008), 40.

[34] Department-Related Parliamentary Standing Committee on Science and Technology and Environment and Forests, *Two Hundred And Third Report on the National Green Tribunal Bill, 2009* (New Delhi: Rajya Sabha Secretariat, 2009), available at http://164.100.47.5/newcommittee/reports/EnglishCommittees/Committee%20on%20S%20and%20T,%20Env.%20and%20Forests/For%20Net.htm, accessed on 15 September 2020. The committee was chaired by T. Subbarami Reddy, Member of Parliament.

[35] The NGT Act, 2010 (No. 19 of 2010), gazetted on 2 June 2010.

[36] Website of the NGT, available at https://greentribunal.gov.in/, accessed on 2 February 2020.

[37] Jayashree Nandi, 'Benches Shut, NGT Forcing Petitioners to Come to Delhi', *Times of India*, 26 April 2018, available at https://timesofindia.indiatimes.com/home/environment/developmental-issues/benches-shut-ngt-forcing-petitioners-to-come-to-delhi/articleshow/63918018.cms on, accessed on 15 September 2020.

Index of Laws, Legal Cases and Government and Parliamentary Committee Reports

LAWS

ACTS

Air (Prevention and Control of Pollution) Act, 1981, 124, 321

Atomic Energy Act, 1962, 281n44

Biological Diversity (BDA) Act, 2002, 46, 59, 201, 201n2, 218–221

Coal Bearing Areas (Acquisition and Development) Act, 1957, 281n44, 282, 288

Compensatory Afforestation Fund Act, 2016, 70–71, 89

Damodar Valley Corporation Act, 1948, 281n44

Draft Model Bill for the Conservation, Protection and Regulation of Groundwater, 2011, 241n26

Environment Protection Act (EPA), 1986, 6, 17, 59, 74, 78–81, 83, 104, 110, 124, 139, 150n72, 158–163, 171–179, 181, 183, 185, 188–189, 191, 196, 240, 240n20, 241, 243, 318, 321–322

Factories Act, 1948, 121

Forest (Conservation) Act, 1980, 5, 66, 72, 89, 94, 94n15, 95–97, 103, 106, 108, 110, 113, 115, 321

Goa Ground Water Regulation Act, 2002, 242

Himachal Pradesh Private Forests Act, 1954, 90n4

Indian Easement Act, 1882, 235n3, 239

Indian Forest Act (IFA), 1927, 89–90, 90n3, 90n4, 91–93, 93n14, 94, 97, , 106, 110, 113, 262, 321, 327

Inter-State Water Disputes Act of 1956, 253n57, 265

Indian Tramways Act, 1886, 281n44

LEGAL CASES

Supreme Court

Supreme Court judgment dated 22 February 2017 in Writ Petition (Civil) No. 375 of 2012 (*Paryavaran Suraksha Samiti v. Union of India*), 145n59

Supreme Court judgment dated 27 February 2012 in Writ Petition (Civil) 512 of 2002 (*In RE: Networking of Rivers*) Writ Petition (Civil) 668 of 2002, 13n31

Supreme Court judgment dated 28 August 1996 in Writ Petition (Civil) 914 of 991 (*Vellore Citizens Welfare Forum v. Union of India & Ors*), 35n15, 55n75, 123n6

Supreme Court judgment dated 29 July 1980 in Special Leave Petition (Criminal) No. 2856 of 1979 (*Municipal Council, Ratlam v. Vardhichand*), 40n35

Supreme Court judgment dated 29 March 2019 in Civil Appeal No. 12251 of 2018 (*Hanuman Laxman Aroskar v. Union of India*), 24n56

Supreme Court judgment dated 7 February 2018 in Civil Appeal No. 32138 of 2015 (*Goa Foundation v. M/s Sesa Sterlite Ltd. & Ors*), 123n3

Supreme Court of India judgment dated 22 September 2004 in Appeal (Criminal.) 1350 of 2003 (*Kachrulal Bhagirath Agrawal & Ors v. State of Maharashtra & Ors*), 37n23, 42n39

Supreme Court of India judgment dated 4 April 2013 in Writ Petition (Civil) No. 180 of 2011 (*Orissa Mining Corporation v. Ministry of Environment & Ors*), 55n76, 58n83

Supreme Court of India judgment dated 6 July 2011 in Civil Original Jurisdiction Interlocutory Application (I.A.) Nos. 1868, 2091, 2225–2227, 2380, 2568 and 2937 (*Lafarge Umiam Mining Pvt. Ltd*) in Writ Petition (Civil) No. 202 of 1995 (*T.N. Godavarman Thirumulkpad v. Union of India & Ors*), 320n1

Supreme Court of India judgment dated 6 July 2011 in Interlocutory Application (I.A.) Nos. 1868, 2091, 2225–2227, 2380, 2568 and 2937 (*Lafarge Umiam Mining Pvt. Ltd*) in Writ Petition (Civil) No. 202 of 1995 (*T.N. Godavarman Thirumulkpad v. Union of India & Ors*), 166n18, 320n1

Supreme Court of India order dated 29 November 2018 in 105699/2018 in Original Suit. No. 4/2007 (*State of Orissa v. The State of Andhra Pradesh & Ors*) related to the Polavarm Multipurpose Project, 57n80

Supreme Court order dated 9 May 2002 in I.A. No. 295 in Writ Petition (Civil) 202 of 1995 with I.A. No. 171 of 1996, 110

Supreme Court order in Writ Petition (Civil) 337 of 1995 (*The Centre for Environmental Law [CEL], WWF v. Union of India and Ors.*), 203n4, 222–234

Supreme Court Special Leave Petition Nos. 000474–000476 of 2018 (*Narbheshanker Prabhashanker Khetia & Ors v. Union of India & Ors*), 285n50

Supreme Court Writ Petition (Civil) No. 13029 of 1985 (*M. C. Mehta v. Union of India & Ors*), 74n21, 131n26, 132n27, 133n29, 134n33, 134n34, 150n72

High Courts

High Court of Delhi judgment dated 10 February 2010 in Writ Petition (Civil) No. 4572 of 2007 (*Manushi Sangathan v. Govt. of Delhi & Ors*), 316*n*49

High Court of Delhi order dated 25 August 1999 in Writ Petition (Civil) No. 2145 of 1999 (*Centre for Public Interest Litigation, Delhi v. Union of India*), 128*n*18

High Court of Delhi order dated 9 January 2019 in Writ Petition (Civil) 13521 of 2018 & CM No. 52715/2018 (*Social Action for Forest and Environment v. Union of India & Anr*), 141*n*49

High Court of Delhi Writ Petition (Civil) No. 9340 of 2009 and CM Appl Nos. 7127 of 2009, 1249 of 2009 (*Utkarsh Mandal vs Union of India*) dated 26 November 2009, 167*n*25

High Court of Gujarat at Ahmedabad judgment dated 11 November 2017 in Special Civil Application No. 20362 of 2015 with Special Civil Application No. 12012 of 2014 to Special Civil Application No. 12023 of 2014, 283*n*48

High Court of Gujarat judgment dated 2 March 2000 in Special Civil Application No. 8529 of 1999 (*Centre for Social Justice [Jan Vikas] v. Union of India*), 57*n*81

High Court of Karnataka judgment dated 9 April 1991 in Writ Petition No. 23138 of 1980 (*V. Lakshmipathy and Others v. State of Karnataka and Others*), 34*n*10

High Court of Karnataka Writ Petition No. 17841 of 2018 (*Dattatraya. T. Devare and Other v. State of Karnataka & Ors*), 107*n*57

High Court of Kerala judgment dated 12 July 1999 in O.P. No. 24160 of 1998 (K. Ramakrishnan and Anr. v. State of Kerala and Ors), 39n29

High Court of Kerala judgment dated 16 December 2003 in Writ Petition (Civil) No. 34292 of 2003 (*Perumatty Grama Panchayat v. State of Kerala*), 53*n*68

High Court of Madhya Pradesh at Indore, Writ Petition (Civil) 217 of 2012 (*Kamalchand v. State of M.P. and Others*), 275*n*28

High Court of Madhya Pradesh at Jabalpur, Writ Petition No. 1359 of 2009 (*Narmada Bachao Andolan v. the State of Madhya Pradesh*), 275*n*27

High Court of Madhya Pradesh judgment dated 2 February 2016 in Writ Petition No. 1263 of 2012 (*Indore Hygiene Product & Ors v. Indore Development Authority & Ors*), 275*n*29

High Court of Madhya Pradesh judgment dated 9 December 1985 in Civil Revision Petition No. 1326 of 1984 (*Krishna Gopal v. State of M.P*) (1986), 38*n*28

High Court of Uttarakhand judgment dated 21 December 2018 in Writ Petition (M/S) No. 3437 of 2016 (*Divya Pharmacy v. Union of India*), 217*n*33

NATIONAL GREEN TRIBUNAL

National Green Tribunal judgment dated 16 October 2015 in O.A. No. 17 of 2014(CZ) (*Bio Diversity Management Committee v. Union of India & Ors*), 215*n*29

National Green Tribunal judgment dated 17 July 2014 in Application No. 41 of 2013 West Zone (*Rajendrasinh Mansinh Kashtrya v. Gujarat Pollution Control Board & Ors*), 154*n*84

National Green Tribunal judgment dated 18 February 2014 in Application No. 87 of 2013 West Zone (*Ramubhai Kariyabhai Patel v. Union of India & Ors*), 184*n*73

National Green Tribunal judgment dated 19 January 2014 in Appeal No. 68 of 2012 (*State Pollution Control Board, Odisha v. M/s Swastik Ispat Pvt. Ltd. & Ors*), 144*n*56

National Green Tribunal judgment dated 19 January 2014 in Appeal No. 69 of 2012 (*State Pollution Control Board, Odisha & Ors v. M/s Patnaik Steel & Alloys Ltd & Ors*), 144*n*56

National Green Tribunal judgment dated 21 August 2017 in O.A. No. 15 of 2015 Eastern Zone (*Tseten Lepcha v. Union of India & Ors*), 195*n*103

National Green Tribunal judgment dated 22 December 2016 in O.A. No. 199 of 2014 (*Almitra H. Patel & Ors v. Union of India & Ors*), 181*n*67

National Green Tribunal judgment dated 23 April 2013 in Application No. 82 of 2013 (*Aditya N. Prasad v. Union of India & Ors*), 107*n*56

National Green Tribunal judgment dated 26 August 2016 in O.A. No. 318 of 2013 (*Rajendra Singh Bhandari v. State of Uttarakhand and Ors*), 136*n*40

National Green Tribunal judgment dated 26 February 2016 Appeal No. 04 of 2014 Eastern Zone (*Themrei Tuithung & Ors. v. of Manipur & Ors*), 114*n*71

National Green Tribunal judgment dated 28 November 2018 in Appeal No. 87 of 2018 (*Vedanta Ltd v. State of Tamil Nadu & Ors*), 156*n*88

National Green Tribunal judgment dated 30 March 2012 in Appeal No. 8 of 2011 (*Prafulla Samantaray & Anr v. Union of India & Ors*), 268*n*11

National Green Tribunal judgment dated 4 May 2016 in Appeal No. 28 of 2013 (*Paryawaran Sanrakshan Sangarsh Samiti Lippa v. Union of India & Ors*), 117*n*80

National Green Tribunal judgment dated 6 February 2014 in Application No. 237 (THC)/2013 (CWPIL No. 15 of 2010) (*Court on Its Own Motion v. State of Himachal Pradesh & Ors*), 317*n*50

National Green Tribunal judgment dated 7 May 2015 in O.A. No. 521 of 2014 (*Om Dutt Singh & Anr v. State of Uttar Pradesh & Ors*), 55*n*77

National Green Tribunal judgment dated 7 November 2012 in Appeal No. 7 of 2012 (*Vimal Bhai & Ors v. Union of India & Ors*), 96*n*21, 113*n*70

Other Judicial Forums

Central Information Commission judgment dated 13 May 2019 in CIC/ SA/A/2016/000209 (*Kanchi Kohli v. PIO, M/o Environment & Forest*), 326*n*11

Darjeeling Chief Judicial Magistrate C.R. Case 48 of 2008, 219*n*37

Gujarat Pollution Appellate Authority Appeal No. 2 of 2014 (*Rajendrasinh Mansinh Kashtrya v. Gujarat Pollution Control Board & Ors*), 154*n*82

INTERNATIONAL CASES

High Court of New Zealand, Wellington Registry Civ 2015-485-919 [2017] NZHC 733 (*Sarah Thomson and The Minister for Climate Change Issues*), 313*n*42

Lahore High Court Lahore Judicial Department Case No. Writ Petition No. 25501 of 2015 (*Asghar Leghai v. Federation of Pakistan etc.*), 315*n*47

The England and Wales Court of Appeal on 15 March 1958 in *Attorney General v. PYA Quarries Ltd.* ([1958] EWCA Civ 1), 36*n*21

The United Kingdom House of Lords on 17 July 1868 in *Rylands v. Fletcher* ([1868] UKHL 1), 51*n*63

OFFICIAL REPORTS

JUDICIAL COMMITTEE REPORTS

CEC Report (Interim) regarding the Closure of the Saw Mills and Other Wood Based Industries in the State of Uttar Pradesh pursuant to The Hon'ble Supreme Court's Order Dated 1September 2006 in I.A. No. 1399 and I.A. No. 1569 with I.A. No. 946, Dated 10 October 2006, 112*n*67

Recommendations of the Central Empowered Committee in Interlocutory Application No.566 of 2000 in Writ Petition (Civil) 202 of 1995, 9 August 2002, New Delhi, 98*n*28, 102*n*40

Report of the CEC about Receipts and Utilisation of Funds and Related Issues (2nd Report)—I.A. 827, New Delhi, 12 October 2004, 104*n*49

Report of the CEC regarding Non-utilisation of Funds Received towards Net Present Value (NPV), Compensatory Afforestation, etc., and the Proposed Mechanism for Its Utilisation, New Delhi, 10 January 2008, 104*n*48

PARLIAMENTARY COMMITTEE REPORTS

Committee on Estimates. *Performance of the National Action Plan on Climate Change (NAPCC): Thirtieth Report*. New Delhi: Lok Sabha Secretariat, 2018, 308*n*24, 308*n*26, 310*n*31

Department-Related Parliamentary Standing Committee on Science and Technology and Environment and Forests. *Status of Forests in India: Three Hundred Twenty Fourth Report.* New Delhi: Rajya Sabha Secretariat, 2019, 91n5

Department-Related Parliamentary Standing Committee on Science and Technology and Environment and Forests. *One Hundred and Ninety Fourth Report on the Compensatory Afforestation Fund Bill, 2008.* New Delhi: Rajya Sabha Secretariat, 2008, 105n51

Department-Related Parliamentary Standing Committee on Science and Technology and Environment and Forests. *One Hundred and Ninety Second Report on the Functioning of Central Pollution Control Board.* New Delhi: Rajya Sabha Secretariat, 2008, 147, 147n64

Department-Related Parliamentary Standing Committee on Science and Technology and Environment and Forests. *Two Hundred and Third Report on the National Green Tribunal Bill.* New Delhi, 2009, 344n34

Planning Commission of India. *Evaluation Study on Functioning of State Pollution Control Boards (PEO Study 180).* New Delhi (undated), 147n63

Public Accounts Committee. *Central Pollution Control Board: Audit Review.* New Delhi: Lok Sabha Secretariat (chaired by Shri Bhagwan Shankar Rawat, Member of Parliament), 1994, 121n1

Standing Committee on Rural Development. *Thirty-Ninth Report on the Land Acquisition (Amendment) Bill, 2007.* New Delhi: Lok Sabha Secretariat, 2007, 277n36

GOVERNMENT REPORTS

Cabinet Secretariat. *Report of the Committee on Allocation of Natural Resources.* New Delhi, 2011, 52n67, 331n27

Cabinet Secretariat. *Report on Reforming Investment Approval and Implementation Procedure—Part II.* New Delhi, 2002, 164n13

Law Commission of India. *One Hundred and Eighty-Sixth Report on Proposal to Constitute Environment Courts.* New Delhi, 2003, 15n37, 333n31

Ministry of Environment and Forests Report of the Committee to Evolve Road Map on Management of Wastes in India, Government of India, New Delhi (2010), 180n63

Ministry of Environment, Forests and Climate Change Committee to Review the Issues relating to the Coastal Regulation Zone Notification, 2011 (2014), 325, 326n14

Ministry of Environment, Forests and Climate Change High Level Committee to Review Various Acts (2014), 49*n*56

Ministry of Personnel, Public Grievances and Pensions report: *Second Administrative Reforms Commission: Fifteenth Report (State and District Administration)*. New Delhi, 2009, 67*n*14.

Ministry of Water Resources, River Development and Ganga Rejuvenation report: *A 21st Century Institutional Architecture for India's Water Reforms: Report submitted by the Committee on Restructuring the CWC and CGWB*. New Delhi, 2016, 81*n*32, 82*n*34, 236*n*7, 239*n*17, 244*n*29

Niti Ayog report on *Sustainable Development Goals (SDGs), Targets, CSS, Interventions, Nodal and other Ministries (2017)*, 55*n*74.

Planning Commission High Level Committee on National Mineral Policy (2006), 330–331.

Planning Commission of India. *Report of the PPP Sub-Group on Social Sector: Public Private Partnership*. New Delhi, 2004, 327*n*17

Planning Commission of India. *Report of the Working group of Planning Commission of India on 'Leasing of Degraded Forest Lands'*. New Delhi, 1998, 327*n*16

POLICIES

Draft National Forest Policy, 2018, 328

National Conservation Strategy and Policy Statement on Environment and Development, 1992, 65*n*10

National Environment Policy, 2006, 49–50, 54, 159, 329–330

National Forest Policy, 1952, 118–119

National Forest Policy, 1988, 117–118

National Water Policy, 2002, 46, 50, 241, 241*n*25, 257, 264–265

National Water Policy, 2012, 46, 264–265, 333

National Wildlife Action Plan (2002–2016), 193

National Wildlife Action Plan (2017–2031), 229

General Index